AF352816

Hydrodynamics and Heat Mass Transfer in Two-Phase Dispersed Flows in Pipes or Ducts

Hydrodynamics and Heat Mass Transfer in Two-Phase Dispersed Flows in Pipes or Ducts

Editors

Maksim Pakhomov
Pavel Lobanov

MDPI • Basel • Beijing • Wuhan • Barcelona • Belgrade • Manchester • Tokyo • Cluj • Tianjin

Editors

Maksim Pakhomov
Laboratory of Thermal and
Gas Dynamics
Institute of Thermophysics
Russian Academy of Sciences
Novosibirsk
Russia

Pavel Lobanov
Laboratory of Physical
Hydrodynamics
Institute of Thermophysics
Russian Academy of Sciences
Novosibirsk
Russia

Editorial Office
MDPI
St. Alban-Anlage 66
4052 Basel, Switzerland

This is a reprint of articles from the Special Issue published online in the open access journal *Water* (ISSN 2073-4441) (available at: www.mdpi.com/journal/water/special_issues/JL55W26E61).

For citation purposes, cite each article independently as indicated on the article page online and as indicated below:

LastName, A.A.; LastName, B.B.; LastName, C.C. Article Title. *Journal Name* **Year**, *Volume Number*, Page Range.

ISBN 978-3-0365-8197-2 (Hbk)
ISBN 978-3-0365-8196-5 (PDF)

Cover image courtesy of Makism Pakhomov

Contents

Preface to "Hydrodynamics and Heat Mass Transfer in Two-Phase Dispersed Flows in Pipes or Ducts"

Two-phase gas–liquid and gas-dispersed flows are frequently encountered in the energy, nuclear, chemical, geothermal, oil and gas and refrigeration industries. Two-phase gas–liquid flows can occur in various forms, such as flows transitioning from pure liquid to vapor as a result of external heating, separated flows behind a flow's sudden expansion or constriction, dispersed two-phase flows where the dispersed phase is present in the form of liquid droplets, or gas bubbles in a continuous carrier fluid phase (i.e. gas or liquid). Typically, such flows are turbulent with a considerable interfacial interaction between the carrier fluid and the dispersed phases. The interfacial heat and mass transfer is very important in the modeling of such flows. The variety of flow regimes significantly complicates the theoretical prediction of hydrodynamics of the two-phase flow. It requires the application of numerous hypotheses, assumptions, and approximations. Often the complexity of flow structures makes it impossible to theoretically describe its behavior, and so empirical data are applied instead. The correct simulation of two-phase gas–liquid flows is of great importance for safety's sake and the prediction of energy equipment elements.

This Special issue discusses papers related to various aspects of the hydrodynamics and heat mass transfer in two-phase gas–liquid and gas-dispersed flows in pipes or ducts. They can be used as well as in basic sciences and vast applications. Authors from Russia, China, the U.S.A., the U.K., India, and Singapore published their recent contributions for this Special Issue. We are happy to see that all papers present findings characterized as unconventional, innovative, and methodologically new. We hope that the readers of the journal Water can enjoy and learn about experimental and numerical study of two-phase flows using the published material and share the results with the other researchers. In total, 11 papers were presented in this Special Issue (2 of them are reviews and 9 papers are the article) and 3 papers from this list are feature papers.

The main goals of this Special Issue are to present the recent advances in experimental and numerical studies of hydrodynamics and heat mass transfer in two-phase gas–liquid and gas-dispersed flows in pipes or ducts for engineering and natural systems that can be very useful for researchers and engineers in heat transfer and heat exchangers communities.

Maksim Pakhomov and Pavel Lobanov
Editors

Editorial

Hydrodynamics and Heat Mass Transfer in Two-Phase Dispersed Flows in Pipes or Ducts

Maksim A. Pakhomov [1,*] and Pavel D. Lobanov [2]

1 Laboratory of Thermal and Gas Dynamics, Kutateladze Institute of Thermophysics, Siberian Branch of Russian Academy of Sciences, Acad. Lavrent'ev Avenue, 1, 630090 Novosibirsk, Russia
2 Laboratory of Physical Hydrodynamics, Kutateladze Institute of Thermophysics, Siberian Branch of Russian Academy of Sciences, Acad. Lavrent'ev Avenue, 1, 630090 Novosibirsk, Russia; lobanov@itp.nsc.ru
* Correspondence: pakhomov@ngs.ru or pakhomov@itp.nsc.ru

Citation: Pakhomov, M.A.; Lobanov, P.D. Hydrodynamics and Heat Mass Transfer in Two-Phase Dispersed Flows in Pipes or Ducts. *Water* 2023, 15, 1969. https://doi.org/10.3390/w15111969

Received: 11 May 2023
Accepted: 15 May 2023
Published: 23 May 2023

Two-phase gas-liquid and gas-dispersed flows are frequently encountered in energy, nuclear, chemical, geothermal, oil and gas and refrigeration industries. Two-phase gas-liquid flows can occur in various forms, such as flows transitioning from pure liquid to vapor as a result of external heating; separated flows behind a flow sudden expansion or constriction; and dispersed two-phase flows, where the dispersed phase is present in the form of liquid droplets or gas bubbles in a continuous carrier fluid phase (i.e., gas or liquid). Typically, such flows are turbulent with a considerable interfacial interaction between the carrier fluid and the dispersed phases. The interfacial heat and mass transfer is very important in the modeling of such flows. The variety of flow regimes significantly complicates the theoretical prediction of the hydrodynamics of the two-phase flow. It requires the application of numerous hypotheses, assumptions, and approximations. Often, the complexity of the flow structure makes it impossible to theoretically describe its behavior, and so empirical data are applied instead. The correct simulation of two-phase gas-liquid flows is of great importance for safety and the prediction of energy equipment elements.

Therefore, understanding and controlling such flows demands study of the transport phenomenon in different macro- and microscales. Currently, the fast development of computer facilities allows numerical or experimental investigations of two-phase gas–liquid and gas-dispersed flows to be carried out in complicated flows of geometry; this can be considered an effective method to solve the abovementioned tasks.

This Special Issue presents papers related to various aspects of hydrodynamics and heat mass transfer in two-phase gas–liquid and gas-dispersed flows in pipes or ducts. These findings can also be used in basic sciences and in a vast range of applications. We consider the main goals to have been successfully reached. Authors from Russia, China, the USA, the UK, India and Singapore have published their recent contributions in this Special Issue. Here, we present 11 papers (2 are reviews and 9 are articles). Three papers from this list are feature papers in this Special Issue. We are happy to see that all papers present findings characterized as unconventional, innovative, and methodologically new. We hope that the readers of *Water* can enjoy and learn about the experimental and numerical study of two-phase flows using the published material and share their results with other researchers.

The main goals of this Special Issue were to present recent advances in experimental and numerical studies of hydrodynamics and heat mass transfer in two-phase gas–liquid and gas-dispersed flows in pipes or ducts for engineering and natural systems. These findings can be very useful for researchers and engineers in heat transfer and heat exchanger communities. The following paragraphs provide a short overview of the published articles.

Review paper [1] presents a comparative analysis of the capabilities of various experimental techniques for studying droplet entrainment in annular gas–liquid flows. Entrainment occurs due to the break-up of small-scale ripple waves propagating on top of large-scale disturbance waves. This process must be studied in two planes simultaneously;

the instantaneous film thickness must be measured over an area, together with the parameters of the entrained droplets. This goal cannot be achieved with conductance probes, optical visualization, or Planar LIF. Re-emission or absorption techniques accompanied with side-view visualization are suitable for this purpose. To reduce the errors caused by optical distortions, usage of X-ray or near-infrared radiation is recommended.

A review of the experimental and numerical methods used for the investigation of two-phase annular flow in vertical pipes, mainly in the last two decades, is provided by [2]. The variety of measured methods and numerical simulations are compared to highlight their advantages and challenges. Experimental techniques were summarized and their advantages and limitations were shown to readers. The advantages and deficiencies of the numerical model are presented and discussed. Many previous theoretical or empirical models considered the gas flow as one of the dominating mechanisms; they did not consider the effect of gas superficial velocity in the formulas. These models did not predict the annular flows well without the gas stream in the core zone of the pipe.

The authors of [3] proposed a method for modifying the inner surface of a mini-channel by applying a laser pulse on the outer wall of the channel. The boiling of R-125 in the vertical channel with a diameter of 1.1 mm and length of 50 mm was studied at varied pressures and flow rates in channels with modified surfaces. A significant increase in the heat transfer coefficient and critical heat flux was found. The increase in pressure leads to a reduction in the effect of wall modification on the heat transfer coefficient. The results of this study can be used for the development of advanced heat exchange devices based on mini-channels.

The influence of the discrete representation of the continuous bubble size distribution on the flow structure inside the bubble column, as well as comparison between the polydisperse and monodisperse approaches, is shown in [4]. The authors use the Euler–Euler approach to model the bubble flow dynamics. The turbulence of the carrier phase was modeled by the k-ω SST model, considering the effect of additional turbulence production due to the presence of bubbles. The population balance model was used to model the bubble size distribution. The main aim of the paper was to propose and assess the most compact discrete representation of the bubble size distribution capable of predicting complex flow structure inside the bubble column reactor. The authors have shown that the new hybrid discrete representation with only four classes of bubbles can capture the flow behavior inside the column; additionally, with the increasing bubble size, the monodisperse approach can be sufficient to predict important multiphase parameters, such as the volume fraction of the dispersed phase or the specific interfacial area density.

An experimental study of the movement of bubbles after injection from a single capillary in an inclined pipe was performed [5]. The high-speed shadowgraph method and image analysis procedure are utilized. The dependencies of the velocity and average size of gas bubbles on the angle of inclination of the pipe at various distances from the place of gas injection are obtained. A map of regime parameters has been constructed, according to which various flow regimes are classified. The regime parameters were obtained, under which chains of bubbles were formed; no coalescence of the bubbles occurred. This demonstrates that the increase in the gas flow rate leads to the coalescence of bubbles and the destruction of bubble chains.

A Reynolds-averaged Navier–Stokes (RANS) Euler–Euler simulation of flow, turbulence and heat transfer of droplet-laden flow in a ribbed duct was carried out in [6]. The carrier phase turbulence was predicted using the elliptical blending second-moment closure by observing the presence of droplets. The authors show that the mean and fluctuational flow structures undergo significant modification in comparison with the straight duct. The significant augmentation in the level of turbulent kinetic energy (up to 2 times) and heat transfer (almost 2.5 times) in the ribbed duct two-phase mist flow, in comparison with the single-phase rubbed flow, was predicted.

In the experimental and numerical studies of swirling flows in an aerial vortex reactor [7], it was found that, similar to the case of two rotating liquids, a strongly swirling jet is formed at the reactor axis and the entire flow takes on the structure of a miniature gas–liquid tornado. The aerating gas only interacts with the liquid through the free surface. It was also revealed that the radial velocity component slips at the interface. Moreover, despite the difference in density being more than three orders of magnitude, the spiral air flow converging to the reactor axis forms a divergent vortex motion of the liquid medium. This feature causes an intensification of the interphase mass transfer due to the high-speed motion of the aerating gas.

The authors of [8] combine laser treatment and hot wire chemical vapor deposition (HW CVD) for the change wettability of single-crystal silicon surface. The wetting properties varied from supehydrophilic to superhydrophobic. The influence of contact angle on the spreading of falling water droplets for a Weber number of up to 40 was measured. A new approach for the generalization of experimental data was suggested. The authors of [9] investigated the effect of gravitational forces on the rate of evaporation of a water droplet deposited on a structured surface. The comparisons were made between the dynamics of evaporation of a sessile and suspended droplet. This was carried out by using a bifilar surface with a seat and an experimental setup with a rotary mechanism. The authors showed and hypothesized about the differences in the dynamics of evaporation for sessile and suspended droplets.

The authors of [10] utilized the HW CVD method to the fabrication of highly efficient hydrophobic separation membranes by depositing fluoropolymer coatings with different wetting properties. The separation of emulsions of water and commercial crude oil was studied. It was shown that membrane-wetting properties affect the rate and efficiency of the separation. The incensement in water contact angle and decrement in the pore size of membranes lead to higher separation efficiency. The optimal parameters for the use of membranes for separating emulsions were found. The highest efficiency is achieved when separating membranes with a superhydrophobic coating (water contact angle = 170°) and the minimum pore size (40 μm). It was shown that the proposed membranes can easily be washed and reused with close-to-original properties.

The effect of wall proximity and surface tension on the single and rising bubble path in still water close to the wall was numerically studied in [11] using OpenFOAM based on the Volume of Fluid (VOF) method. The authors found that bubble motion and rising close to the wall increases the drag and leads to an early transition from rectilinear to the planar zigzagging regime. The bouncing motion of gas bubbles in pure water was studied. The authors revealed the bubble motion in the spanwise direction is insignificant and the bouncing motion is two-dimensional.

The advanced underpinning physics of the mechanism is summarized in several groups, including wavy liquid film, droplet behavior, entrainment and local void fraction. Unsteady annular two-phase flow has not been widely experimentally and numerically studied and is not well understood. The current understanding of unstable annular flow is generally derived from the steady-state case. Therefore, the investigation of unsteady annular flow and the transient region of a stable flow is recommended with particular focus on the wavy liquid film, gas core and entrained droplets.

All of the abovementioned reviews and research papers have demonstrated the importance and usefulness of two-phase gas-liquid and gas-dispersed flow studies, using both measurements and numerical simulations for science and a range of applications.

Author Contributions: Conceptualization, M.A.P. and P.D.L.; writing—original draft preparation, M.A.P. and P.D.L.; writing—review and editing. All authors have read and agreed to the published version of the manuscript.

Funding: Authors are grateful to the state contract with IT SB RAS (Projects 121032200034-4 and 121031800217-8).

Institutional Review Board Statement: Not applicable.

Informed Consent Statement: Not applicable.

Data Availability Statement: Not applicable.

Acknowledgments: We are grateful to all authors for submitting their contributions to the present Special Issue and for its successful completion. We deeply appreciate the *Water* journal reviewers and Academic Editors for the encouragement and thorough work, which are truly valuable for us to improve the quality and impact of all these submitted papers. Finally, we sincerely thank the editorial staff of *Water* for their efforts and hard work during the editing and publication of this Special Issue.

Conflicts of Interest: The authors declare no conflict of interest.

References

1. Cherdantsev, A.V. Experimental investigation of mechanisms of droplet entrainment in annular gas-liquid flows: A review. *Water* **2022**, *14*, 3892. [CrossRef]
2. Xue, Y.; Stewart, C.; Kelly, D.; Campbell, D.; Gormley, M. Two-phase annular flow in vertical pipes: A critical review of current research techniques and progress. *Water* **2022**, *14*, 3496. [CrossRef]
3. Belyaev, A.V.; Dedov, A.V.; Sidel'nikov, N.E.; Jiang, P.; Varava, A.N.; Xu, R. Flow boiling heat transfer intensification due to inner surface modification in circular mini-channel. *Water* **2022**, *14*, 4054. [CrossRef]
4. Chernyshev, A.; Schmidt, A.; Chernysheva, V. An impact of the discrete representation of the bubble size distribution function on the flow structure in a bubble column reactor. *Water* **2023**, *15*, 778. [CrossRef]
5. Gorelikova, A.E.; Randin, V.V.; Chinak, A.V.; Kashinsky, O.N. The effect of the angle of pipe inclination on the average size and velocity of gas bubbles injected from a capillary into a liquid. *Water* **2023**, *15*, 560. [CrossRef]
6. Pakhomov, M.A.; Terekhov, V.I. RANS modeling of turbulent flow and heat transfer in a droplet-laden mist flow through a ribbed duct. *Water* **2022**, *14*, 3829. [CrossRef]
7. Sharifullin, B.R.; Skripkin, S.G.; Naumov, I.V.; Zuo, Z.; Li, B.; Shtern, V.N. Intense vortex motion in a two-phase bioreactor. *Water* **2023**, *15*, 94. [CrossRef]
8. Starinskiy, S.; Starinskaya, E.; Miskiv, N.; Rodionov, A.; Ronshin, F.; Safonov, A.; Lei, M.-K.; Terekhov, V. Spreading of impacting water droplet on surface with fixed microstructure and different wetting from superhydrophilicity to superhydrophobicity. *Water* **2023**, *15*, 719. [CrossRef]
9. Starinskaya, E.; Miskiv, N.; Terekhov, V.; Safonov, A.; Li, Y.; Lei, M.-K.; Starinskiy, S. Evaporation dynamics of sessile and suspended almost-spherical droplets from a biphilic surface. *Water* **2023**, *15*, 273. [CrossRef]
10. Melnik, A.; Bogoslovtseva, A.; Petrova, A.; Safonov, A.; Markides, C.N. Oil–water separation on hydrophobic and superhydrophobic membranes made of stainless steel meshes with fluoropolymer coatings. *Water* **2023**, *15*, 1346. [CrossRef]
11. Mundhra, R.; Lakkaraju, R.; Das, P.K.; Pakhomov, M.A.; Lobanov, P.D. Effect of wall proximity and surface tension on a single bubble rising near a vertical wall. *Water* **2023**, *15*, 1567. [CrossRef]

 *water*

Review

Experimental Investigation of Mechanisms of Droplet Entrainment in Annular Gas-Liquid Flows: A Review

Andrey V. Cherdantsev [1,2]

1   Coherent and Nonlinear Acoustics Laboratory, Geophysical Acoustics Department, Geophysical Research
    Division, Institute of Applied Physics, 46 Ulyanov Str., Nizhniy Novgorod 603950, Russia;
    cherdantsev@itp.nsc.ru
2   Laboratory of Transfer Processes in Multiphase Systems, Kutateladze Institute of Thermophysics,
    1 Lavrentiev Ave., Novosibirsk 630090, Russia

Abstract: Entrainment of liquid from the film surface by high-velocity gas stream strongly affects mass, momentum and heat transfer in annular flow. The construction of basic assumptions for simplified physical models of the flow, as well as validation of numerical models, requires detailed experimental investigation of droplet entrainment process and the preceding stages of film surface evolution. The present paper analyzes the achievements and perspectives of application of various experimental approaches to qualitative and quantitative characterization of droplet entrainment. Optical visualization in at least two planes simultaneously may provide enough information on transitional liquid structures and detaching droplets, given that the side-view image is not obscured by the wall film. A planar LIF technique is not suitable for this purpose, since real objects are hidden by curved agitated interface and replaced by optical artifacts. To characterize the waves evolving into the transitional liquid structures, film thickness measurements in the plane of the wall are necessary. Such measurements can be achieved by intensity-based optical techniques, such as Brightness-Based LIF, near-infrared or X-ray attenuation techniques, combined with the side-view observations.

Keywords: annular flow; disturbance waves; ripples; droplet entrainment; experimental methods

Citation: Cherdantsev, A.V.
Experimental Investigation of
Mechanisms of Droplet Entrainment
in Annular Gas-Liquid Flows: A
Review. *Water* **2022**, *14*, 3892.
https://doi.org/10.3390/
w14233892

Academic Editors: Maksim
Pakhomov and Pavel Lobanov

Received: 31 October 2022
Accepted: 27 November 2022
Published: 29 November 2022

Publisher's Note: MDPI stays neutral
with regard to jurisdictional claims in
published maps and institutional affiliations.

1. Introduction

At high gas content, gas-liquid flow follows an annular pattern: liquid film is formed on the duct walls and sheared by high-velocity gas stream in the duct core. Interaction of the liquid surface to strong gas shear produces waves of different types and scales. At large gas and liquid flow rates, both the liquid and gas dispersed phase appears in such a flow in the form of liquid droplets torn from the film surface and gas bubbles entrapped by liquid film. Due to its high speed, intense mixing, and large interfacial area, annular flow is widely used in heat exchangers, cooling systems, and chemical reactors. It occurs in oil-and-gas production and transportation [1,2], gas purification, and propulsion engines. It may also occur as a result of liquid boiling/evaporation or steam condensation in an originally single-phase flow, as happens in nuclear industry [3] or solar energy plants [4].

Droplet entrainment is perhaps the most complex phenomenon among those taking place in annular flow. The flow rate of liquid travelling as droplets may make up to 80% of the total liquid flow rates even in the absence of heat flux [5]. As a result, the wall film gets thinner, which affects the rate of heat removal from the heated duct walls. Liquid in the entrained droplets is transported along the pipe with the speed close to that of the gas stream (for comparison, the typical speed of the largest waves is an order of magnitude smaller). The gas stream loses its energy on acceleration of entrained droplets; thus, the pressure drop in the flow increases. Mass and heat exchange between the wall film and the gas core is also intensified. Entrained droplets may hit the film surface creating either a large number of small-size droplets ("secondary entrainment"), or a large number of small bubbles entrapped in the liquid film. Entrapped bubbles, as well as the perturbation of the film surface created by impacting droplets may serve as nucleation sites for formation of dry spots on heated walls. The presence of droplets and bubbles increases the interfacial area thus enhancing heat transfer and chemical reactions between the phases. The droplets also affect the turbulence in the gas phase: small droplets attenuate the turbulence and large droplets augment it, see [6].

It is not surprising that understanding the entrainment physics is of high importance for industrial applications. Numerous experimental studies were devoted to measuring the properties of droplets in annular flow. The droplet flow rate was measured in an integral manner, using sampling probes [7], a film extraction technique [8], and tracer additives [9]. The size and velocity distributions of entrained droplets were measured by the diffraction technique [10], Doppler anemometry [11], and interferometry [12]. To study the droplets, the film is usually extracted from the pipe walls to provide a clear view.

At present, the prediction of entrainment is mainly made in form of correlations, providing an empirical generalization of droplet parameters [13,14]. There exist numerical models of entrainment [15–18], but the dynamics of the modeled flow vary greatly in different computational setups used by different authors. The construction of simplified physically-based models of entrainment is possible, but it critically depends on basic assumptions employed in the models. Previous attempts to create such models [19–21] required strong empirical corrections, effectively reducing each model into another empirical correlation (see [22]).

To construct the basic assumptions for the simplified physical models, and to validate the numerical models of the process, it is necessary to study the very process of entrainment (i.e., the physical events leading to detachment of droplets) on both qualitative and quantitative levels. One needs to understand the whole process of deformation of the film surface leading to the detachment of droplets; the parameters of the evolving film perturbations and deformed liquid structures must be measured at each stage of their evolution, together with the parameters of the droplets created in the process of the interface destruction. Such studies are relatively rare in the field, and some observations and their interpretations in different studies contradict each other.

In the present paper we analyze the available experimental results on the process of droplet entrainment from film surface. It should be noted that secondary entrainment due to droplet impact, bubble burst, or break-up of a liquid membrane in transitional slug/churn flow are ignored here, though the main methodical conclusions of the present paper are also applicable for these processes. The paper is structured in relation to the methods of investigation. However, the description of each technique also presents the method's contribution to the current understanding of entrainment physics and the method's capability to further elucidate the entrainment process. Thus, the name of each chapter consists of two parts, with the second part reflecting the method's role in the development of our understanding.

2. Non-Optical Techniques: From Wave Shape to First Entrainment Hypotheses

The main number of techniques suitable for studying the entrainment process can be considered "optical", even if they employ radiation with a wavelength out of the visible spectrum. Nonetheless, a large number of studies were carried out based on non-optical techniques, and some of these studies have exerted their influence on the present field of knowledge.

Both conductance and capacitance methods are based on measuring the conductivity/capacity, respectively, between the electrodes contacting liquid. The electrodes may be flush-mounted into the duct wall or introduced into the flow ("parallel-wire" or "twin-wire" probes). In the former configuration [23–25], the dependence of the probe signal on film thickness gets saturated at a certain thickness value, defined mainly by the probe size, i.e., the distance between the flush-mounted electrodes. The signal of the probe is averaged over this distance, so the spatial resolution of the probe is limited by the probe size. Because of that, the shape of the waves with a longitudinal size comparable to the probe size will be distorted in the probe record, not to mention the detection of any fine liquid structures on top of large waves. Reducing the probe size might improve the spatial resolution, but it will also reduce the range of measurable film thicknesses. The twin-wire probes [26–28] have linear calibration dependence and generally better spatial resolution compared to flush-mounted probes. On the other hand, they might perturb the flow; besides, a capillary meniscus can be formed around the wires affecting the measurements.

Such probes were intensively used to obtain temporal records of film thickness and study the wave structure. It was established that the appearance of entrained droplets (detected based on visual observations or sampling probe studies) requires the presence of large-amplitude liquid structures on the film surface. These structures are known as "disturbance waves". Due to the limitations of the technique described above, the disturbance waves in such records have a smooth shape with a steep front slope and a single well-pronounced crest (Figure 1). Based on this shape, two hypotheses were proposed in [29]: shearing-off the disturbance wave crests and wave undercutting (Figure 2). Though these hypotheses did not receive any direct experimental confirmation, they remain the most popular in the literature; the former hypothesis is used as the basic assumption in majority of the aforementioned simplified physical models.

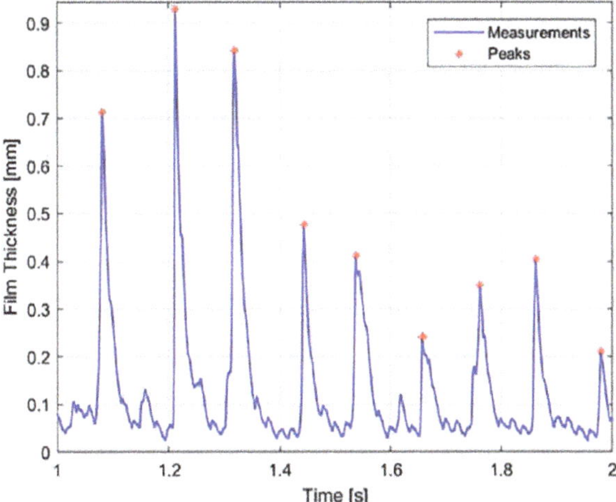

Figure 1. A temporal record of film thickness obtained by a flush-mounted conductance probe [25].

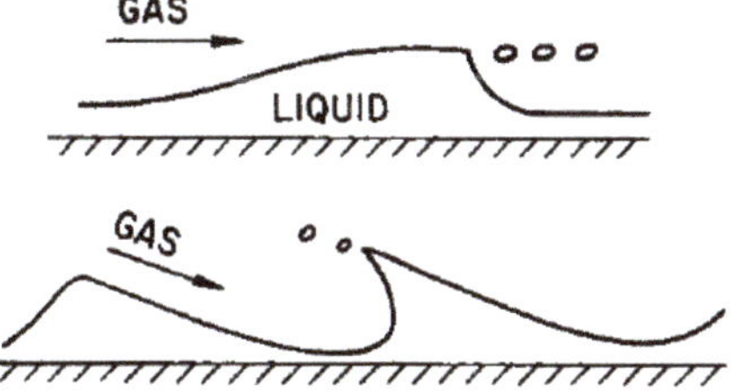

Figure 2. Entrainment hypotheses on shearing-off the disturbance wave crest (**top**) and disturbance wave undercutting (**bottom**) [29].

The non-optical techniques have wider capabilities than merely taking film thickness records in one point. It is possible to create multi-component probes, obtaining the records simultaneously in many points, obtaining the film thickness records resolved along one [30] or two [31,32] spatial coordinates. E.g., in [30] it was found that the waves on the base film are produced at the rear slope of the disturbance waves, and in [31] it was shown that the disturbance waves, even if they are closed around the pipe circumference, show strong amplitude variation in circumferential direction and also large temporal variation as they propagate downstream. Nonetheless, such approaches are limited by the wall film measurements and are hardly suitable for studying the droplets simultaneously with the waves. The measurements in the central part of the duct can be made with tomographic modifications of conductance/capacitance techniques, namely, wire-mesh sensor [33] and electric capacitance tomography [34], respectively. Such approaches allow one to reconstruct instantaneous distributions of the liquid phase in one cross-section of the duct. However, these measurements are not resolved in longitudinal coordinate; the results are also averaged over the distance between the sensor layers; the accuracy of the reconstruction is also quite limited by the large distance between neighboring wires or by the number and size of the capacitance plates.

3. Shadow Visualization: From Simplicity to Complexity

Optical techniques are convenient for observing a large spatial domain. Thus, it is easy to observe a complex object evolving in time and shifting in space; multiple objects of a different nature (e.g., waves and droplets) can be studied simultaneously. In the further analysis, we consider only the approaches employing domain-based measurements and ignore the techniques limited to pointwise applications such as laser focus displacement [35], total internal reflection [36], chromatic confocal imaging [37], etc. Moreover, in this section we only discuss the direct optical visualization approach, which normally consists in imaging the flow illuminated by white light.

Obviously, one camera can see one spatial plane. Three spatial coordinates would define three main working planes. We denote the axis parallel to the flow direction—longitudinal coordinate—as x. The transverse coordinate parallel to the wall over which the film flows is denoted as y. The transverse coordinate orthogonal to the wall and the film surface is denoted as z. In pipes, y-coordinate can be referred to as the circumferential or azimuthal coordinate and can be also replaced by the azimuthal angle in polar coordinates with the origin at the pipe axis. Respectively, the z-coordinate can be referred to as the radial coordinate, and x as the axial coordinate.

The simplest way to visualize flow in pipes is to obtain backlit images in x-y planes. The lamp is placed on one side of the pipe, and the camera "sees" this light passing through the flow. In the very first studies it was observed that the disturbance waves have a "milky" appearance due to strong agitation of their surface. This alone contradicts the "smooth" profiles of disturbance waves obtained by low-resolution conductance probes as mentioned above, and to the related hypotheses.

For annular flow inside pipes the whole pipe wall is covered by agitated liquid film. Therefore, it is quite difficult to organize a clear view of the objects in the gas core through this interface. On the other hand, it is generally agreed that the disturbance waves are qualitatively the same in ducts of different size, shape and orientation, despite the related quantitative difference in wave parameters. The first comprehensive study of the interrelation between the roughness of disturbance wave surface and liquid entrainment was made in [38] on a liquid film flowing along the bottom of a horizontal rectangular duct by viewing the film from the top (i.e., in the x-y plane). It was found that there exist smaller-scale horseshoe-shaped "ripples" on top of disturbance waves (Figure 3). Under the action of gas shear, these ripples got stretched like soap bubbles and broken into a number of droplets. This scenario is more complex than those proposed in [29]. Moreover, it is more plausible: shorter structures are easier to deform and break by the gas stream than the whole crest of a disturbance wave, which normally has a large longitudinal size of the order of 1 cm. Some more recent visualization studies in pipes (e.g., [18,39]) generally follow the pattern proposed in [38].

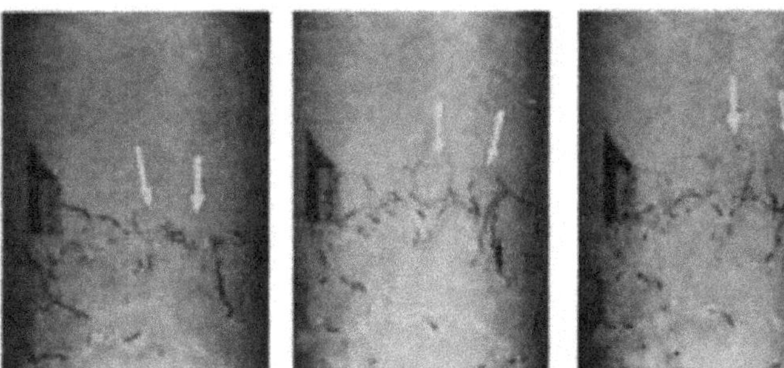

Figure 3. Visualization of entrainment due to break-up of ripples on top of a disturbance wave in a horizontal rectangular duct in the x-y plane [38].

The same process was observed later in [40] in the y-z plane. This camera was oriented along the pipe axis through a specially organized viewing window. Although in this configuration it is impossible to distinguish between the disturbance waves and ripples on top of them, the deformed liquid structures being shattered into droplets can be clearly seen. Using an analogy with secondary atomization of droplets in a gas stream, the entrainment events were separated into two types: bag break-up and ligament break-up (Figure 4). The former (Figure 4a) is very likely the same as the process observed in [38]: part of a wave is blown into a liquid bag; at some point, this bag gets torn into many droplets. During the latter (Figure 4b), an isolated liquid jet ("ligament") is formed from the film surface; this ligament is then broken into a small number of relatively large droplets. The break-up of a ligament is reminiscent of Rayleigh–Plateau instability. Later, similar observations were made in [41,42].

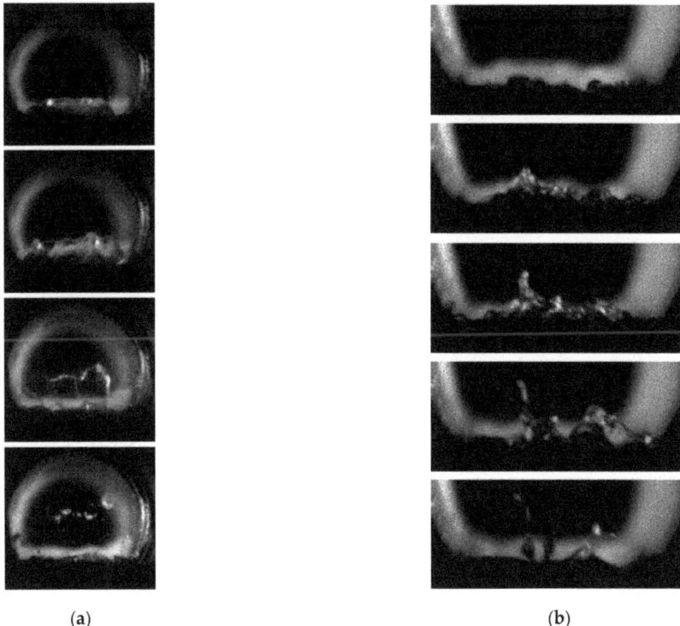

(**a**)   (**b**)

Figure 4. Visualization of entrainment due bag break-up (**a**) and ligament break-up (**b**) in a horizontal pipe in the y-z plane [42].

The direct visualization of the entrainment process in the x-z plane was made in [43]. In this work, there was clear optical access to the breaking wave and droplets, since the film was flowing on the outer surface of a cylinder (namely, part of a rectangular rod bundle mimicking nuclear installations). Again, bag and ligament break-up can be observed (Figure 5). It is also quite clear that only a small fraction of a disturbance wave—namely, a short ripple on its surface—directly participates in the entrainment process.

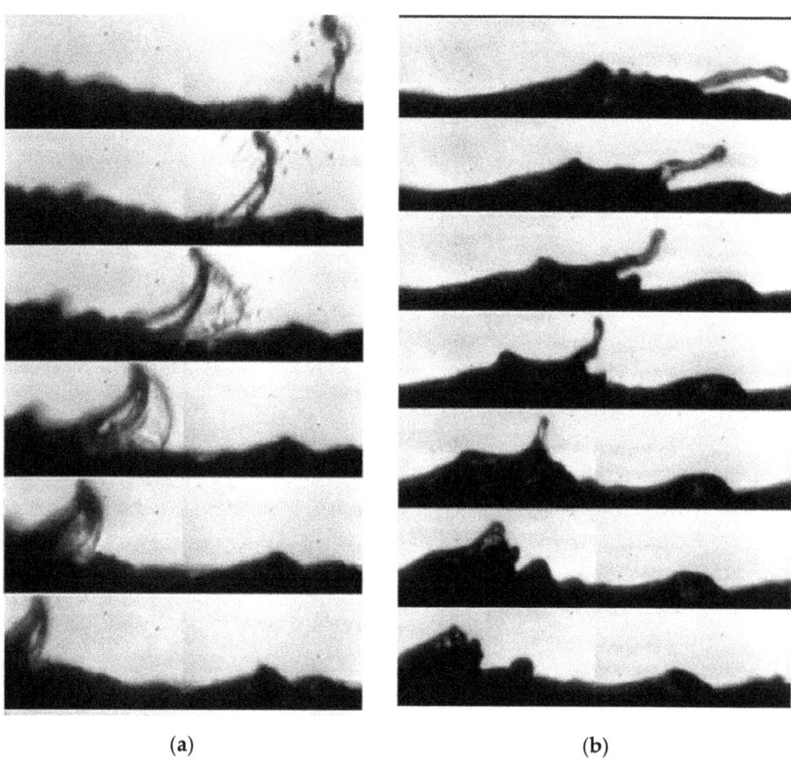

(a)   (b)

Figure 5. Visualization of entrainment due to bag break-up (**a**) and ligament break-up (**b**) on the outer surface of a vertical cylinder in the x-z plane [43].

It is also educational to see the entrainment visualization presented in the works [44,45], where laboratory-modeled sea waves were studied in a large (20 m by 2 m by 2 m) tank sheared by strong wind. The visualization was carried out in both the x-z and x-y planes, and the same events of bag and ligament break-up were observed (Figure 6). In this case it is especially clear that the wind does not tear off the crests of the large-scale waves, which are orders of magnitude higher and longer than the disturbance waves in annular flow. On the opposite, the typical transverse size of the bags created on the surface of deep gas-sheared liquid layer, is of the order of 1 cm. The transverse and longitudinal size of the ripples broken into droplets in the annular flow is of the same order of magnitude [46].

For both the thin-film flow [43] and the deep-layer flow [45], it is reported that the liquid bag consists of a thin central film and a thicker rim around it. The central film is broken first into small droplets; afterwards, the rim is broken into larger droplets.

The entrainment events presented in Figures 3–6 are essentially three-dimensional. The only expected kind of symmetry is the bilateral symmetry of a liquid bag, which is also not necessarily true. Quantitative optical investigations require simultaneous imaging of the entrainment events in at least two planes.

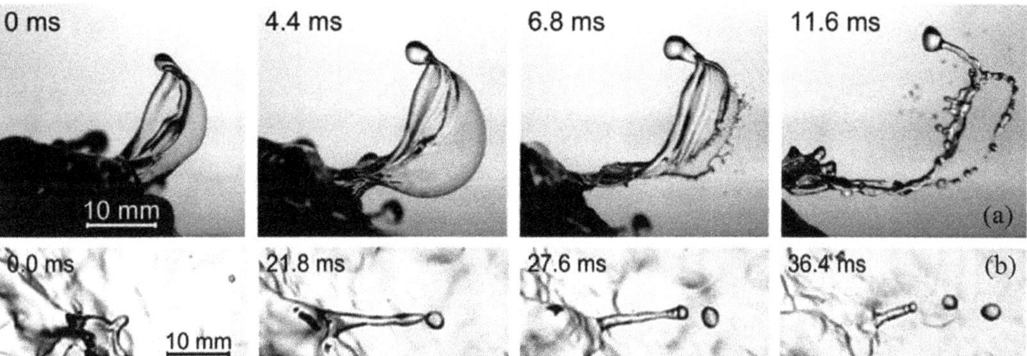

Figure 6. Visualization of entrainment: (**a**) due to bag break-up in the *x-z* plane; (**b**) ligament break-up in the *x-y* plane on surface of a 2 m deep gas-sheared liquid layer [44].

Such investigation can provide the dynamic characteristics of the transitional liquid structure (i.e., bag or ligament), together with the characteristics of the entrained droplets. However, the important characteristics of liquid film, including the parameters of the disturbance wave and three-dimensional evolution of the ripples that serve as the basis for formation of the transitional liquid structures, cannot be extracted merely from shadow visualization. An additional technique applicable for film thickness measurements and compatible with optical visualization is required.

4. Planar Laser-Induced Fluorescence (PLIF) Technique: Deceptive Clarity

The planar LIF method is based on direct imaging of one section of a liquid film (doped with fluorescent dye) illuminated by a laser sheet. The sheet is usually oriented in the *x-z* plane, and the camera views the fluorescing liquid along *y*-axis. For annular flow, this method is mainly applied in circular pipes [47–51]. In some of these papers [49,51], the authors claim that entrainment events can be observed directly using PLIF technique. Sometimes, such images are interpreted based on seeming similarity to previously reported entrainment mechanisms [49], and sometimes, the interpretation involves the construction of new hypotheses [51].

Figure 7 shows a number of such events presented in [49]. In this work, a downward air-water annular flow was studied in a 32.4 mm diameter pipe. The bright areas in the images are presumably occupied by liquid. The pipe wall is on the right-hand side of each sub-image, and the interface is on the left. The flow direction is from top to bottom. The events presented in Figure 7a,b were interpreted as shearing-off disturbance wave crests and disturbance wave undercutting, respectively, since they are reminiscent of the images shown in Figure 2. Figure 7c shows a structure reminiscent of a liquid ligament (see Figures 4b, 5b and 6b). Figure 7d-e and Figure 7f are interpreted, respectively, as bursting of a large bubble and droplet impingement, discussed in [29] and before.

In [51], upward annular flow was studied in a 25 mm pipe. Figure 8b shows an example sequence of PLIF-images. In this image, the pipe wall is on the left, and the flow goes upwards. In total, five types of entrainment events were allegedly observed. These events are not directly related to previously proposed hypotheses; instead, new hypotheses are constructed. All the event types are alike; they are all related to the partial disappearance of a disturbance wave image. The difference between the types is related to which part of a wave disappears: the whole small-scale wave (type 1); the front part of a large wave (type 2); the rear part of a large wave (type 3, see Figure 8); the top part of a large wave (type 4); part of wave containing a bubble image (type 5).

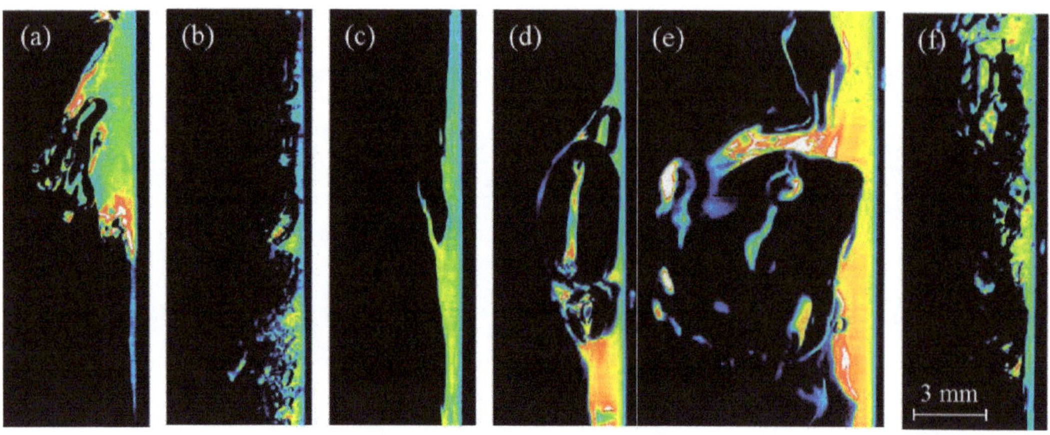

Figure 7. PLIF-images of downward annular flow [49]. The presented events are interpreted as: shearing-off disturbance wave crests (**a**); disturbance wave undercutting (**b**); liquid ligament (**c**); bubble burst (**d**,**e**); droplet impingement (**f**).

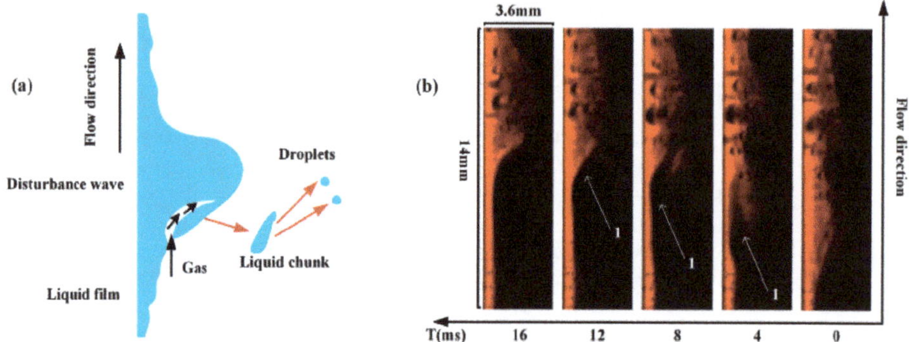

Figure 8. PLIF images in upward annular flow [51]. The event is interpreted as entrainment due to "shearing of the ending of the disturbance wave". (**a**) A sketch illustrating the proposed interpretation; (**b**) A sequence of raw PLIF-images with part of a wave image disappearing.

The images presented above may seem to be a direct visualization of the entrainment process, but this interpretation is doubtful from both hydrodynamics and optics points of view. Let us first list the issues related to hydrodynamics. The events shown in Figure 7a,b exactly correspond to hypotheses based on speculation [29] and not observed with plain visualization. The ligament formation (Figure 7c) was observed indeed in visualization studies, but in all cases the ligaments are oriented along the flow (see Figure 5b, Figure 6b, and Figure 13b), whilst in Figure 7c it is oriented against the flow. All the entrainment "types" suggested in [51] assume that a very strong and narrow gas jet literally cuts through the liquid, "slicing" it in chunks. Such behavior does not look plausible, especially with a low gas speed of 6 m/s. Besides, the probability of all entrainment events in [51] is reported to decrease quickly with increasing gas speed, which also contradicts all measurements of entrainment intensity (see, e.g., [13,14]).

From an optics point of view, it is not likely that the process of entrainment can be directly observed in the PLIF-images without any distortion. Analysis of optical distortions related to PLIF-imaging in pipes was carried out in [52,53], though these papers mainly dealt with the error of film thickness measurements. At present, no analysis considering optical distortions for PLIF-visualization of the entrainment phenomena is presented in the literature. For that reason, such analysis will be carried out in the present review.

Let us start with a simple case of annular film with nearly uniform thickness. An example of a raw PLIF-image for such case (flow of liquid film falling under the action of gravity along the inner walls of a vertical pipe) is shown in Figure 9. A typical PLIF-image consists of a bright stripe near the pipe wall and a dark area above it. There also might appear a thin bright line above the film image far in the dark area (so-called "ghost image"). It should be noted that there appear to be some "barbs" on the film surface (at $x = 6$ mm and $x = 20$ mm), even in absence of gas flow and entrainment. A straightforward interpretation of PLIF-images is that the bright stripe shows the film profile and the dark area shows the gas above this film in the illuminated pipe section.

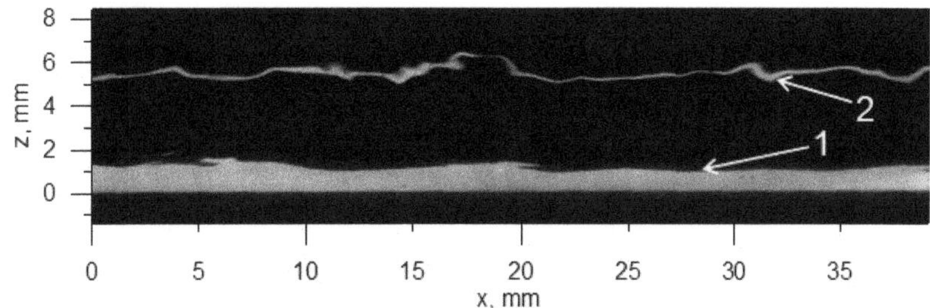

Figure 9. Raw PLIF-image of a falling film flow in a vertical 32.4 mm pipe; taken from the experimental data set [52]. (1) PLIF-image of liquid film (both true and false image, see below); (2) "Ghost" image of the film seen through the pipe wall.

To understand how correct this interpretation is, let us consider the application of the PLIF method to a waveless liquid film with a uniform thickness of 5 mm in a 50 mm diameter pipe (the numbers are arbitrary and do not quantitatively change the conclusions). A portion of such a pipe section is sketched in Figure 10; the interface is shown by thick blue line, and the pipe inner wall by thick black line. One longitudinal section of the film is illuminated by a laser sheet (shown by the pale green rectangle). The camera is viewing the pipe from the right. For simplicity, we assume that the angle between the camera axis and the laser sheet plane is 90°, and the rays going to camera are parallel. In addition, we neglect the refraction on the pipe inner wall. This assumption implies the same refraction coefficient for the liquid and pipe wall (e.g., if the pipe is made of FEP, see [49]), and a flat vertical outer wall of a pipe on the right-hand side (i.e., if the pipe is placed into an optical box, see [49]). The ray tracing is very simple here, based merely on Snell's law, Fresnel equations, equality of the angles of incidence and reflection, and the geometry of the film and pipe. Namely, the ray path is defined by the angle α between the vector connecting the origin to a particular point at the interface, and the z-axis. The angle of incidence is then defined as $\pi/2 - \alpha$ if this angle exceeds the angle of total internal reflection or TIR-angle (48.6° for air-water system), the ray will not pass through the interface.

As illustrated by the rays drawn in Figure 10, the real picture is quite different from the straightforward one. The rays passing through the illuminated section of the liquid film (shown by red lines in Figure 10 between $z = -20$ mm and $z = -25$ mm) indeed collect the fluorescence emitted by the laser-illuminated liquid film and show the real profile of film thickness (zone A). The rays slightly above this section (shown by magenta lines, between $z = -20$ mm and $z = -18.5$ mm) are totally internally reflected from the interface. After reflection, these rays also pass through the illuminated section of the film, so they also deliver fluorescence to the camera image, creating a false film image (see also [53]). The false image width is about 30% of that of the real image for a film of constant thickness (zone B). The false image is also deformed (namely, non-uniformly compressed along z) and turned upside down. Thus, the border of the bright stripe in the PLIF image does not correspond to the real interface, even if the 30% increase is taken into account. The real interface is inside the bright stripe. The brightness of real and false images is approximately the same, so the position of the true interface cannot be detected based on brightness values. Although sometimes a thin brighter or darker line can be seen where the true interface is located, its presence is not guaranteed [52].

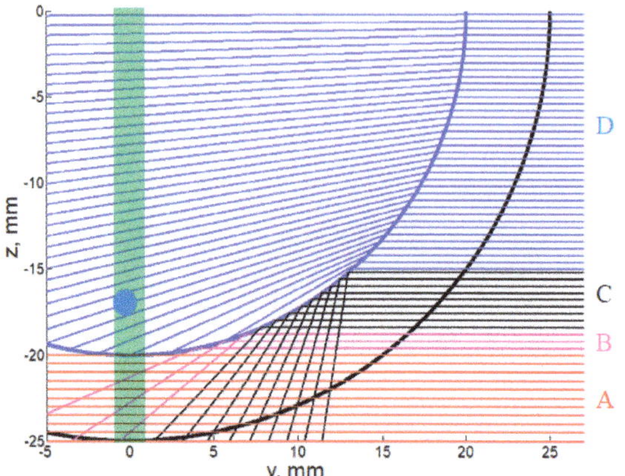

Figure 10. Rays leading to the camera in the PLIF method. The calculations are made within the present review, as a development of the analysis first presented in [53].

The rays passing above the false image (shown by black lines, between $z = -18.5$ mm and $z = -15$ mm) are also totally internally reflected at the interface but they do not pass through the illuminated section of the film, so the pixels collecting fluorescence along these rays remain dark (zone C). These rays do not enter into the gas phase, so the image collected by them has no relation to the gas section above the film. Thus, any object above the film surface in the illuminated section (like an entrained droplet drawn in Figure 10 or a deformed liquid structure) cannot be seen by the camera in areas B and C.

Such objects can be observed in zone D by the rays passing even higher (the blue lines in Figure 10, for $z > -15$ mm). These rays do penetrate into the gas phase, though the reflection coefficient may be still high closely to zone C. The transmitted rays are refracted and hit the interface at different points. Those rays which pass through the illuminated section of the film (at $z \sim -13$–14 mm for the circular interface) would create the ghost image of the film (as the one shown in Figure 9). Additionally, the interpretation of such an image as a reflection of the fluorescence from the farther wall of the pipe, proposed in [52], is incorrect: it is just fluorescence seen by the rays refracted at the closer wall. If some objects like droplets are in the laser sheet above the film, they absorb the light and emit fluorescence, so they can be seen in zone D. However, these objects will be seen far from the film image, separated by the false image zone B and the blind zone C; their images will be distorted by the curved interface even in the case of a smooth film of constant thickness. Moreover, in the flow conditions where droplets or complex liquid structures may appear, the whole film surface would be agitated with short-length three-dimensional waves. In this case, even the ghost image disappears from the PLIF-images due to strong distortions. Each interface perturbation on the optical path would work as a lens massively distorting the images of droplets, especially if the "focus distance" of such a "lens" is small compared to the distance from a droplet to the "lens". Even worse, the laser sheet itself may be refracted at the agitated interface.

Given the above, none of the interpretations provided for Figure 7 is correct. The bright spots near the apparent interface (Figure 7a,b,f) are not related to droplets which cannot be seen in such a system. If the elongated structures connected to the interface, interpreted as liquid ligaments (specified for Figure 7c, but also observable for Figure 7a,b,f), were real, they would have not been visible in PLIF-images, being also hidden from view by an inclined interface. Only the bubble shown in Figure 7d,e is possibly a real large bubble. Most likely, this bubble is located between the laser sheet and the camera, judging by the bright longitudinal line in its center. A more detailed description of how to distinguish the real y-position of a bubble relative to the laser sheet can be seen in Section 3.4 in [52]. It is still unclear whether that bubble has actually burst or its upper border (i.e., thin liquid film covering the bubble) has just disappeared from view.

To properly interpret the images shown in Figures 7 and 8, one needs to answer two questions:

(1)   How a bright isolated area may appear in PLIF-images?
(2)   How a part of a bright area connected to the film image may disappear?

The simple modeling shows that both questions have the same answer and that this answer is related to the presence of three-dimensional waves on the film surface between the laser sheet and the camera, mainly in zones B and C shown in Figure 10. If all the waves were perfectly two-dimensional, i.e., if the cross-section of the interface was always circular (with only the radius varying due to passage of waves), the PLIF-image would have consisted only of the true and false film images, with an approximately constant ratio of their sizes. However, for three-dimensional waves, the position of the interface will show oscillations along the azimuthal coordinate; the local azimuthal slope and, hence, the incidence angle, will also undergo strong variations.

Figure 11a shows a fragment of uniform annular film with imposed small azimuthal perturbation of the film surface. The perturbation is selected in the form of one period of cosine function with amplitude of 0.5 mm (10% of film thickness) and azimuthal size of 6 mm. The position of this perturbation is characterized by the y-coordinate of its center, L. The perturbation does not affect the rays obtaining the true film image (see ray 1 passing at $z = -20.5$ mm), and the rays hitting the interface between the laser sheet and the perturbation (ray 2 at $z = -19.5$ mm). However, the rays reflected from the perturbed area may not reach the illuminated section of the film due local change of the interface slope (ray 3 at $z = -18.5$ mm). The rays reflected near the outer edge or outside the perturbation (ray 4 at $z = -17.5$ mm), may again hit the illuminated section and create a bright patch in the false film image.

In general, a ray may create a bright spot in the false film image region if the conditions below are satisfied:

(1)   The reflection coefficient of a ray hitting the interface is close or equal to unity. This means that the angle of incidence, taking into account the local change of interface slope due to the perturbation, is close to the TIR-angle or exceeds it.
(2)   The reflected ray crosses the laser-illuminated section of the film.
(3)   There is no interface on the reflected ray's optical path between the reflection point and the illuminated section.

Technically, there could appear bright spots in more complicated cases. For example, the reflected ray hitting the interface (condition 3 is broken) may get totally reflected again into the illuminated section. Alternatively, the ray hitting a local steep slope may pass through the interface (condition 1 is broken) and be refracted to penetrate again into the illuminated film section. Such cases were neglected in the present analysis: if any of the conditions (1–3) is not satisfied, the corresponding camera pixel level is set to zero; if all the conditions are satisfied, it is set to unity.

Figure 11b shows how the raw PLIF brightness profile would look at different y-positions of the perturbation. If $L = 0$, the upper edge of the false film image does not change; however, its bottom part will be darkened by the right edge of the perturbation. It should be noted that in this case the true film thickness in the illuminated section is 10% higher, but it is not detected by PLIF. In the medium range of L (5–8 mm), part of the false image will be darkened, reducing the apparent film thickness. There will also appear an additional bright spot above the film image, creating an illusion of a droplet above the film surface or a liquid ligament/overturning wave if this bright area is still connected to the true/false film image at neighboring x. This spot may be also shifted far from the area of the false film image, strengthening the illusion of a droplet flying above the interface. If the perturbation is located outside the false film area ($L \geq 10$ mm), it does not affect the false image.

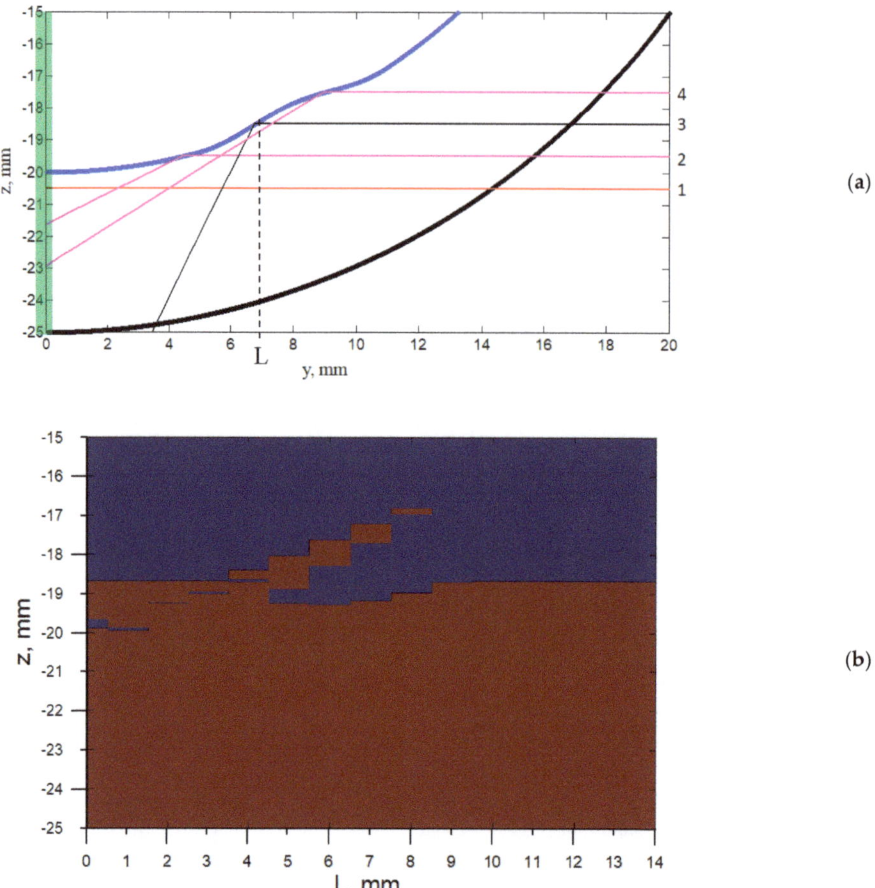

Figure 11. (a) The shape of annular liquid film with a small circumferential perturbation and the rays' reflection at different heights. (b) Modeled PLIF-image for different y-positions of the perturbation. The calculations are made within the present review, as a development of the analysis first presented in [53].

Thus, the presence of even a small and smooth perturbation effectively "shatters" the "mirror" surface creating the false film image. Parts of this surface stop acquiring fluorescence due to altering the path of reflected light, which leads to temporal disappearance of the false film image and, in the case of severe perturbations, part of the true film image as well. This effect can make an illusion of disappearance of part of a wave, which can be further erroneously interpreted as liquid entrainment (as it was done in [51]). Furthermore, some "shards" of the "shattered mirror" may still collect fluorescence, creating isolated bright spots in PLIF image. These spots can be erroneously interpreted as liquid droplets or structures connected to the film (liquid ligaments, overturning waves) and attributed to entrainment events, as it was done in [49].

In real annular flow with entrainment, the film surface is strongly agitated by the gas shear, being covered with three-dimensional ripples on top of disturbance waves and on the base film. These waves may have a variety of amplitude and transverse size values, introducing more severe perturbations of the film surface compared to the one analyzed above. The disturbance waves themselves are not uniform in amplitude across the pipe circumference [31]. Their fronts (as well as those of ripples) may be curved or slanted in the x-y plane [46]; as a result, additional azimuthal slopes may appear in the y-z cross-section studied by PLIF.

To summarize, optical artifacts created by the "mirror" perturbations severely distort the PLIF images, making the PLIF technique inappropriate for studying the entrainment process. It should be also noted that the shadow visualization in the x-z plane inside circular pipes [54–56] is inapplicable to studying entrainment for similar reasons: neither droplets nor three-dimensional liquid structures in the investigated x-z plane will be seen due to the total internal reflection of light at the interface between the investigated pipe section and the camera. Besides, only the lowest film thickness value will be measurable in a given cross-section of the pipe.

5. Brightness-Based Laser-Induced Fluorescence (BBLIF) Technique: Enhancing the Visualization Studies

The BBLIF technique also employs the principle of laser-induced fluorescence, but the film thickness is recovered in a different manner compared to the PLIF method. Namely, both the camera and the laser are aimed at the same fragment of the x-y plane. Each pixel of the camera collects the fluorescence along a ray approximately normal to the wall and passing through the film. This value of fluorescence intensity is converted into local film thickness [46]. A matrix of film thickness, $h(x,y)$, can be obtained at each time instant.

Application of this technique allows one to clarify the spatiotemporal dynamics and three-dimensional shape of the ripples on liquid film. Experiments in downward flow in a vertical pipe have shown [57] that all the ripples are generated at the rear slopes of disturbance waves. Depending on the relative x-coordinate of the point of inception, a ripple may either lag behind the "parent" disturbance wave and travel with low speed over the base film ("slow ripples") or accelerate and travel with high speed over the disturbance wave ("fast ripples"). The same spatiotemporal evolution of ripples was observed in a horizontal rectangular duct [46]. At low liquid and high gas flow rates, where entrainment is not observed, fast ripples do not exist: only the "primary" waves generating slow "secondary" waves can be seen, and the crests of the primary waves remain smooth. This observation confirms that ripples on top of disturbance waves are necessary for liquid entrainment. The fast ripples often disappear from view when they reach the front of the disturbance wave or even before that. This happens due to shattering of the fast ripple into droplets as first described in [38]. Moreover, the entrained droplets are seen in the BBLIF data since they also contain the fluorophore; the spatiotemporal trajectory of such a droplet starts where the trajectory of the fast ripple disappears [58]. Figure 12 shows examples of the spatiotemporal evolution of disturbance waves, fast and slow ripples, and the entrained droplets.

Experiments [46] show that the fast ripples are three-dimensional and have horseshoe-shaped fronts in the x-y plane. These ripples are placed in a staggered order in several rows (the number of rows is mainly defined by the ratio of the longitudinal size of a disturbance wave to that of a fast ripple). The transitional liquid structures are distinguishable in BBLIF data since they also contain fluorophore. It was observed that the bag break-up occurs due to the deformation and breaking of the whole front of a fast ripple (Figure 13a). At the same time, ligament break-up occurs at the junctions of the side edges of neighboring fast ripples, where a longitudinally-oriented liquid hump is formed. The gas shear strips the liquid from this hump and forms a ligament oriented in the flow direction (Figure 13b). Indeed, the ligaments are most frequently observed between fast ripples.

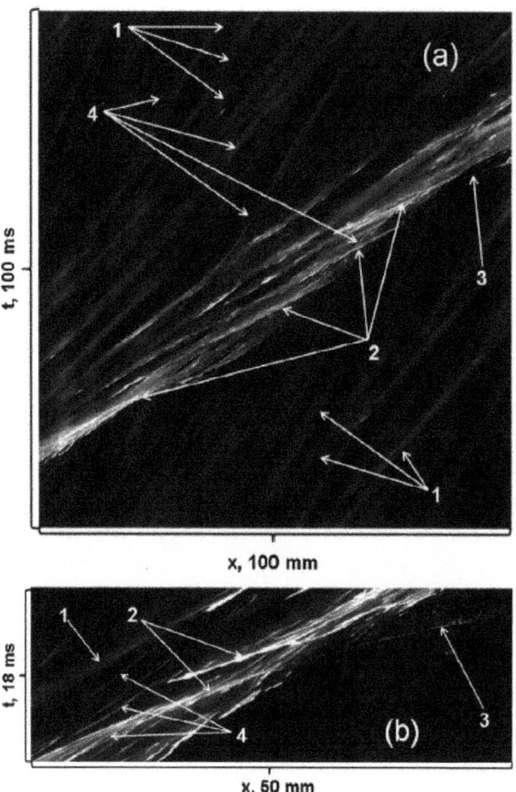

Figure 12. Spatiotemporal evolution of disturbance waves and ripples on gas-sheared liquid film in a horizontal rectangular duct [46]. Local brightness of the image corresponds to local instantaneous film thickness. A disturbance wave is seen as an inclined bright non-uniform stripe. The numbers denote: slow ripples (1), fast ripples (2), entrained droplet (3), entrapped bubbles (4). (**a**) A large fragment of h(x,t) matrix showing a disturbance wave and its surroundings; (**b**) A smaller magnified fragment showing the structure of a disturbance wave.

Both the longitudinal and transverse size of fast ripples decreases as the gas speed is increased [46]. Thus, the amount of entrainment events of both types per unit surface area is expected to grow with gas speed. The relative contribution of the bag break-up mechanism into total liquid entrainment is expected to get smaller compared to that of the ligament break-up at high gas flow rates. The reason is that each broken ripple becomes narrower; hence, less liquid is torn from film surface via bag break-up. At the same time, the number of junctions where ligaments are formed increase with gas speed.

For a complete description of the entrainment events, the BBLIF can be combined with simultaneous visual observations in the *x-z* plane. An additional camera looking from the side can be employed and synchronized with the BBLIF camera. No additional light source is required for the second camera since it may also use the same laser illumination. Within this approach, all instantaneous coordinates of the entrained droplets may be obtained, together with the shape of the transitional structures and overturning waves. A detailed description of such a stereoscopic approach is given in [59]. Though that paper mainly deals with impact of previously entrained droplets on the film surface, the same data can be used to investigate the entrainment events. Figure 14 shows an example of an entrainment event seen by two cameras simultaneously.

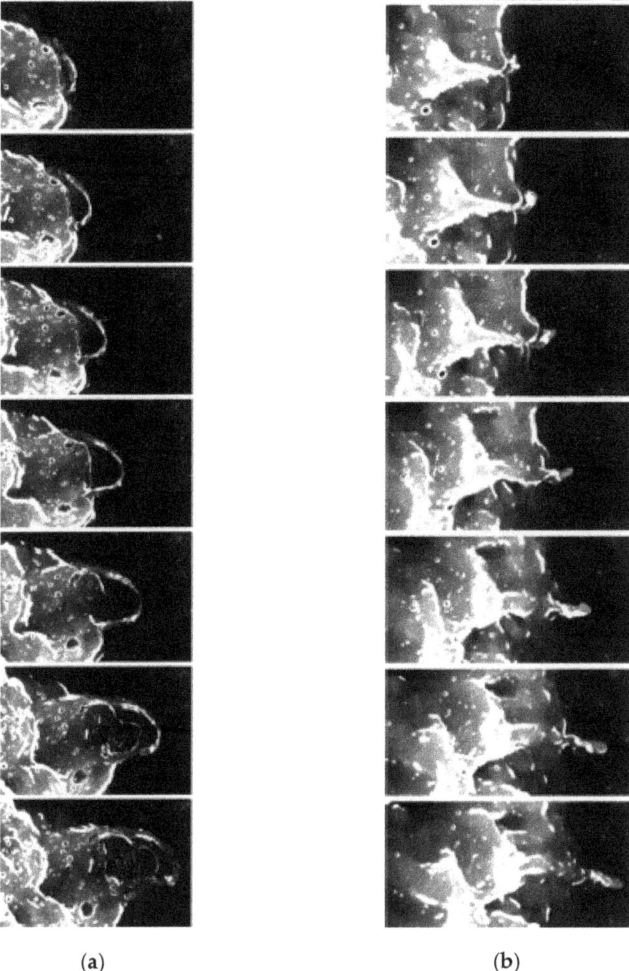

(a)   (b)

Figure 13. Bag break-up (**a**) and ligament break-up (**b**) imaged in the *x-y* plane by BBLIF technique. Gas-sheared liquid film in a horizontal rectangular duct [46]. Image size is 20 mm × 13 mm (**a**) and 20 mm × 10 mm (**b**). Time step is 0.5 ms.

The BBLIF technique has its own optical artifacts. At steep interface slopes, approaching and exceeding the TIR-angle, defined approximately as $dh/dx \geq 1$ and/or $dh/dy \geq 1$, the coefficient of the reflection of light from the interface is of order of unity, which is unaccounted in the film thickness calculation process. This reflection leads to the appearance of narrow but high-amplitude non-physical peaks of film thickness exactly on the steep slopes (see, e.g., [52]). This phenomenon does not affect the film thickness measurements on the base film, or rear slopes of disturbance waves, or even in the fast ripple area before the intensive deformation into a liquid bag or ligament begins, so it is possible to quantitatively investigate the initial stages of the entrainment process. After the deformation starts, all kinds of interface slopes, including those over 90^0 with overturning, are possible.

The real height of the deformed fast ripples and more complex structures may be extracted from the images obtained by the second camera in a stereoscopic approach as described above. It must be noted that the optical distortions at steep slopes are inherent not only to BBLIF, but to the majority of intensity-based optical techniques, including light-absorption techniques [60,61], pigment luminance technique [62], reflection-based [63], and Schlieren-based methods [64–66].

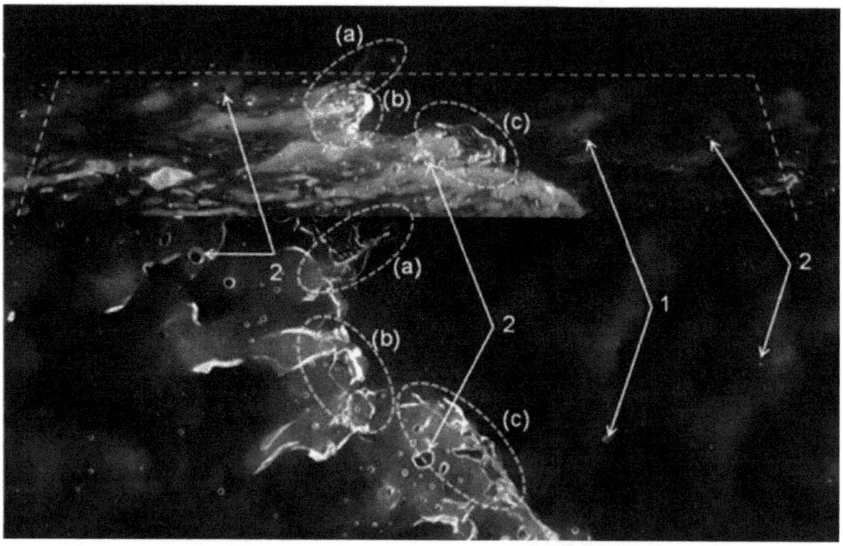

Figure 14. An example of simultaneous visualization of entrainment events in the x-z plane (**top image**) and in the x-y plane by the BBLIF-technique (**bottom image**). Entrained droplet (1) and entrapped bubbles (2) are shown for both images. Fast ripples (a–c) are shown in both images; ripple (a) is being shattered into droplets, and ripples (b,c) are in the process of growth and overturning [59]. The imaged area is 52 mm by 20 mm. The dashed trapezium shows the borders of the image viewed in the x-y plane.

6. X-ray Technique: Towards Distortion-Free Experiments

The optical distortions described above appear due to the strong difference in refractive indices of liquid and gas, which manifest themselves at steep local slopes of the interface. The only way to safely avoid such issues is to eliminate this difference. To the best of our knowledge, it cannot be solved by the selection of liquids and gas, since the difference remains large enough for any combination of gas and liquid. However, it can be eliminated by using non-optical radiation, namely, X-rays, as the main measurement tool. For X-rays, the refractive index in both liquid and gas is very close to unity. At the same time, the attenuation rate of the radiation intensity is different in gas and liquid media, so the spatial distribution of the two phases can be distinguished by comparing the intensity of rays passing through the media.

The pioneering X-ray visualization studies of annular flow [67,68] have shown the ability of this technique to overcome the optical obscuring of the objects in the gas core by agitated liquid film on pipe walls. In particular, complex liquid structures ("wisps") were observed in the gas core at large liquid flow rates. Most likely, such structures appear due to the same bag and ligament break-up mechanisms. However, in this case bags and ligaments are not entirely scattered into droplets; instead, large amorphous chunks of liquid are detached from the film surface and carried by the gas stream.

A recent review of applications of the X-ray tomography technique to various multiphase flows can be found in [69]. A number of researchers [70–72] used X-ray radiography to obtain cross-sectional void fraction distributions in stratified and annular gas-liquid flow in pipes. Usually, two source-detector systems oriented along the y and z axes were used simultaneously to reconstruct liquid phase distribution. In [73], the system rotated at low speed to obtain time-averaged circumferential and longitudinal film thickness profiles before and after an orifice in the pipe. However, studies of high-speed transient processes such as liquid entrainment are rare. In [74], a broad-band X-ray source was used for observation of the dynamics of an annular liquid sheet, detached liquid droplets, and entrapped air bubbles in an atomizer (Figure 15). In this work, outstanding spatial and temporal resolution was achieved. Nonetheless, the attenuation rate depends strongly on the wavelength, so the quantitative reconstruction of liquid phase distribution is impossible using a broadband X-ray source; a monochromatic X-ray source is required instead.

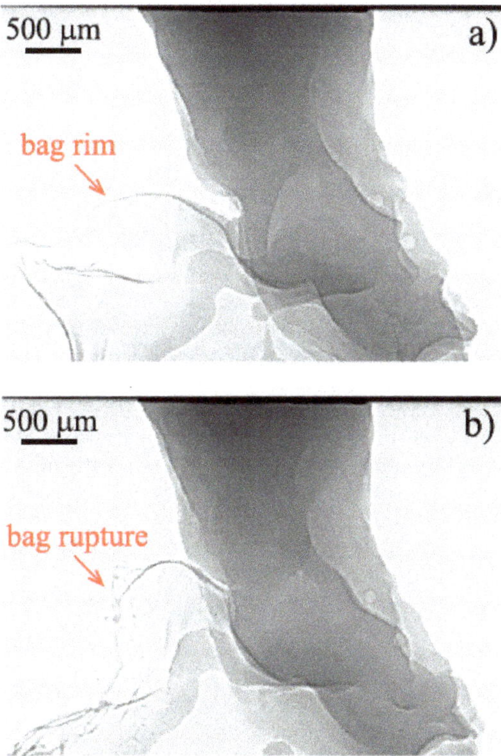

Figure 15. Deformation of annular liquid sheet with droplet entrainment in a two-fluid coaxial atomizer studied with broadband X-ray radiography [74]. Subfigures (**a**,**b**) show consecutive frames separated by 1 ms time interval.

An alternative way to avoid strong distortions is to use near-infrared radiation (NIR); in this range, the liquid refractive index is also reduced, though it does not exactly reach unity. This technique is also based on light attenuation by liquid and can be used to reconstruct instantaneous liquid distribution in the flow (see, e.g., [75,76]).

Application of X-ray or NIR attenuation techniques to entrainment studies have the same requirements as those for visible-light techniques, namely: obtaining instantaneous images of the flow in at least two planes; at least the measurements in the x-y plane must allow one to quantitatively reconstruct the local instantaneous film thickness; temporal and spatial resolution of the method must be less than the smallest temporal and spatial scale of the phenomenon under study. The advantages of these approaches are the possibility to investigate entrainment for annular flow in pipes and less risk of distortions produced by refraction and reflection.

7. Discussion and Conclusions

The physical mechanism of liquid entrainment from disturbance waves in annular gas-liquid flow is a long-discussed matter. A number of hypotheses concerning this mechanism were proposed based on experimental observations. The hypotheses on shearing-off the disturbance wave crests and undercutting of disturbance waves are based on the concept of smooth disturbance waves, which, in turn, stems from film thickness records obtained by low-resolution measurement techniques. Direct visualization studies of the entrainment process carried out in three different planes and BBLIF studies of spatiotemporal evolution of film thickness show that the surface of disturbance waves is unstable and covered with ripples of much smaller length but comparable amplitude. These ripples are generated at the rear slopes of disturbance waves and travel with high speed towards the fronts of disturbance waves. The ripples are broken into droplets by two possible mechanisms known as bag break-up and ligament break-up. The bag break-up occurs when a three-dimensional ripple wave grows and its whole front is stretched and blown up by the gas stream. The ligament break-up occurs at the junctions of fast ripples where liquid is stripped by the gas shear and a forward-oriented jet is formed. Similar processes are found on the surface of a deep layer of liquid when sea storm conditions are modeled experimentally. The typical scale of the bags on deep water is quite close to that on thin films.

Comprehensive studies require reconstruction of instantaneous film thickness in the plane of the wall to analyze the wave processes leading to entrainment, as well as detecting the position and size of entrained droplets in the same plane. Simultaneously, side-view visualization is needed to investigate the parameters of bags, ligaments, and entrained droplets, to finally gain three-dimensional characterization of these objects. With visible-light techniques, proper side-view visualization is possible only when there is no liquid film between the studied object and the camera, e.g., when liquid film flows on a single flat plane or on a convex surface. In annular pipe flows, a curved and agitated film surface on the side walls creates strong optical distortions, especially affecting the PLIF-technique. Such a challenge may be overcome by X-ray or near-infrared attenuation techniques, though no studies of entrainment with an appropriate problem statement have been carried out yet in annular flow. A brief summary of the applicability of experimental methods to the investigation of fast ripples and transitional structures is given in Table 1.

Table 1. Summary of limitations and capabilities of experimental techniques related to investigation of entrainment process.

Experimental Technique	Fast Ripples		Bags, Ligaments, and Droplets	
	Domain	Distortions	x-y Plane	x-z Plane
Non-optical techniques *	x-y	Smoothened out due to coarse resolution	No	No
Shadow visualization	x	Underestimation due to 3D waves	Yes	Yes
Planar Laser-Induced Fluorescence	x or y	Overestimation due to mirror effect; underestimation due to 3D waves	No	No
Brightness-Based Laser-Induced Fluorescence	x-y	Local overestimation at steep slopes	Yes	Compatible with shadow visualization or broad-band X-ray
X-ray and NIR techniques	x-y	No	Yes	Compatible with shadow visualization or broad-band X-ray

Note: * Conductance, capacitance, etc.

Finally, it must be noted that qualitatively the mechanisms of liquid entrainment are more or less clear at present. On the other hand, the quantitative data on the entrainment process are scarce: properties of fast ripples [46] and liquid bags [45] were measured only separately, and in very different conditions. The future studies of the entrainment process should report a quantitative analysis of the whole process, starting from the formation of fast ripples, formation and break-up of bags and ligaments, and creation of droplets.

Funding: The main analysis was carried out in the Institute of Applied Physics under support of Russian Science Foundation (project 21-19-00755). Analysis of PLIF distortions was carried out under the State Contract of the Kutateladze Institute of Thermophysics (Registration number 121031100246-5).

Conflicts of Interest: The author declares no conflict of interest.

References

1. Van Nimwegen, A.T.; Portela, L.M.; Henkes, R.A.W.M. Modelling of upwards gas-liquid annular and churn flow with surfactants in vertical pipes. *Int. J. Multiph. Flow* **2018**, *105*, 1–14. [CrossRef]
2. Tuttle, S.G.; Fisher, B.T.; Kessler, D.A.; Pfützner, C.J.; Jacob, R.J.; Skiba, A.W. Petroleum wellhead burning: A review of the basic science for burn efficiency prediction. *Fuel* **2021**, *303*, 121279. [CrossRef]
3. Le Corre, J.M. Phenomenological model of disturbance waves in annular two-phase flow. *Int. J. Multiph. Flow* **2022**, *151*, 104057. [CrossRef]
4. Sandá, A.; Moya, S.L.; Valenzuela, L. Modelling and simulation tools for direct steam generation in parabolic-trough solar collectors: A review. *Renew. Sust. Energy Rev.* **2019**, *113*, 109226. [CrossRef]
5. Sawant, P.; Ishii, M.; Mori, M. Droplet entrainment correlation in vertical upward co-current annular two-phase flow. *Nucl. Eng. Des.* **2008**, *238*, 1342–1352. [CrossRef]
6. Azzopardi, B.J. Turbulence modification in annular gas/liquid flow. *Int. J. Multiph. Flow* **1999**, *25*, 945–955. [CrossRef]
7. Karami, H.; Pereyra, E.; Torres, C.F.; Sarica, C. Droplet entrainment analysis of three-phase low liquid loading flow. *Int. J. Multiph. Flow* **2017**, *89*, 45–56. [CrossRef]
8. Lopes de Bertodano, M.; Assad, A.; Beus, S.G. Experiments for entrainment rate of droplets in the annular regime. *Int. J. Multiph. Flow* **2001**, *27*, 685–699. [CrossRef]
9. Pitton, E.; Ciandri, P.; Margarone, M.; Andreussi, P. An experimental study of stratified–dispersed flow in horizontal pipes. *Int. J. Multiph. Flow* **2014**, *67*, 92–103. [CrossRef]
10. Azzopardi, B.J.; Zaidi, S.H. Determination of entrained fraction in vertical annular gas/liquid flow. *J. Fluids Eng.* **2000**, *122*, 146–150. [CrossRef]
11. van't Westende, J.M.C.; Kemp, H.K.; Belt, R.J.; Portela, L.M.; Mudde, R.F.; Oliemans, R.V.A. On the role of droplets in cocurrent annular and churn-annular pipe flow. *Int. J. Multiph. Flow* **2007**, *33*, 595–615. [CrossRef]
12. van Eckeveld, A.C.; Gotfredsen, E.; Westerweel, J.; Poelma, C. Annular two-phase flow in vertical smooth and corrugated pipes. *Int. J. Multiph. Flow* **2018**, *109*, 150–163. [CrossRef]
13. Cioncolini, A.; Thome, J.R. Entrained liquid fraction prediction in adiabatic and evaporating annular two-phase flow. *Nucl. Eng. Des.* **2012**, *243*, 200–213. [CrossRef]
14. Aliyu, A.M.; Almabrok, A.A.; Baba, Y.D.; Archibong, A.E.; Lao, L.; Yeung, H.; Kim, K.C. Prediction of entrained droplet fraction in co-current annular gas–liquid flow in vertical pipes. *Exp. Therm. Fluid Sci.* **2017**, *85*, 287–304. [CrossRef]
15. Han, H.; Gabriel, K. A numerical study of entrainment mechanism in axisymmetric annular gas-liquid flow. *J. Fluids Eng.* **2007**, *129*, 293–301. [CrossRef]
16. Kumar, P.; Das, A.K.; Mitra, S.K. Physical understanding of gas-liquid annular flow and its transition to dispersed droplets. *Phys. Fluids* **2016**, *28*, 072101. [CrossRef]
17. Sato, Y.; Niceno, B. Large eddy simulation of upward co-current annular boiling flow using an interface tracking method. *Nucl. Eng. Des.* **2017**, *321*, 69–81. [CrossRef]
18. Zhou, R.; Xia, T.; Wei, A.; Zhang, X. Investigation on liquid nitrogen and vaporous nitrogen countercurrent flow considering droplet entrainment. *Cryogenics* **2020**, *109*, 103125. [CrossRef]
19. Holowach, M.J.; Hochreiter, L.E.; Cheung, F.B. A model for droplet entrainment in heated annular flow. *Int. J. Heat Fluid Flow* **2002**, *23*, 807–822. [CrossRef]
20. Ryu, S.H.; Park, G.C. A droplet entrainment model based on the force balance of an interfacial wave in two-phase annular flow. *Nucl. Eng. Des.* **2011**, *241*, 3890–3897. [CrossRef]
21. Liu, L.; Bai, B. Generalization of droplet entrainment rate correlation for annular flow considering disturbance wave properties. *Chem. Eng. Sci.* **2017**, *164*, 279–291. [CrossRef]
22. Cherdantsev, A.V. Overview of physical models of liquid entrainment in annular gas-liquid flow. *AIP Conf. Proc.* **2018**, *1939*, 020006. [CrossRef]
23. Chu, K.J.; Dukler, A.E. Statistical characteristics of thin, wavy films: Part II. Studies of the substrate and its wave structure. *AIChE J.* **1974**, *20*, 695–706. [CrossRef]
24. Zhao, Y.; Markides, C.N.; Matar, O.K.; Hewitt, G.F. Disturbance wave development in two-phase gas–liquid upwards vertical annular flow. *Int. J. Multiph. Flow* **2013**, *55*, 111–129. [CrossRef]

25. Rivera, Y.; Muñoz-Cobo, J.L.; Cuadros, J.L.; Berna, C.; Escrivá, A. Experimental study of the effects produced by the changes of the liquid and gas superficial velocities and the surface tension on the interfacial waves and the film thickness in annular concurrent upward vertical flows. *Exp. Therm. Fluid Sci.* **2021**, *120*, 110224. [CrossRef]
26. Miya, M.; Woodmansee, D.E.; Hanratty, T.J. A model for roll waves in gas-liquid flow. *Chem. Eng. Sci.* **1971**, *26*, 1915–1931. [CrossRef]
27. Han, H.; Zhu, Z.; Gabriel, K. A study on the effect of gas flow rate on the wave characteristics in two-phase gas–liquid annular flow. *Nucl. Eng. Des.* **2006**, *236*, 2580–2588. [CrossRef]
28. Wang, G.; Dang, Z.; Ishii, M. Wave structure and velocity in vertical upward annular two-phase flow. *Exp. Therm. Fluid Sci.* **2021**, *120*, 110205. [CrossRef]
29. Ishii, M.; Grolmes, M.A. Inception criteria for droplet entrainment in two-phase concurrent film flow. *AIChE J.* **1975**, *21*, 308–318. [CrossRef]
30. Sekoguchi, K.; Takeishi, M.; Ishimatsu, T. Interfacial structure in vertical upward annular flow. *Phys. Chem. Hydrodyn.* **1985**, *6*, 239–255.
31. Belt, R.J.; Van't Westende, J.M.C.; Prasser, H.M.; Portela, L.M. Time and spatially resolved measurements of interfacial waves in vertical annular flow. *Int. J. Multiph. Flow* **2010**, *36*, 570–587. [CrossRef]
32. Fershtman, A.; Barnea, D.; Shemer, L. Wave identification in upward annular flow-a focus on ripple characterization. *Int. J. Multiph. Flow* **2021**, *137*, 103560. [CrossRef]
33. Vieira, R.E.; Parsi, M.; Torres, C.F.; McLaury, B.S.; Shirazi, S.A.; Schleicher, E.; Hampel, U. Experimental characterization of vertical gas–liquid pipe flow for annular and liquid loading conditions using dual wire-mesh sensor. *Exp. Therm. Fluid Sci.* **2015**, *64*, 81–93. [CrossRef]
34. Abdulkadir, M.; Hernandez-Perez, V.; Kwatia, C.A.; Azzopardi, B.J. Interrogating flow development and phase distribution in vertical and horizontal pipes using advanced instrumentation. *Chem. Eng. Sci.* **2018**, *186*, 152–167. [CrossRef]
35. Hazuku, T.; Takamasa, T.; Matsumoto, Y. Experimental study on axial development of liquid film in vertical upward annular two-phase flow. *Int. J. Multiph. Flow* **2008**, *34*, 111–127. [CrossRef]
36. Hurlburt, E.T.; Newell, T.A. Optical measurement of liquid film thickness and wave velocity in liquid film flows. *Exp. Fluids* **1996**, *21*, 357–362. [CrossRef]
37. Zhou, D.W.; Gambaryan-Roisman, T.; Stephan, P. Measurement of water falling film thickness to flat plate using confocal chromatic sensoring technique. *Exp. Therm. Fluid Sci.* **2009**, *33*, 273–283. [CrossRef]
38. Woodmansee, D.E.; Hanratty, T.J. Mechanism for the removal of droplets from a liquid surface by a parallel air flow. *Chem. Eng. Sci.* **1969**, *24*, 299–307. [CrossRef]
39. Zhang, Z.; Wang, Z.; Liu, H.; Gao, Y.; Li, H.; Sun, B. Experimental study on bubble and droplet entrainment in vertical churn and annular flows and their relationship. *Chem. Eng. Sci.* **2019**, *206*, 387–400. [CrossRef]
40. Azzopardi, B.J. *Mechanisms of Entrainment in Annular Two Phase Flow*; AERE-R 11068; UKAEA Atomic Energy Research Establishment: Harwell, UK, 1983.
41. Badie, S.; Lawrence, C.J.; Hewitt, G.F. Axial viewing studies of horizontal gas–liquid flows with low liquid loading. *Int. J. Multiph. Flow* **2001**, *27*, 1259–1269. [CrossRef]
42. Lecoeur, N. Interfacial Behaviour in Stratified and Stratifying Annular Flows. Ph.D. Thesis, Imperial College, London, UK, 2013. [CrossRef]
43. Pham, S.H.; Kawara, Z.; Yokomine, T.; Kunugi, T. Detailed observations of wavy interface behaviors of annular two-phase flow on rod bundle geometry. *Int. J. Multiph. Flow* **2014**, *59*, 135–144. [CrossRef]
44. Troitskaya, Y.; Kandaurov, A.; Ermakova, O.; Kozlov, D.; Sergeev, D.; Zilitinkevich, S. Bag-breakup fragmentation as the dominant mechanism of sea-spray production in high winds. *Sci. Rep.* **2017**, *7*, 1614. [CrossRef] [PubMed]
45. Troitskaya, Y.; Kandaurov, A.; Ermakova, O.; Kozlov, D.; Sergeev, D.; Zilitinkevich, S. The "bag breakup" spume droplet generation mechanism at high winds. Part I: Spray generation function. *J. Phys. Oceanogr.* **2018**, *48*, 2167–2188. [CrossRef]
46. Cherdantsev, A.V.; Hann, D.B.; Azzopardi, B.J. Study of gas-sheared liquid film in horizontal rectangular duct using high-speed LIF technique: Three-dimensional wavy structure and its relation to liquid entrainment. *Int. J. Multiph. Flow* **2014**, *67*, 52–64. [CrossRef]
47. Schubring, D.; Ashwood, A.C.; Shedd, T.A.; Hurlburt, E.T. Planar laser-induced fluorescence (PLIF) measurements of liquid film thickness in annular flow. Part I: Methods and data. *Int. J. Multiph. Flow* **2010**, *36*, 815–824. [CrossRef]
48. Farias, P.S.C.; Martins, F.J.W.A.; Sampaio, L.E.B.; Serfaty, R.; Azevedo, L.F.A. Liquid film characterization in horizontal, annular, two-phase, gas–liquid flow using time-resolved laser-induced fluorescence. *Exp. Fluids* **2012**, *52*, 633–645. [CrossRef]
49. Zadrazil, I.; Matar, O.K.; Markides, C.N. An experimental characterization of downwards gas–liquid annular flow by laser-induced fluorescence: Flow regimes and film statistics. *Int. J. Multiph. Flow* **2014**, *60*, 87–102. [CrossRef]
50. Xue, T.; Li, Z.; Li, C.; Wu, B. Measurement of thickness of annular liquid films based on distortion correction of laser-induced fluorescence imaging. *Rev. Sci. Instrum.* **2019**, *90*, 033103. [CrossRef]
51. Liu, J.; Xue, T. Experimental investigation of liquid entrainment in vertical upward annular flow based on fluorescence imaging. *Prog. Nucl. Energy* **2022**, *152*, 104383. [CrossRef]
52. Cherdantsev, A.V.; An, J.S.; Charogiannis, A.; Markides, C.N. Simultaneous application of two laser-induced fluorescence approaches for film thickness measurements in annular gas-liquid flows. *Int. J. Multiph. Flow* **2019**, *119*, 237–258. [CrossRef]

53. Häber, T.; Gebretsadik, M.; Bockhorn, H.; Zarzalis, N. The effect of total reflection in PLIF imaging of annular thin films. *Int. J. Multiph. Flow* **2015**, *76*, 64–72. [CrossRef]

54. Pan, L.M.; He, H.; Ju, P.; Hibiki, T.; Ishii, M. Experimental study and modeling of disturbance wave height of vertical annular flow. *Int. J. Heat Mass Transf.* **2015**, *89*, 165–175. [CrossRef]

55. Isaenkov, S.V.; Cherdantsev, A.V.; Vozhakov, I.S.; Cherdantsev, M.V.; Arkhipov, D.G.; Markovich, D.M. Study of primary instability of thick liquid films under strong gas shear. *Int. J. Multiph. Flow* **2019**, *111*, 62–81. [CrossRef]

56. Lin, R.; Wang, K.; Liu, L.; Zhang, Y.; Dong, S. Study on the characteristics of interfacial waves in annular flow by image analysis. *Chem. Eng. Sci.* **2020**, *212*, 115336. [CrossRef]

57. Alekseenko, S.; Antipin, V.; Cherdantsev, A.; Kharlamov, S.; Markovich, D. Two-wave structure of liquid film and wave interrelation in annular gas-liquid flow with and without entrainment. *Phys. Fluids* **2009**, *21*, 061701. [CrossRef]

58. Alekseenko, S.V.; Cherdantsev, A.V.; Markovich, D.M.; Rabusov, A.V. Investigation of droplets entrainment and deposition in annular flow using LIF technique. *At. Sprays* **2014**, *24*, 193–222. [CrossRef]

59. Cherdantsev, A.V.; Sinha, A.; Hann, D.B. Studying the impacts of droplets depositing from the gas core onto a gas-sheared liquid film with stereoscopic BBLIF technique. *Int. J. Multiph. Flow* **2022**, *150*, 104033. [CrossRef]

60. Mendez, M.A.; Scheid, B.; Buchlin, J.M. Low Kapitza falling liquid films. *Chem. Eng. Sci.* **2017**, *170*, 122–138. [CrossRef]

61. Yang, H.; Guo, Y.; Li, C.; Tao, J.; Zhang, Y.; Wu, W.; Su, M. Development of a two-line DLAS sensor for liquid film measurement. *Spectrochim. Acta A Mol. Biomol. Spectrosc.* **2020**, *224*, 117420. [CrossRef]

62. Ohba, K.; Nagae, K. Characteristics and behavior of the interfacial wave on the liquid film in a vertically upward air-water two-phase annular flow. *Nucl. Eng. Des.* **1993**, *141*, 17–25. [CrossRef]

63. Åkesjö, A.; Gourdon, M.; Vamling, L.; Innings, F.; Sasic, S. Hydrodynamics of vertical falling films in a large-scale pilot unit–a combined experimental and numerical study. *Int. J. Multiph. Flow* **2017**, *95*, 188–198. [CrossRef]

64. Vinnichenko, N.A.; Pushtaev, A.V.; Plaksina, Y.Y.; Uvarov, A.V. Measurements of liquid surface relief with moon-glade background oriented Schlieren technique. *Exp. Therm. Fluid Sci.* **2020**, *114*, 110051. [CrossRef]

65. Kofman, N.; Mergui, S.; Ruyer-Quil, C. Characteristics of solitary waves on a falling liquid film sheared by a turbulent counter-current gas flow. *Int. J. Multiph. Flow* **2017**, *95*, 22–34. [CrossRef]

66. Cherdantsev, A.V.; Gavrilov, N.V.; Ermanyuk, E.V. Study of initial stage of entry of a solid sphere into shallow liquid with Synthetic Schlieren technique. *Exp. Therm. Fluid Sci.* **2021**, *125*, 110375. [CrossRef]

67. Bennett, B.A.W.; Hewitt, G.F.; Kearsey, H.A.; Keeys, R.K.F.; Lacey, P.M.C. Paper 5: Flow visualization studies of boiling at high pressure. In *Proceedings of the Institution of Mechanical Engineers, Conference Proceedings*; SAGE: London, UK, 1965; Volume 180, pp. 260–283.

68. Hewitt, G.F.; Roberts, D.N. *Studies of Two-Phase Flow Patterns by Simultaneous X-ray and Flash Photography*; AERE-M 2159; Atomic Energy Research Establishment: Harwell, UK, 1969.

69. Aliseda, A.; Heindel, T.J. X-ray flow visualization in multiphase flows. *Ann. Rev. Fluid Mech.* **2021**, *53*, 543–567. [CrossRef]

70. Hu, B.; Langsholt, M.; Liu, L.; Andersson, P.; Lawrence, C. Flow structure and phase distribution in stratified and slug flows measured by X-ray tomography. *Int. J. Multiph. Flow* **2014**, *67*, 162–179. [CrossRef]

71. Skjæraasen, O.; Kesana, N.R. X-ray measurements of thin liquid films in gas–liquid pipe flow. *Int. J. Multiph. Flow* **2020**, *131*, 103391. [CrossRef]

72. Bieberle, A.; Shabestary, A.M.; Geissler, T.; Boden, S.; Beyer, M.; Hampel, U. Flow morphology and heat transfer analysis during high-pressure steam condensation in an inclined tube part I: Experimental investigations. *Nucl. Eng. Des.* **2020**, *361*, 110553. [CrossRef]

73. Porombka, P.; Boden, S.; Lucas, D.; Hampel, U. Horizontal annular flow through orifice studied by X-ray microtomography. *Exp. Fluids* **2021**, *62*, 5. [CrossRef]

74. Machicoane, N.; Bothell, J.K.; Li, D.; Morgan, T.B.; Heindel, T.J.; Kastengren, A.L.; Aliseda, A. Synchrotron radiography characterization of the liquid core dynamics in a canonical two-fluid coaxial atomizer. *Int. J. Multiph. Flow* **2019**, *115*, 1–8. [CrossRef]

75. Dupont, J.; Mignot, G.; Prasser, H.-M. Two-dimensional mapping of falling water film thickness with near-infrared attenuation. *Exp. Fluids* **2015**, *56*, 90. [CrossRef]

76. Wang, C.; Zhao, N.; Fang, L.; Zhang, T.; Feng, Y. Void fraction measurement using NIR technology for horizontal wet-gas annular flow. *Exp. Therm. Fluid Sci.* **2016**, *76*, 98–108. [CrossRef]

Review

Two-Phase Annular Flow in Vertical Pipes: A Critical Review of Current Research Techniques and Progress

Yunpeng Xue [1,*], Colin Stewart [2], David Kelly [2], David Campbell [2] and Michael Gormley [2,*]

1   Singapore-ETH Centre, Future Resilient Systems, ETH Zurich, CREATE Campus, 1 CREATE Way, #06-01 CREATE Tower, Singapore 138602, Singapore
2   Institute for Sustainable Building Design, Heriot-Watt University, Edinburgh EH14 4AS, UK
*   Correspondence: yunpeng.xue@sec.ethz.ch (Y.X.); m.gormley@hw.ac.uk (M.G.)

Abstract: Two-phase annular flow in vertical pipes is one of the most common and important flow regimes in fluid mechanics, particularly in the field of building drainage systems where discharges to the vertical pipe are random and the flow is unsteady. With the development of experimental techniques and analytical methods, the understanding of the fundamental mechanism of the annular two-phase flow has been significantly advanced, such as liquid film development, evolution of the disturbance wave, and droplet entrainment mechanism. Despite the hundreds of papers published so far, the mechanism of annular flow remains incompletely understood. Therefore, this paper summarizes the research on two-phase annular flow in vertical pipes mainly in the last two decades. The review is mainly divided into two parts, i.e., the investigation methodologies and the advancement of knowledge. Different experimental techniques and numerical simulations are compared to highlight their advantages and challenges. Advanced underpinning physics of the mechanism is summarized in several groups including the wavy liquid film, droplet behaviour, entrainment and void fraction. Challenges and recommendations are summarized based on the literature cited in this review.

Keywords: two-phase flow; annular flow; experimental techniques; annular pipe flow; building drainage

Citation: Xue, Y.; Stewart, C.; Kelly, D.; Campbell, D.; Gormley, M. Two-Phase Annular Flow in Vertical Pipes: A Critical Review of Current Research Techniques and Progress. *Water* **2022**, *14*, 3496. https://doi.org/10.3390/w14213496

Academic Editors: Maksim Pakhomov, Pavel Lobanov and Chin H. Wu

Received: 13 September 2022
Accepted: 23 October 2022
Published: 1 November 2022

Publisher's Note: MDPI stays neutral with regard to jurisdictional claims in published maps and institutional affiliations.

1. Introduction

The building drainage system (BDS) provides a means to safely remove human waste from a building. It consists of vertical 'stack' pipes and horizontal 'branch' pipes connecting appliances to the vertical stack. The importance of this system in securing public health for occupants has been highlighted recently in the COVID-19 pandemic. Recent work has shown that cross-transmission of pathogens such as SARS and SARS-CoV-2 responsible for the COVID-19 disease, and laboratory surrogate pathogens such as *pseudomonas putida* (bacterium) and PMMoV (virus), have been shown to travel on airflows inside the BDS which is naturally subject to pressure gradients in the system [1–5]. Understanding two-phase flow in BDS is therefore a public health as well as a fluid mechanics issue.

The flow regime in a drainage system is generated by the unsteady transient flow of a free-falling annular flow, which is a simplified version of a two-phase annular flow with zero or natural airflow and can facilitate the transport of pathogens under certain conditions. An incomplete understanding of the complex two-phase annular flow in vertical pipes limits the extent of current design standards and hence the system quality. Indeed, the vertical annular flow pattern can be found in many important industrial facilities, such as nuclear reactors, refrigeration equipment, heat exchangers and condensers, transportation of oil and natural gas, etc. As the most important two-phase flow regime, annular flow has been widely investigated due to its large involvement in industrial processes and the significant complexity of the flow mechanism.

Professor Hewitt and Professor Azzopardi contributed to the field of annular two-phase flow through their series of studies and significant advancement of knowledge.

Hewitt published the fundamental understanding of the annular flow and summarised the literature in 1970, including theoretical models, interfacial waves, and entrained droplets [6]. Azzopardi published a comprehensive review of the literature up to 1997, focused on the droplets in the flow, i.e., entrainment, size, motion, and re-deposition onto the wall film [7]. Berna et al. summarised the research on annular flow with a particular focus on droplet entrainment [8,9]. Relevant review work focusing on the flow maps for two-phase flows in vertical pipe and annuli was reported by Wu et al. in 2017 [10]. However, due to its nature of complexity, it is always challenging to capture detailed and accurate information about the flow at high temporal and spatial resolutions. The incompletely understood annular two-phase flow requires further advancement in experimental techniques, numerical modelling and a better understanding of the mechanism.

Therefore, this paper reviews the investigations on the annular two-phase flow in vertical pipes with a particular focus on the investigation methodologies and advances in understanding the mechanism. Experimental techniques used in previous investigations and numerical modelling are reviewed in the first part of the paper. The challenges of these electrical, optical or mechanical techniques are highlighted at the end of this part. The advanced understanding of the flow characteristics including the wavy liquid film, disturbance waves, droplet behaviours and droplet entrainment is summarized and compared in the second part followed by a conclusion of current challenges and further recommendations.

2. Investigation Methodologies

Due to the complexity of the annular flow mechanism, experimental techniques have been widely used to obtain detailed information on the flow properties of the annular two-phase flow. Pressure and temperature profiles can be collected using pressure sensors and temperature sensors. However, it is always challenging to obtain accurate information about the flow, such as fluid/gas velocity, film thickness, droplet entrainment, drop size, drop velocity, void fraction, disturbance waves, etc. Some experimental techniques used to measure the void fraction and flow regime [11], and film thickness [12,13] have been summarized. The following section summarises and evaluates the experimental techniques used in previous research, focusing on the flow regime, annular thin film and entrained droplets, followed by a summary of the numerical investigations in the field.

2.1. Visualization and Photography

Due to the development of visual/optical techniques and digital image processing technology, optical methods have shown great advantages in studying gas-liquid two-phase flow, from which qualitative and quantitative results can be obtained. Flow visualization is a useful intuitive method to study flow in transparent pipes, particularly in identifying the flow regime map. Qiao et al. [14] conducted a detailed flow visualization study to characterize the flow regimes in a vertical downward pipe. Flow regime maps for each inlet, including bubbly, slug, churn-turbulent, and annular flow, were developed (as shown in Figure 1). Nimwegen et al. [15–17] successfully recorded the visualized flow pattern in a vertical pipe with surfactants, which caused the formation of foam. The videos clearly show the significant impact of the surfactants on flow behaviour. To identify the flow regime, similar visualization results have been observed by different groups of researchers [18–38].

Film properties, such as thickness and disturbance wave, can be also captured from the visualization results. Pan et al. [39,40] conducted flow visualization experiments in an air-water two-phase annular flow. High-speed videos of the vertical upward flow were recorded to capture the liquid film properties and disturbance wave data using Matlab code, based on which a prediction model of gas-liquid interfacial shear stress for vertical annular flow was proposed. To minimise the issue of the different refractive indices, a transparent square box was used with water filled inside. Schubring et al. [41], Lin et al. [42,43], Moreira et al. [44] and Barbosa et al. [45] obtained the liquid-film thickness and disturbance-wave characterization from high-speed imaging. Barbosa et al. conducted a series of experiments using a test section with a specially constructed transparent liquid inlet. High-speed video recordings clearly showed the process of wave formation and the wave frequencies and typical velocities were also obtained from the recorded videos.

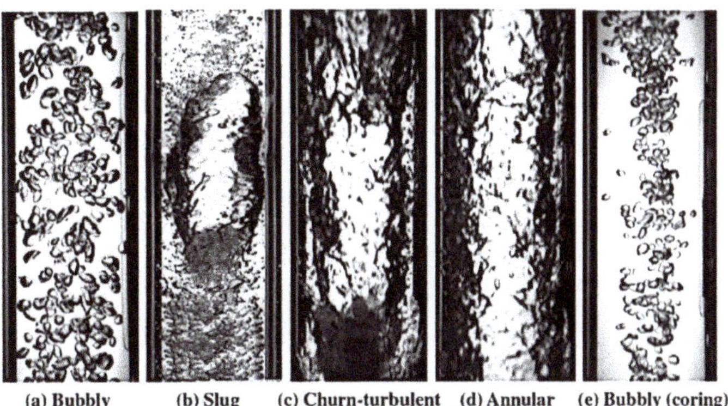

(a) Bubbly	**(b) Slug**	**(c) Churn-turbulent**	**(d) Annular**	**(e) Bubbly (coring)**

Figure 1. The visualized two-phase flow regime map, reprinted with permission from [14] 2017, Elsevier.

2.2. Laser-Induced Fluorescence (LIF)

To obtain detailed information on the annular film, Laser Induced Fluorescence (LIF) technique has shown broad application prospects due to its high temporal and spatial resolutions. The fluorescence of a certain wavelength can be excited by the laser, which is then captured by a high-speed camera, and the film thickness can be obtained from the recorded images. This enables a measurement of the film thickness and real-time visualization of liquid film flow. A typical experimental setup for a planer laser-induced fluorescence (PLIF) measurement of the liquid film thickness and the visualized liquid film is shown in Figure 2. Häber et al. [46] successfully collected images of the illuminated film and employed a ray-tracing technique to analyse the wavy annular film. They analyzed the errors in the measurement and reported that the uniform film was widened by about 30% due to the deflected fluorescence. Similarly, Eckeveld et al. [47] used a novel implementation of PLIF to measure the film thickness, which had shadows on the laser sheet and appeared as dark lines in the recorded images, preventing strong reflections in the measurements.

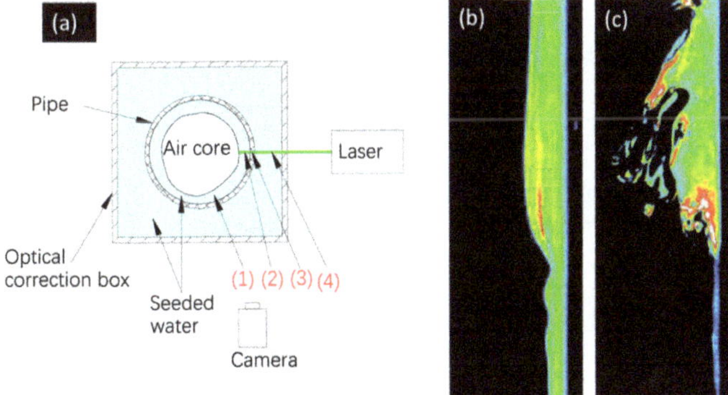

Figure 2. (**a**), A typical experimental setup for PIV/LIF measurement, (1)—(4) are the annular film in the front view, the targeting annular film full of tracer particles/dye that is illuminated by a laser sheet, the transparent wall, and water in the optical correction box, (**b**), a visualized falling film, and (**c**), a disturbance wave. (**b,c**) are adapted with permission from [48], 2014, Elsevier.

Schubring et al. [49,50] also obtained visualization results of the liquid film using PLIF in a vertical annular flow and proposed an improved visualization algorithm for PLIF measurement [51]. Xue et al. [52,53] combined PLIF with high-speed photography to study the liquid film thickness and droplet entrainment. They further [54,55] established a distortion model and proposed a new distortion correction method for the PLIF measurement of the film thickness with a virtual dual-view vision sensor. The method was found capable of measuring circumferential film thickness and distribution characteristics. Vasques et al. [56] used the Brightness Based Laser Induced Fluorescence technique (BBLIF) to study the interfacial wave structure of the liquid film in both upward and downward annular gas-liquid flows. Cherdantsev et al. [13] and Fan et al. [57] also used the BBLIF in the investigation of the co-current downward annular gas-liquid flows as it can be easily resolved in two spatial coordinates and to reconstruct the 3-D shape of the interface wave. Other work using the LIF technique to successfully obtained the film properties in gas-liquid annular flow is seen in [19,48,58–64].

2.3. Particle Image Velocimetry (PIV)

Particle image velocimetry (PIV) and Particle tracking velocimetry (PTV) have been used to obtain the velocity profiles of the liquid film. Illuminated by the laser sheet, the tracer particles can be detected to calculate the velocity vectors. Figure 3 is an example of the liquid film velocity vectors obtained from PIV/PTV measurement by Zadrazil et al. [64,65]. Adomeit and Renz [19] and Ashwood et al. [66] also obtained averaged velocity distribution in a thin annular film using PIV, but the quality of the velocity vector fields was not good. Charogiannis et al. [67] successfully employed simultaneous PIV and LIF measurements in an annular flow to capture the instantaneous velocity vector field and identify the annular film thickness. The main challenges in the application of PIV to obtain the velocity profiles of the liquid film are large velocity gradients in the thin film, strong fluctuation of the free surface, the presence of droplets, and the disturbance waves that can be 10 times the average base film thickness. We have recently conducted PIV measurements of the flow velocity in a gas-liquid annular flow in a vertical pipe. Post-processing of the velocity data and image intensity profile ensures the identification of the interface and hence can provide a statistical understanding of the film thickness and velocity.

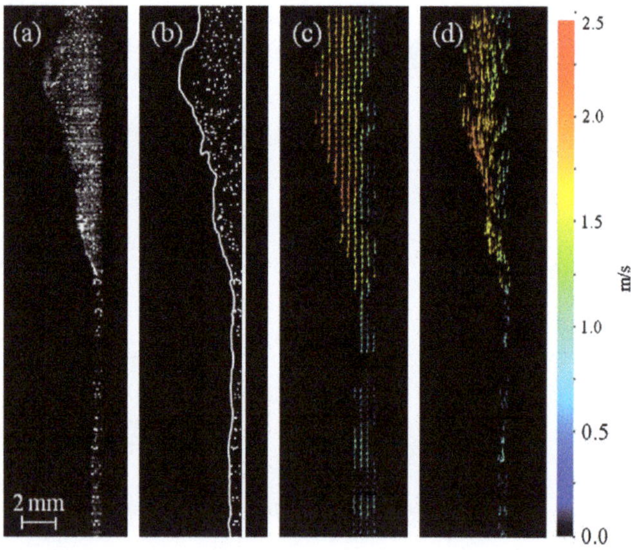

Figure 3. (**a**). A typical raw PIV image; (**b**), a processed raw image; (**c**), a velocity vector field and (**d**), a PTV velocity vector field [64].

2.4. Laser Focus Displacement Meter (LFD)

Annular film properties can be also detected by other techniques, such as laser focus displacement meter, ultrasonic flow meter and near-infrared sensor. Hazuku et al. [68] used a laser focus displacement meter (LFD) to measure the film thickness in an annular flow. Okawa et al. [69] performed film thickness measurements using a laser focus displacement meter, which focuses on a target adopted in automatic focusing. The LFD technique was found capable of accurately measuring film thickness and improving the spatial resolution up to 0.2 μm and time resolution up to 1 kHz. Other similar optical techniques tested in the measurement of the film thickness are the total internal reflection (TIR) method [70,71], the pigment luminance (PLM) method [72] and chromatic confocal imaging (CCI) [73].

2.5. Ultrasonic Flow Meter and Near-Infrared Sensor

Liang et al. [74] measured the film thickness and velocity using ultrasound Doppler velocimetry. Important assumptions include a uniform circumferential liquid film with no waves and no entrained gas. Hence, the measurement accuracy was not as good as other techniques. Wang et al. [75] developed a new ultrasonic echo resonance main frequency (UERMF) measuring system to measure the film thickness and reported an agreement between the conductance probe and the UERMF with an error of less than ±10%. Al-Aufi et al. [76] performed film thickness measurements using ultrasonic pulse-echo techniques and demonstrated the potential of using different signal processing methods. Challenges in the measurement were noticed when the gas-liquid interface experienced large waves. Similarly, near-infrared light sensors have also been used to measure the film thickness based on the different absorption characteristics of gas and liquid [77–81]. However, due to the significant impact of the entrained liquid droplets on the absorption coefficient, light guide pipes were designed and inserted into the annular flow, which ensured the light was only absorbed by the liquid film.

2.6. Conductance Sensor

As one of the widely used sensors, the conductance sensor can measure the film thickness and void fraction in the two-phase annular flow based on the relationship between the fluid fraction and its conductivity. Coney [82] described the theoretical behaviour of flat electrodes wetted by a liquid layer in 1973 and Hewitt [83] reviewed the application of the conductance probe technique up to 1978. Damsohn and Prasser [84] designed a novel flat high-speed liquid film sensor with a high spatial resolution based on the electrical conductance method to measure the thickness of the dynamic liquid films in a two-phase flow, which could be implemented into the annular flow. Some typical conductance probes are shown in Figure 4. The inserted parallel wires probe can capture the local film thickness at the measurement point with a high temporal resolution, while the embedded electrodes are adopted to reduce the disturbance to the film and to measure the average local film thickness.

Wang et al. [85] employed a conductance probe with adjustable insertion depth to measure the film thickness in 2018, and the measurement error was less than 1%. Polansky and Wang [86] applied POD (Proper Orthogonal Decomposition) on a large number of tomograms for the identification of flow patterns in a gas-liquid annular flow. The images were obtained using electrical impedance tomography via a series of electrodes mounted non-intrusively (no induced disturbance to the flow) but invasively (direct contact with the fluid) on the pipe. Measurement of the liquid film thickness by inserting electrodes or parallel wire probes into the pipes is widely used in the field due to its simplicity and low cost [23,24,87–109]. With no induced disturbance to the flow, the embedded electrodes can be grouped as parallel strip electrodes, concentric circle electrodes, circular rod electrodes, ring-shaped electrodes, and so on [12,36,39,40,110–128].

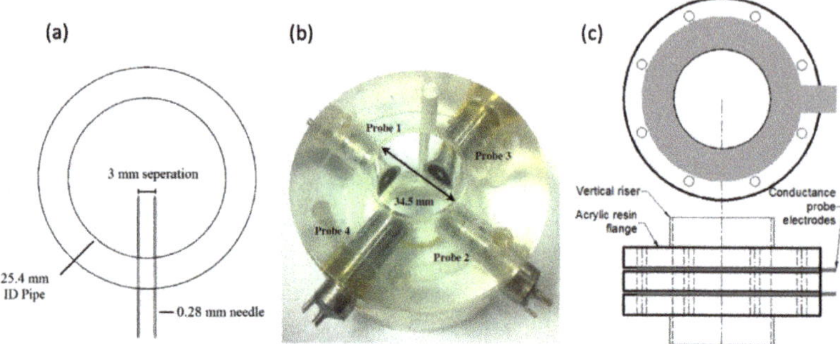

Figure 4. Some typical conductance probes. (**a**), Parallel-wire probes [109], 2021 (**b**), embedded concentric conductance probes [118], 2013 and(**c**), ring-shaped probes [129], 2019. Figures are adapted with permission from the references, Elsevier.

The conductance probe is capable of measuring the void fraction of the annular flow. Fossa [130] used ring-shaped electrodes and plate electrodes to measure the conductance of the mixtures in pipes and hence to determine the liquid fraction of gas-liquid flow. They further optimised the impedance probes for void fraction measurement in annular, stratified, and dispersed flows [131]. Yang et al. [132] also used conductance probes (3 ring-type impedance meters and 3 arc-type impedance meters) to measure the average void fraction of the two-phase flow.

2.7. Capacitance Sensor

In two-phase flow, the permittivity of different fluids is generally different which allows the measurement of the void fraction of the flow components using a capacitance sensor. different types of these capacitance sensors, such as concave [133–135], parallel [136], ring [74,137–139], and helical [140] sensors, were successfully used in the measurements of the void fraction/gas holdup in the two-phase flows. The selection of this sensor is because it is simple, easy to implement, and relatively low cost. A typical concave capacitance sensor has two concave electrodes mounted on the tube circumference opposite to each other between which the capacitance of the flow is correlated to the local void fraction as shown in Figure 5. A parallel capacitance sensor has several pairs of plates mounted on the outer surface of an insulating pipeline [136]. It can measure the capacitance between all possible combination pairs of the electrodes and then estimate the void fraction based on the image reconstruction algorithm. A ring-type capacitance sensor has two ring electrodes separated in the axial direction of the tube [137] and a helical plate capacitance sensor has two helical electrodes [140].

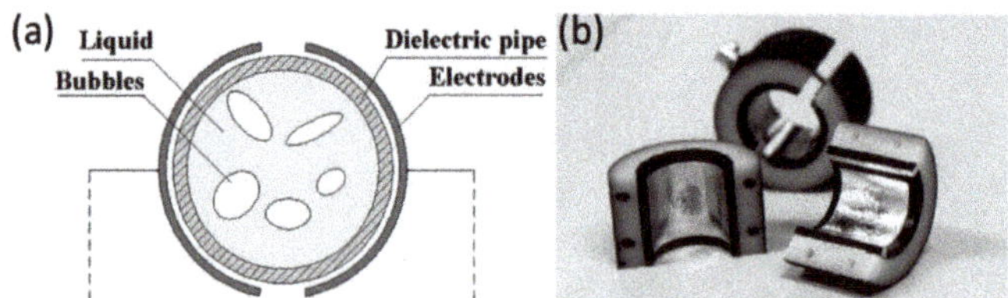

Figure 5. (**a**), The structure of a typical concave capacitance sensor, adapted with permission from [134], 2004, Elsevier and (**b**), an actual concave capacitance sensor, adapted with permission from [135], 2014, Elsevier.

Atkinson and Huang [141] developed mathematical models for the capacitance sensors used in the measurement of the liquid annulus thickness. Ahmed [139] compared the concave and ring-type capacitance sensors in the void fraction measurement and flow pattern identification in an air/oil two-phase flow experimentally and theoretically. It was reported that the ring-type sensors were more sensitive to the void-fraction signal and had better agreement with the theoretical predictions. Elkow and Rezkallah [140] compared the concave and helical capacitance sensors together with quick-closing valves and a gamma densitometer in the measurement of the void fraction in gas-liquid flows. The fluid temperature was found to have a strong impact on the measured capacitance. Kerpel et al. [142,143] used a concave capacitive void fraction sensor to determine the void fraction and flow regime of a two-phase flow in a small tube. Annular flow, slug flow and intermittent flow were set in the experiment and an alternative calibration technique based on the statistical parameters of the measurement was proposed.

2.8. Wire-Mesh Sensor (WMS)

Phase fraction of two-phase flow (for instance, a droplet in the air core of annular flow or bubbles in bubbly flow) can be also measured using a wire-mess sensor (WMS), as shown in Figure 6, which has two sets of wire electrodes with a small axial displacement between them and being perpendicular to each other. In this way, a matrix-like arrangement of the measuring points (crossing points) is constructed. Based on the different electrical properties of the flow phases, the instantaneous flow conditions at the measurement points can be collected, from which the fractions of the flow phases can be distinguished.

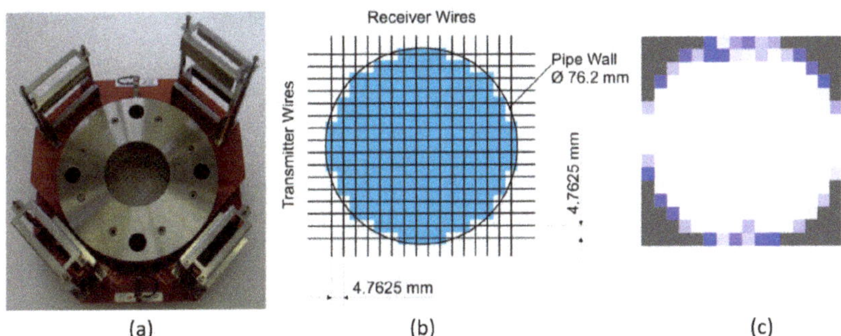

Figure 6. (**a**), An actual wire mesh sensor, adapted with permission from [144]. 2015, Elsevier, (**b**), wire distribution of the 16 × 16 sensors, adapted with permission from [145]. 2015, Elsevier, (**c**), a typical measurement frame showing the liquid phase (blue) and gas phase (white), adapted with permission from [145]. 2015, Elsevier.

The wire-mesh sensor was first used by Johnson [146] to measure the fraction of water in oil, based on the different conductivities. Prasser et al. [147] further developed an electrode-mesh conductivity sensor, which was the basic form of current WMS, to measure the void fraction distributions of a gas-liquid flow. The wire mesh sensor has been widely used in the measurement of multi-phase flow properties, such as void fraction distribution, flow velocity, bubble size, and film thickness [18,148–152]. Two review articles summarised the application of different types of WMS for the measurement of flow properties in two-phase flows and their associated uncertainties [153,154].

Lucas et al. [155] conducted a comprehensive investigation of the flow properties of an air-water flow in a vertical pipe for a wide range of flow rates (including annular flow). Using the wire-mesh sensor technology, they collected different flow properties, including the gas volume fraction, bubble size, and gas velocity. Vieira et al. [144,145] used a dual WMS with a sampling frequency of 10 kHz to detect the local instantaneous cross-section distribution of the gas core in a vertical pipe. The dual sensor enables simultaneous measurements of the local void fraction at two positions along the flow direction and the flow velocity can be captured from the correlation of the two signals. Silva et al. [156]

introduced a novel wire-mesh sensor based on fluid permittivity in a silicone oil bubbly flow. Sensor and measuring electronics were evaluated showing good stability and accuracy in the capacitance measurement.

2.9. Radiative Imaging

As a contactless measurement method, neutron imaging is an attractive option in two-phase flow investigation for the high spatial resolution of the flow structure, such as void fraction and film thickness [157–159]. In applications of gamma-densitometry [140,160–163], the accurate correlation between the loss of radiation intensity and its void fraction provided information on the prevailing flow regime. Banowski et al. [149], Zboray et al. [164] and Misawa et al. [150] used X-ray CT (Computer Tomography) in annular flow and obtained good time series of void fractions.

2.10. Film Extraction

The wavy liquid film is always a challenge in investigating the entrained droplets, especially when optical techniques are employed. Therefore, film extraction is generally needed to remove the unsteady optical distortion due to the liquid film and to accurately measure the property of the entrained drops [165]. Figure 7 presents two examples of the film removal configuration in the investigation of the entrained drops in annular flow. The liquid film can be extracted through the porous wall with further purge gas injection at the window or removed from the annular flow through a slightly smaller circular slit. Measurement results of the droplet profiles showed a negligible impact on the film extraction [166].

The extraction of the liquid film was not only required in the visualization study [165] but also needed in laser Doppler anemometry (LDA) [26,167,168] and laser diffraction [99,166,169] measurements. Film extraction can be also used to obtain the void fraction using the flow rate of the separated liquid and gas flow. Bertodano et al. [170], Okawa et al. [171], and Sawant et al. [172] used a film double extraction unit to separate the gas flow and the liquid flow. With single-phase flow meters, the flow rate of the separated liquid/gas flow can be determined which can be used to further calculate the void fraction. However, the droplets entrained in the gas core have to be removed as well, which may otherwise result in non-negligible errors.

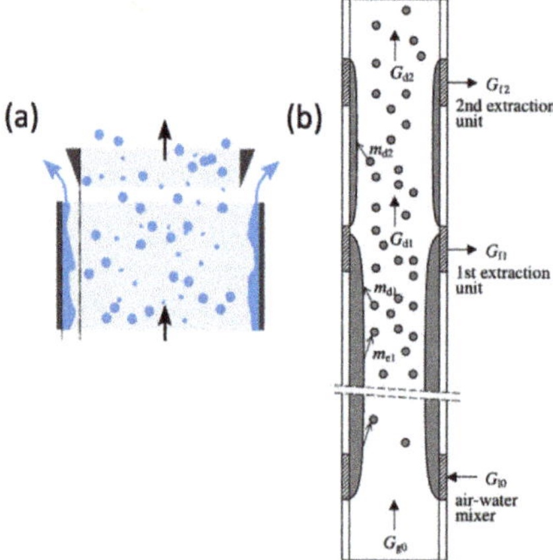

Figure 7. (**a**), Removal of the liquid film by a circular slit with a wedge-shaped edge, adapted with permission from [47], 2018, Elsevier and (**b**), the extraction of the liquid film by a porous wall for the laser diffraction measurement, adapted with permission from [171], 2005, Elsevier.

2.11. Shadow Photography and Laser Diffraction

With the fluctuating film removed, shadow photography and laser diffraction have shown their capability to measure the droplet size in the gas core. Fore et al. [165] recorded images of the entrain drops in a nitrogen/water annular flow, which had the liquid film extracted through a porous wall (Figure 8). The size of the drops was measured directly from images and statistics analysis provided information on the drop size distribution. When a laser beam passes through a dispersed droplet, the angular variation in the intensity of scattered light is determined by the drop size. In such a way, the distribution of the entrained droplets can be obtained from laser diffraction results. Simmons and Hanratty [166] measured the drop size in air-water annular flow at atmospheric pressure using the laser diffraction technique (Malvern Spraytec R 5008 instrument, Malvern Panalytical Ltd, Malvern, UK). This optical technique has been successfully used by many researchers [47,99,100,169,173–179] in obtaining the droplet size distribution.

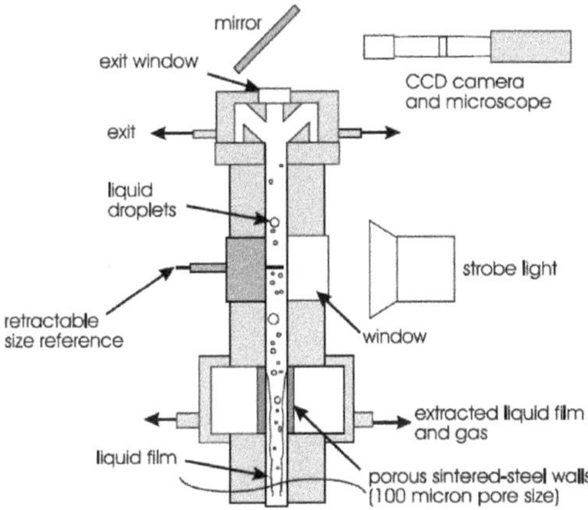

Figure 8. Optical setup of the droplet size measurement, adapted with permission from [165], 2002, Elsevier.

2.12. Laser Doppler Anemometry (LDA)

Another optical technique for the droplet velocity measurement used is the laser Doppler anemometry/phase Doppler anemometry (LDA/PDA) [18,176,179,180]. The velocity can be obtained from the light scattered by the droplets and the Doppler shift, which is related to the velocity component perpendicular to the bisector of the two laser beams. Trabold et al. [161] conducted local measurements in a droplet-laden vapour core in an annular flow, using a gamma densitometer, hot-film anemometer, and LDA. Van't Westende [167,168] and Zhang et al. [26] used phase Doppler anemometry to measure the size and velocity of the droplets at 11 radial positions of an annular flow.

2.13. Other Mechanical Methods

Barbosa et al. [181] used an isokinetic probe, which was similar to a Pitot tube, to collect the entrained drop and the gas flow of an annular flow, which were then separated for the void fraction measurement. Oliveira et al. [182] achieved accurate measurement of flow rate in gas-liquid flows using a venturi coupled to a void fraction sensor. A quick closing valve (QCV) is another mechanical method to provide an exact void fraction measurement by isolating a section of flow in the conduit and the void fraction can be directly determined [140,183–194]. However, it is not a practical method to determine the void fraction for continuous flow as it disrupts the flow. Hurlburt and Hanratty [195] developed an immersion sampling system to measure the droplet diameter in an air-water annular flow. A high-speed shutter apparatus was used to capture the drops in the small cavity filled with high-viscosity oil. The droplet images were processed to obtain drop diameters.

2.14. Numerical Simulation

With the development of computing power, numerical simulation (CFD) has shown its capability to investigate the flow properties of annular two-phase flow. However, due to its complexity, i.e., multi-phase, atomisation, deposition, droplets, bubbles, film, and waves, an accurate and detailed simulation of the annular two-phase flow attracts great interest. A common shortcoming of current CFD models is the treatment method used for the dynamic gas-liquid interface configuration [196]. Because of the challenges in determining the interface dynamics, a simple wave shape or even a smooth gas-liquid interface was used in many of the solution models. Han and Gabriel [196–199] modelled the gas core flow in a typical annular flow with the simplified flow in the liquid film region. Van der Meulen [18] simulated the droplet behaviour and droplet trajectory in gas-liquid flows in vertical pipes using Star-CD. They also quantified the deposition of the drops by diffusion or direct impaction mechanisms. Xie et al. [200] numerically studied the three-dimensional droplet impacts on a liquid film in a vertical annular flow (Figure 9).

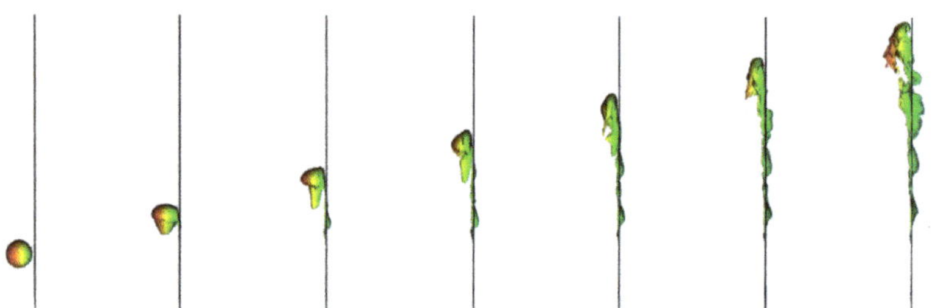

Figure 9. Simulation of the droplet deposition process in an upwards annular two-phase flow (side view) [200].

Alipchenkov et al. [201] used a population balance model (PNM) for droplet size distribution, which was based on conservation equations. Liu and Li [202,203] used a CFD-PBM (population balance model) model to investigate the droplet's size distribution in an annular flow system as well as liquid roll waves directly determined. Van't Westende [167] performed quasi-1D and 3D large eddy simulation (LES) simulations to compute the pressure gradient and deposition of a dispersed phase in an upward airflow. The PDA measurement results were used in the simulations, mimicking the atomisation process of an actual annular dispersed flow as realistically as possible. Adineh et al. [163] compared the experimental and numerical results of the void fraction inside a vertical pipe with the numerical modelling done using the MCNP (Monte Carlo N-Particle Transport) code. Saxena and Prasser [204] tested different turbulence models to predict the void fraction and capture the liquid film flow properties.

Kishore and Jayanti [205] developed a finite volume method-based model to investigate steady state, 2-D annular flow in a rough-walled duct. It was shown that the simplified model gave reasonable predictions of the flow parameters under equilibrium and non-equilibrium conditions. Kiran et al. [185] tested the VOF (volume of fluid) model and two turbulence models (realizable k-ε and Shear Stress Transport (SST) k-ω models) in the simulation of annular flows at high gas velocity and an average error of 20% was reported. Hassani et al. [206] simulated the two-phase flow using the VOF method and piecewise interface calculation (PLIC) algorithm to track the interface. Fan et al. [57] developed new turbulence damping models to overcome the under-prediction of the turbulence level in the numerical study using the volume of fluid (VOF) method.

It should be noted that some assumptions are generally used in the simulations, such as fully developed and stable annular flow, isothermal flow, no mass exchange between the two phases, uniform distribution of the liquid film, no entrapped gas bubble in the liquid film and uniformly dispersed drops in the gas core. These assumptions can significantly simplify the numerical simulation process but may lead to some non-negligible errors and hence have to be carefully validated.

2.15. Challenges of Current Experimental Techniques

Figure 10 simply summarises the research methodologies used in the investigation of the annular two-phase flow in vertical pipes. The non-invasive and non-intrusive optical measurements, i.e., flow visualization, planar laser-induced fluorescence (PLIF), particle image velocimetry (PIV), particle tracking velocimetry (PTV), the laser Doppler anemometry/phase Doppler anemometry (LDA/PDA), and laser diffraction are getting more attractive due to the main advantage of no disturbance to the flow, high temporal and spatial resolutions, and capability to capture the unsteady flow structure. However, the optical transparency of the pipe or a window that enables the imaging of the flow is essential in the application. Optical distortion of the transparent wall is another limitation that needs to be minimised by a specially designed configuration or corrected in the data post-processing. The main challenge in capturing the wavy film characteristics in annular two-phase flow, particularly the statistical results (for example PIV), is the strong fluctuation of the liquid film thickness. Some other measurement techniques, such as laser focus displacement meter and ultrasonic flow meter, also have limited application due to the complex unsteady flow characteristics in annular two-phase flow.

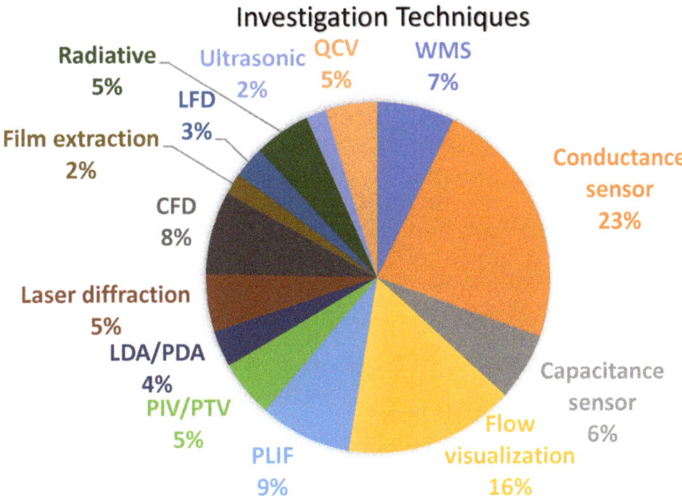

Figure 10. A summary of the major research techniques used in the investigation of the annular two-phase flow in the recent two decades.

The electrical sensors, i.e., the wire mesh sensor (WMS), conductance sensor, and capacitance sensor, are still widely used due to their simplicity, reliability, and high temporal resolution. The main limitations of these electrical sensing technologies are the invasive and intrusive nature of the wires, the requirement of the fluid electrical properties, limited spatial resolution or single-point measurement, and relatively high measurement uncertainty compared to other techniques. As a wire mess sensor needs to be positioned within the pipe, it has a significant impact on the flow properties. It was reported that the WMS caused velocity alteration due to the induced disturbance and a significant pressure drop [147]. It is not applicable to use the WMS in pipe systems which may contain solid components in the flow, such as the drainage system. The successful application of the WMS is also determined by the different electrical properties of the fluids, i.e., conductivity or permittivity, which are significantly different to distinguish the different phases from the electronic signals [154]. The spatial resolution is determined by the number of wires used in each plane. A high resolution of more wires will increase the occupied area and enhance its impact on the flow. Furthermore, the wires must bear the drag forces in the flow, so the larger diameter of the wire, the greater drag force. Both of the two cases are undesirable. Another issue associated with the spatial resolution is the interaction between electrodes, this implies closer electrodes will cause a greater effect and hence lower the measurement quality. The uncertainty of a WMS in void fraction measurement is generally higher than 10% over a wide flow regime [153]. Therefore, in the two-phase annular flow, the applicability of WMS is limited to cases where the film thickness and droplet size are large enough compared to the mesh pitch. The bulk velocity of a WMS is calculated using the assumption that the interfacial velocity is equal to the void velocity. However, this incorrect assumption indicates the inapplicability of this type of measurement [207].

For a conductance probe, the electrical property of the fluid, particularly the temperature, has a significant impact on the measurement result and accuracy. Therefore, it is important to keep the fluid temperature constant during the experiment and to accurately calibrate the probe at the same temperature. Another limitation of the conductance probe is that it only can measure the local property (film thickness or void fraction) at the measurement point. Additionally, the ring type or embedded probe can only measure the average property at the cross-section. While the void fraction or film thickness is generally not available directly from the conductance probe. The accuracy of the conductance probe is generally about 10% [99,100,108].

3. The Wavy Liquid Film

3.1. Fundamental Understanding of the Liquid Film

For an annular two-phase flow in a vertical pipe, the peripheral liquid film generally includes ripple and disturbance waves and acts as a thin wall for the gaseous core flow with entrained drops, as shown in Figure 11. Annular flow gets stable when the fluid has higher effective viscosity (molecular and turbulent viscosity) in the core region and lower viscosity fluid in the annulus [208]. Several definitions are frequently used in the field to describe the flow, such as instantaneous film thickness, the average thickness of base film/substrate, average film thickness, and local maximum thickness. The liquid film thickness is determined by the piping system configuration and the flow conditions such as the liquid flow rate, fluid (gas/liquid) properties, and flow directions. The average liquid film thickness has been well documented in previous research and new understandings of the mechanism have been reported in the recent two decades due to the advanced experimental methods and techniques.

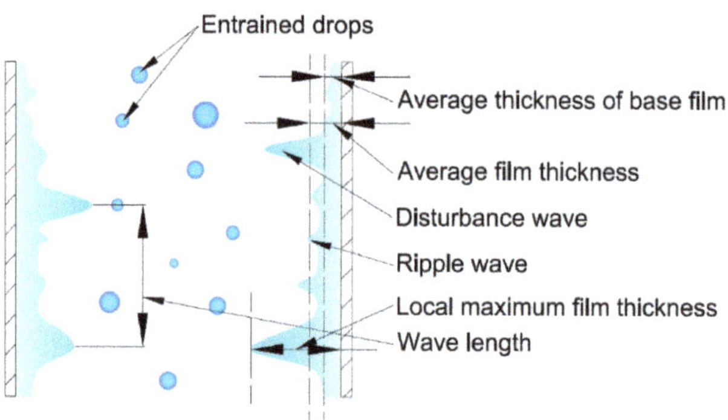

Figure 11. Structure of a typical two-phase annular flow in vertical pipe showing the disturbance waves, ripple waves, liquid film, and entrained droplets.

Film thickness distribution captured directly from LIF results [47,49,56,67] (Figure 12) is found similar across a wide range of flow conditions, although the absolute film thickness changes significantly. The average film thickness and the fluctuation of film thickness decreased with gas flow velocity and increased with the liquid flow velocity and the relationship between the average film thickness and the roughness is determined by the liquid and gas flows [50]. Zhao et al. [118] collected high-frequency film thickness data of a gas-liquid annular flow and found the film variation along the axial direction was not significant (within $\pm 10\%$ of average values). The development of the average thickness was only near the inlet, i.e., up to L/D=20 ($Re_L = 211$) and L/D=25 ($Re_L = 603$). Prior to becoming fully developed, the film decelerates first to a local maximum thickness and then accelerates again to become thinner.

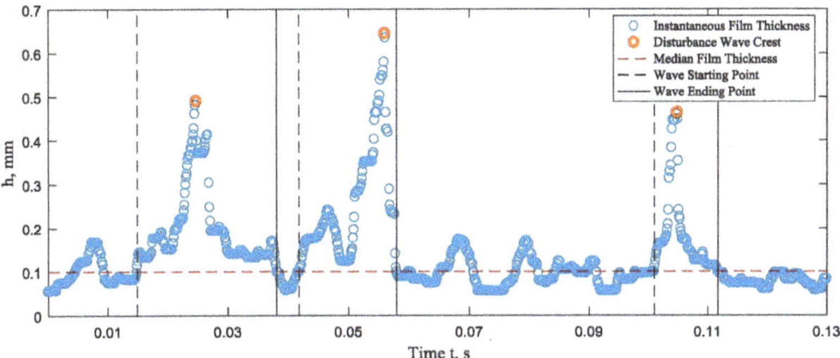

Figure 12. Film thickness distribution conducted from BBLIF [56].

It is well understood that the film thickness generally decreases with the decrease in the liquid superficial velocity and increases in the gas superficial velocity [18,39,48,99,101,111,112,129,148,209]. The liquid film was reported to have 3-D structures with a large height fluctuation in circumferential and axial directions, and a meandering path between the maximum height around the circumference (Figure 13) [54,55]. This fluctuation is mainly caused by the non-uniform generation of the ripple wave and disturbance waves. This difference in film thickness in circumferential directions was also observed from LIF results [59,60] conductive probe measurements [118,123]. It was found that the average length of the disturbance waves was similar to the pipe diameter and independent of the gas/liquid superficial velocities [123].

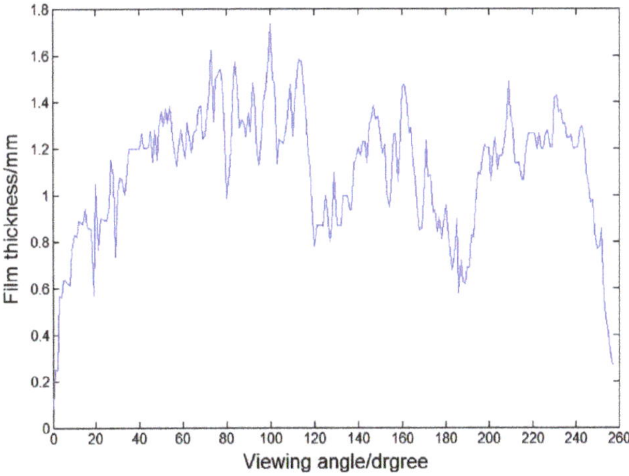

Figure 13. The unsymmetrical circumferential film thickness determined using PLIF, adapted with permission from [54], 2019, John Wiley and Sons.

A good agreement between the Nusselt's predictions and experimental velocity profiles was found at low liquid Reynolds number and significant differences between the measured and Nusselt's predicted profiles were reported in wavy turbulent films (Figure 14), i.e., high Reynolds number [63]. The measured average film thickness data (PLIF) agreed well with previous experimental data and was compared with Nusselt's theory [48]. The visualization results also proved the existence of recirculation zones in front of disturbance waves [64]. Like the velocity profiles within the wavy film, the film thickness was well-described by the Nusselt flow predictions at low Re_L, while with increasing Re_L, the film thickness was increasingly underpredicted by the theory, but with good agreement with Mudawwar and El-Masri's semi-empirical turbulence model [210].

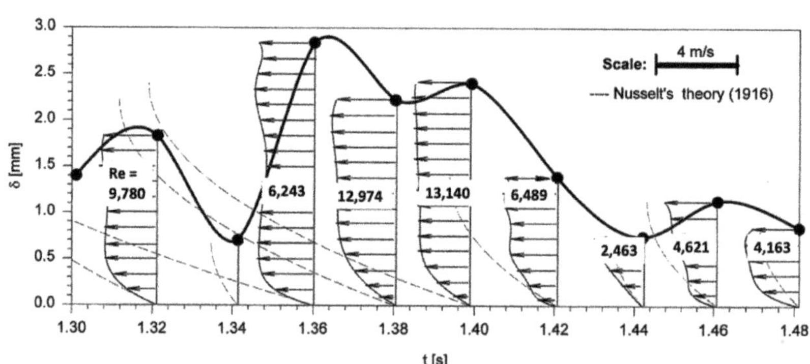

Figure 14. Velocity and thickness profiles of a falling film at a liquid Reynolds number of 5275, adapted with permission from [63], 1998, Elsevier.

Vassallo [211] conducted a near-wall measurement of velocity in the liquid film using a hot-film probe. A modified law of the wall was suggested for annular two-phase flows near the transition regime when the film was thicker. Muñoz-Cobo et al. [125] focused on the effect of the liquid surface tension in vertical annular flow by having different amounts of 1-butanol in the fluid. Reducing the surface tension leads to a reduction in the intermolecular cohesion forces, easier entrainment of the small droplets from the wave peaks and a decrease in the wave amplitude.

3.2. Disturbance Wave Characteristics

Based on the experimental and numerical results, Fan et al. [57] reported the main progress of wave evolution, i.e., generation and development of initial waves, coalescence of initial disturbance waves into large-scale waves, and acceleration of waves with further stable propagation. They also found that the waves generated slow and fast ripples on their rear slopes and droplet entrainment started from the disruption of fast ripples. The disturbance waves were observed only when the liquid film Reynolds numbers exceeded the critical value [118]. Dao and Balakotaiah [105] investigated the occlusion of falling film in a vertical pipe with glycerine. The experimental results reported a good correlation between the liquid Reynolds number, the Kapitza number, and the Bond number. Han et al. [102] studied the effects of gas flow on the disturbance wave in the annular flow. With a constant liquid flow rate, an increase in the gas mass flow rate resulted in a series of changes in the wave characteristics, i.e., decreased wave spacing and increased wave frequency, slightly decreases in wave base height, peak height, and the mean film thickness. They also reported a much more significant increase in the liquid velocity from the base area to the wavy area with an approximate ratio of 1:14.

Alekseenko et al. [58–61] reported quantitative studies of the disturbance wave focusing on its spatial and temporal evolution. Three different regions were defined in the liquid film: the crests of disturbance waves, where the fast ripples existed; the back slopes of disturbance waves, where the slow ripples were generated and their properties gradually changed with increasing distance from the crests; and the base film, where the properties of slow ripples had stabilized values. Rapid changes in the film flow parameters, including the thickness, disturbance wave velocity, and frequency, were found within the first 50 tube diameters [111]. The disturbance waves were found to appear and to achieve the stable circumferential distribution at 5–10 pipe diameters from the injection and this coherence gradually strengthened downstream [118].

The identification of the flow pattern and the pressure gradient was determined by the characteristics and behaviours of the interfacial wave [42,43], and its orientation has significant impacts on the flow identification and pressure gradient. Pressure drop in a downward co-current annular flow measured by Hajiloo et al. [212] suggested that at a fixed gas Reynolds number, a large increase in interfacial friction accompanied a decrease in tube diameter and existing correlations were unsuccessful for the present data.

3.3. Correlations of the Film Thickness

Klyuev and Solov'eva [213] developed a mathematical model for the annular flow, which showed the increase in void fraction resulted in decreases in the average film thickness and the average liquid velocity. Belt et al. [209] improved the Wallis correlation by correcting the film roughness, which was assumed as four times the mean film thickness. The new sand-grain roughness was found proportional to the wave height and can be estimated using the roughness density. The transient behaviour model [214] and critical friction factor model [215,216] were developed to estimate the averaged film thicknesses. The calculated results were agreed to within 20% of the experimental measurements. The liquid film at the top was found significantly different from those at the lower axial positions [148], which had a distinctly different slope from the published correlations and theoretical predictions, and hence suggested a potential change in the film structure in large-scale pipes.

Different correlations of the liquid film thickness based on experimental results and theoretical analysis have been proposed so far, which are summarised in Table 1.

Table 1. Correlations of the average liquid film thickness of annular two-phase flow.

Reference	Correlations of the Liquid Film Thickness
Ishii and Grolmes [217] (1975)	$\delta = 0.347 Re_L^{2/3} \sqrt{\frac{\rho_L}{\tau_i} \frac{\mu_L}{\rho_L}}$
Henstock and Hanratty [218] (1976)	$\delta = \frac{6.59F}{(1+1400F)^{0.5}} D$ $F = \frac{1}{\sqrt{2} Re_G^{0.4}} \frac{Re_L^{0.5}}{Re_G^{0.9}} \frac{\mu_L \rho_G^{0.5}}{\mu_G \rho_L^{0.5}}$
Tatterson et al. [219] (1977)	$\delta = \frac{6.59F}{(1+1400F)^{0.5}} D$ $F = \frac{\gamma(Re_L)}{Re_G^{0.9}} \frac{\mu_L \rho_G^{0.5}}{\mu_G \rho_L^{0.5}}$ $\gamma = \left[\left(0.707 Re_L^{0.5}\right)^{2.5} + \left(0.0379 Re_L^{0.9}\right)^{2.5} \right]^{0.4}$
Hori et al. [220,221] (1978)	$\delta = 0.905 Re_G^{-1.45} Re_L^{0.9} Fr_G^{0.93} Fr_L^{-0.68} \left(\frac{\mu_L}{\mu_{L,ref}}\right)^{1.06} D$
Ambrosini et al. [222] (1991)	$\frac{\rho_L \delta u^*}{\mu_L} = \begin{cases} 0.34 Re_L^{0.6} & Re_L \leq 1000 \\ 0.0512 Re_L^{0,875} & Re_L > 1000 \end{cases}, \quad u^* = \sqrt{\frac{\tau_i}{\rho_L}}$
Fukano and Furukawa [94] (1998)	$\delta = 0.0594 exp\left(-0.34 Fr_G^{0.25} Re_L^{0.19} \chi^{0.6}\right) D$ $\chi = \frac{j_G \rho_G}{j_G \rho_G + j_L \rho_L}$
Okawa et al. [223] (2002)	$\delta \approx \frac{1}{4} \sqrt{\frac{f_w \rho_L}{f_i \rho_G} \frac{(1-E) j_L}{j_G}} D$ $f_i = 0.005 \left(1 + 300 \frac{\delta}{D}\right)$ $f_w = max\left(\frac{16}{Re_L}, 0.005\right)$

Table 1. *Cont.*

Reference	Correlations of the Liquid Film Thickness
MacGillivray [224] (2004)	$\delta = 39\frac{\mu_L}{\rho_L j_L} Re_L^{0.2}\frac{1-a}{a}\left(\frac{\rho_G}{\rho_L}\right)^{0.5}$
Hazuku et al. [68] (2008)	$\delta_{base}\left(g/v_L^2\right)^{1/3} = 0.977 Re_L^{0.143}\tau_i^{*-0.117}$ $\tau_i^* = \frac{\tau_i}{\rho_L g}\left(\frac{g}{v_L^2}\right)^{1/3}$ $\tau_i = \frac{D-2\bar{\delta}}{4}\left(\frac{dp}{dz}\right)_{fric}$ $\tau_i = f_i\rho_G j_G^2/2$
Berna et al. [9] (2014)	$\delta = 7.156 Re_G^{-1.07} Re_L^{0.48}\left(\frac{Fr_G}{Fr_L}\right)^{0.24} D$
Pan et al. [39] (2015)	$\delta = 2.03 Re_L^{0.15} Re_G^{-0.6} D$
Almabrok et al. [148] (2016)	$\delta = 1.4459 Re_L^{0.3051}\left(\frac{g}{v_L^2}\right)^{-\frac{1}{3}}$
Rahman et al. [225] (2017)	$\delta = 1.93\times 10^{-3} Re_G^{-0.246} We_G^{-0.161} Fr_L^{0.15}\left(\frac{\dot{m}_L}{\dot{m}_G}\right)^{0.546}$
Ju et al. [226,227] (2019)	$\delta = 0.071 tanh\left(14.22 We_L^{0.24} We_G^{-0.47} N_{\mu_f}^{0.21}\right) D$ $\delta_{base} = 0.04 tanh\left(4.31 We_L^{0.22} We_G^{-0.44}\right) D$ $N_{\mu_f} = \frac{\mu_L}{\sqrt{\left(\rho_L\sigma\sqrt{\frac{\sigma}{g\Delta\rho}}\right)}}$ $We_L = \frac{\rho_L j_L^2 D}{\sigma}, \; We_G = \frac{\rho_G j_L^2 D}{\sigma}\left(\frac{\Delta\rho}{\rho_G}\right)^{\frac{1}{4}}$
Rivera et al. [125] (2021)	$\delta = 2.35 Re_G^{-1.415} Re_L^{0.414} Ka^{0.781} D$
Rivera et al. [127] (2022)	$\delta = 0.19 Re_L^{0.54}\left(1 - 1.29\times 10^{-5} Re_G^{0.93}\right)\left(\frac{v^2}{g}\right)^{1/3}$
Pan et al. [39]	$\frac{\delta_{DW}}{D} = 1400\left(\frac{u_c}{u_L}\right)^{-\frac{1}{3}}\left[\frac{(\rho_L-\rho_c)g\delta^2}{\sigma}\right]^{\frac{5}{8}}$
Ju et al. [226]	$\delta_{DW} = 0.24 tanh\left(4.22 We_L^{0.16} We_G^{-0.46}\right) D$
Y. Rivera, et al. [125]	$\frac{\delta_{DW}}{D} = 0.554\times 10^{-3} Re_G^{-0.57} Re_L^{0.061} Ka^{1.12}$

3.4. The Void Fraction of Annular Two-Phase Flow

The void fraction is the fraction of the gaseous phase to the total volume of the channel, which is generally between 0.65 and 0.98. Godbole et al. [191] conducted a comprehensive literature review of the void fraction correlations and experimental results in the early years of upward two-phase flow. Most area-averaged void fraction had an increasing trend along the axial direction and decreased after a maximum value of around 80–100 diameters downstream [144,145,148]. However, the decrease in void fraction in the vertical downward annular flow was also observed in some conditions which was a result of the kinematic shock phenomenon. Alves et al. [214] developed a three-field two-phase flow model to simulate the transient annular flow in vertical pipes with a slight tendency of underprediction. Smith et al. [103] proposed a one-dimensional interfacial area transport equation (IATE) using measurements of local void fraction, interfacial area concentration, and interface velocity of an upward annular flow in a large pipe. The dependence of mixture density on the void fraction and correlations based on the slip ratio and drift flux model were analysed [228]. Table 2 summarized the correlations for the void fraction *(a)* annular two-phase flows.

Table 2. Correlations for the void fraction (a) of annular two-phase flows.

Reference	Correlations for the Void Fraction
Tandon et al. [229] (1985)	$$a = \left\{1 - 1.928 Re_L^{-0.315}[F(X_{tt})]^{-1} + 0.9293 Re_L^{-0.63}[F(X_{tt})]^{-2}\right\} \quad 50 < Re_L < 1125$$ $$a = \left\{1 - 0.38 Re_L^{-0.088}[F(X_{tt})]^{-1} + 0.0361 Re_L^{-0.176}[F(X_{tt})]^{-2}\right\} \quad Re_L > 1125$$ $$F(X_{tt}) = 0.15\left[X_{tt}^{-1} + 2.85 X_{tt}^{-0.476}\right]$$ $$X_{tt} = \left(\frac{\mu_L}{\mu_G}\right)^{0.1}\left(\frac{1-x}{x}\right)^{0.9}\left(\frac{\rho_G}{\rho_L}\right)^{0.5}$$
Usui and Sato [93] (1989)	$$(1-a)^{23/7} - 2C_w Fr_L^2\left[1 \pm \frac{C_i}{C_w}\cdot\frac{(1-a)^{16/7}}{a^{5/2}}\cdot\frac{\rho_G}{\rho_L}\left(\frac{j_G}{j_L}\right)^2\right] = 0$$ $$\text{Free falling film, } a = 1 - \left(2C_w Fr_L^2\right)^{7/23}$$
Cioncolini and Thome [230] (2012)	$$a = \frac{hx^n}{1+(h-1)x^n}$$ $$\left(0 < x < 1,\ 10^{-3} < \frac{\rho_G}{\rho_L} < 1,\ 0.7 < \varepsilon < 1\right)$$ $$h = -2.129 + 3.129\left(\frac{\rho_G}{\rho_L}\right)^{-0.2186}$$ $$n = 0.3487 + 0.6513\left(\frac{\rho_G}{\rho_L}\right)^{0.515}$$
Kumar et al. [194] (2017)	$$(1-a)^{23/7} - 2C_w Fr_L^2\left[1 \pm \frac{C_i}{C_w}\cdot\frac{(1-a)^{16/7}}{a^{5/2}}\cdot\frac{\rho_c}{\rho_L}\left(\frac{j_G}{j_L}\right)^2\right] = 0$$ $$\text{Free falling film, } a = 1 - \left(2C_w Fr_L^2\right)^{1/3}$$

4. The Entrained Droplets in the Central Gas Core

4.1. Droplet Behaviour

The surface instability is the reason for droplet formation and its entrainments. In an annular two-phase flow, the top part of the disturbance wave is undercut and forms an open-ended bubble with a filament rim. This bag breaks into many small droplets and the rim's break-up results in a smaller number of larger drops. Some of the droplets can deposit onto the liquid film and hence leads to a decrease in the entrainment. The droplet entrainment was well discussed and summarised by Prof. Azzopardi [7] and supported by recent experimental results as shown in Figure 15 [48]. Kumar et al. [231] numerically observed wave-like protrusion in an orifice, a bag-like break-up in undercutting, and ligament fragmentation that results in dislodge of droplets from the film (Figure 15g). The entrained droplet generally moves with the gas core, with its distribution of size varying in time and space. Tube diameter has been found to have little influence on the sizes of drops [169]. In contrast, the drop size was determined by the liquid flow rate and viscosity, increasing with increases in both [180,232]. The models used to describe the onset of entrained liquid fraction were reported invalid with high viscosity liquid [233] and the droplets in the early stage could be larger than any stable droplet in convection pipe flow [234].

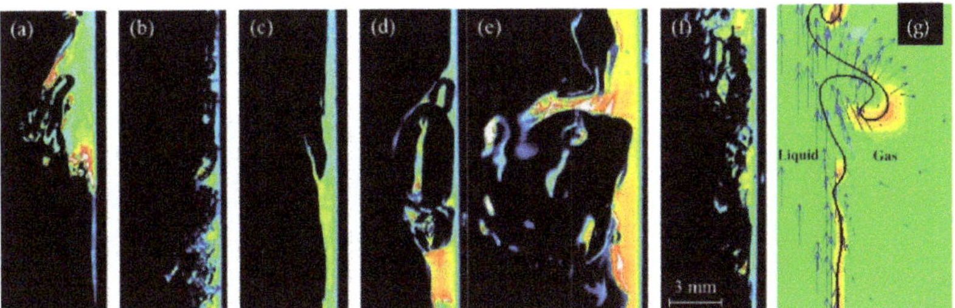

Figure 15. PLIF images showing the droplet entrainment process: (**a**) disturbance wave, (**b**) wave undercut (**c**) ligament break-up, (**d**,**e**) different stages of a bubble burst, (**f**) liquid impingement, adapted with permission from [48], 2014, Elsevier and (**g**) simulation of the droplet formation of undercutting zone, adapted with permission from [231], 2016, Elsevier.

Okawa et al. [171] conducted a series of measurements focusing on the droplet entrainment and deposition of an annular two-phase flow in a small vertical tube. The entrainment rate had a good correlation with the dimensionless number denoting the interfacial shear force and surface tension, while the deposition rate was determined by the droplet concentration in the gas core. It is also reported that at low drop concentrations, the deposition rate varied linearly but was not sensitive at high concentrations [232]. In another work [177], the droplet deposition rate was determined by the deposition constant and the drop velocity fluctuations. Xie et al. [200] studied the three-dimensional droplet deposition process and reported detailed complex interfacial structures during droplet impact.

Zhang et al. [32] measured the size and velocity of the droplets and reported that the ligament break-up of waves produced a large number of drops with high velocity and sphericity. The droplets had a continuous size distribution and the size increased slightly during the accelerated migration in the gaseous core. The droplets experienced axial acceleration and radial deceleration during the radial migration, but the overall distribution of droplet size and velocity remained unchanged. Starting with a small number, the proportion of relatively large drops gradually increased and their impacts on the total momentum could not be neglected. Alamu and Azzopardi [99,100] reported that both wave and droplet dominant frequency increased with an increase in the superficial velocities of both gas and liquid and drops became more elastic as liquid superficial velocity increased.

Trabold and Kumar [161,235] found the vapour turbulence was enhanced by the droplets over the range of drop size and concentration. However, improvement of the experimental technique and reduction in the measurement uncertainties were highly recommended as the uncertainty could be up to 25%. To further reconcile the droplet diameters obtained by various researchers, a definitive study is also highly recommended [7]. Liu and Li [202] showed it was possible to numerically predict the drop size distribution in annular two-phase flow using coalescence and breakup kernels. However, the only good prediction of the large droplets implied that it was necessary to develop a more accurate model. Based on Kataoka's correlation [236], Fore et al. [165] proposed a prediction of the drop size, which is given in Table 3 as well as other recently proposed correlations.

Table 3. Predictions of the drop size in an annular flow.

Reference	Prediction of the Droplet Size
Kocamustafaogullari et al. [234] (1994)	$\frac{d_{32}}{D} = 0.64 C_W^{-4/15} We_m^{-3/5} \left(\frac{Re_G^4}{Re_L}\right)^{4/15} \left(\frac{\rho_G}{\rho_L}\right)^{4/15} \left(\frac{\mu_G}{\mu_L}\right)^{4/15}$ $C_w = 1/35.34 N_\mu^{\frac{4}{5}} \ (N_\mu \leq 1/15) \ C_w = 0.25 \ (N_\mu > 1/15)$
Azzopardi [7] (1997)	$\frac{d_{32}}{D} = 1.91 Re_G^{0.1} We_G^{-0.6} \left(\frac{\rho_G}{\rho_L}\right)^{0.6} + 0.4 \frac{Ej_L}{j_G}$
Fore et al. [165] (2002)	$\frac{d_v}{D} = 0.028 We_G^{-1} Re_L^{-1/6} Re_G^{2/3} \left(\frac{\rho_G}{\rho_L}\right)^{-1/3} \left(\frac{\mu_G}{\mu_L}\right)^{2/3}$
Azzopardi [237] (2006)	$d_{32} = \left[0.069 j_G + 0.0187 \left(\frac{\rho_L j_L}{\rho_G j_G}\right)^2\right] \frac{\sigma}{\rho_G j_G}$
Berna et al. [8] (2015)	$\frac{d_v}{D} = 0.11 We_G^{-0.68} Re_G^{0.33} Re_L^{0.11} \left(\frac{\rho_G}{\rho_L}\right)^{0.31}$
Wang et al. [180] (2020)	$\frac{d_{32}}{D} = 0.022 We_G^{-0.545} We_L^{0.214} Re_L^{-0.249} Re_G^{0.439} \left(\frac{\rho_G}{\rho_L}\right)^{0.117}$

4.2. Correlation of Droplet Entrainment

Droplet entrainment fraction (E) is defined as the ratio of the entrained liquid drop mass flow rate divided by the total liquid mass flow rate and the entrainment rate (ε) is defined as the entrained drop mass rate per unit area of the gas-liquid interface. It has been shown that most of the predictions developed were very restricted in a wide variety of flow conditions. And there has not been a general correlation for the entrainment fraction so far [238]. The correlations of the droplet entrainment rate and entrainment fraction of the annular two-phase flow reported recently are summarised in Tables 4 and 5 in this section.

In our recent work, the droplet entrainment of an annular two-phase flow in a vertical pipe was obtained using the film extraction technique. The entrainment fraction is plotted against various flow rates at different heights, which shows the developing process of the entrained drops to a steady state. The impact of the ventilation on the flow behaviour of the annular flow, i.e., with or without the central gas flow, is examined in form of entrained droplets, which shows a negligible difference in entrainment fraction. These results will be published soon with annular velocity and thickness profiles.

Table 4. Correlations for the droplet entrainment rate in annular flows.

Reference	Correlations of the Droplet Entrainment Rate
Bertodano et al. [170] (1997)	$\frac{\varepsilon D}{\mu_L} = k\left[We_G\left(\frac{\Delta\rho}{\rho_G}\right)^{1/2}\left(Re_{Lf} - Re_{Lfc}\right)\right]^{0.925}\left(\frac{\mu_G}{\mu_L}\right)^{0.26}$ $Re_{Lf} = Re_L(1-E)$
Kataoka et al. [239] (2000)	$\frac{\varepsilon D}{\mu_L} = 6.6 \times 10^{-7} Re_L^{0.74} Re_{Lf}^{0.185} We^{0.925}\left(\frac{\mu_G}{\mu_L}\right)^{0.26}$
Bertodano et al. [240] (2001)	$\frac{\varepsilon D}{\mu_L} = \frac{3.8\times 10^{-6}\sigma}{4}\left(Re_{Lf} - Re_{Lfc}\right)We_G\left(\frac{\Delta\rho}{\rho_G}\right)^{1/2}$
Okawa and Kataoka [241] (2005)	$\varepsilon = \rho_L min\left(0.0038\pi_{e1}, 0.0012\pi_{e1}^{0.5}\right)$ $\pi_{e1} = \frac{f_i\rho_G\left(j_G^2 - j_{Gc}^2\right)\delta}{\sigma}$
Ryu and Park [242] (2011)	$\varepsilon = \rho_L V_e \frac{u_G - u_{DW}}{4h\lambda^2\sqrt{a}}$
Liu and Bai [243] (2017)	$\varepsilon = 4.347 \times 10^{-6}\rho_L Re_L^{0.584}\left(\frac{\rho_L}{\rho_G}\right)^{0.0561}\left(\frac{\tau_i\delta}{\sigma}\right)^{0.391}\left(\frac{D}{\sqrt{\sigma/g\Delta\rho}}\right)^{-0.291}$ $\tau_i = f_i\rho_G j_G^2/2$
Wang et al. [244] (2020)	$\varepsilon = \varepsilon_{max} \times tanh\left(3.56 \times 10^{-6} Re_{la}^{0.47} We^{1.15}\right)$ $\frac{\varepsilon_{max}\sqrt{\sigma/g\Delta\rho}}{\mu_L} = \begin{cases} 0.00515 Re_{la}^{0.69} & (Re_{la} \leq 800) \\ 0.521 & (Re_{la} > 800) \end{cases}$ $Re_{la} = \frac{\rho_L j_L \sqrt{\sigma/g\Delta\rho}}{\mu_L}$

Table 5. Correlations for entrainment fraction in annular two-phase flows.

Reference	Correlations for the Entrainment Fraction
Oliemans et al. [245] (1986)	$\frac{E}{1-E} = 10^{-0.25}\rho_L^{1.08}\rho_G^{0.18}\mu_L^{0.27}\mu_G^{0.28}\sigma^{1.80}D^{1.72}j_L^{0.7}j_G^{1.44}g^{0.46}$
Ishii and Mishima [246] (1989)	$E = tanh\left(7.25 \times 10^{-7} We_G^{1.25} Re_L^{0.25}\right)$
Utsono and Kaminanga [247] (1998)	$E = tanh\left(0.16 Re_L^{0.16} We_G^{0.08} - 1.2\right)$
Petalas and Aziz [248] (2000)	$\frac{E}{1-E} = 0.735\left(\frac{\mu_L^2 j_G^2\rho_G}{\sigma^2\rho_L}\right)^{0.074}\left(\frac{j_L}{j_G}\right)^{0.2}$
Barbosa et al. [181] (2002)	$E = 0.95 + 342.55\sqrt{\frac{\rho_L\dot{m}_L}{\rho_G\dot{m}_G}D^2}$
Pan and Hanratty [249] (2002)	$\frac{E/E_M}{1-E/E_M} = 6 \times 10^{-5}(u_G - u_{Gc})^2\sqrt{\rho_G\rho_L}D/\sigma$ $E_M = 1 - \frac{\dot{m}_{lfc}}{\dot{m}_l}$
Sawant et al. [250] (2008)	$E = \left(1 - \frac{250\ln Re_L - 1265}{Re_L}\right)tanh\left(2.31 \times 10^{-4} Re_L^{-0.35} We_G^{1.25}\right)$
Sawant et al. [172] (2009)	$E = \left[1 - \frac{13N_{\mu f}^{-0.5} + 0.3\left(Re_L - 13N_{\mu f}^{-0.5}\right)^{0.95}}{Re_L}\right]tanh\left(2.31 \times 10^{-4} Re_L^{-0.35} We_G^{1.25}\right)$ $N_{\mu f} = \frac{\mu_L}{\sqrt{\left(\rho_L\sigma\sqrt{\frac{\sigma}{g\Delta\rho}}\right)}}$
Cioncolini and Thome [251] (2010)	$E = \left(1 + 13.18 We_c^{-0.655}\right)^{-10.77}$ $We_c = \frac{\rho_c u_c^2 D_c}{\sigma}$
Cioncolini and Thome [252] (2012)	$E = \left(1 + 279.6 We_c^{-0.8395}\right)^{-2.209}$ $We_c = \frac{\rho_c j_G^2 D_c}{\sigma}$

Table 5. *Cont.*

Reference	Correlations for the Entrainment Fraction
Berna et al. [8] (2015)	$\frac{E}{1-E} = 5.51 \times 10^{-7} We_G^{2.68} Re_G^{-2.62} Re_L^{0.34} \left(\frac{\rho_G}{\rho_L}\right)^{-0.37} \left(\frac{\mu_G}{\mu_L}\right)^{-3.71} C_w^{4.24}$ $C_w = \begin{cases} 0.028 N_\mu^{-0.8} \ (N_\mu \leq 1/15) \\ 0.25 \ (N_\mu > 1/15) \end{cases}$
Aliyu et al. [152] (2017)	$E = \begin{cases} \dfrac{1 \times 10^{-2} We^{0.33} Re_L^{0.27}}{1 + 1 \times 10^{-2} We^{0.33} Re_L^{0.27}} & (j_G > 40 \ m/s) \\ \dfrac{1.25 \times 10^{-3} We^{0.15} Re_G^{0.2} Re_L^{0.23}}{1 + 1.25 \times 10^{-3} We^{0.15} Re_G^{0.2} Re_L^{0.23}} & (j_G \ll 40 \ m/s) \end{cases}$
Zhang et al. [253] (2020)	$E \approx 0.075 \dfrac{D \rho_L \rho_G^2 \delta^2 u_*^8}{j_L \mu_G^{\frac{2}{3}} \sigma^{\frac{7}{3}} u_G^{\frac{5}{3}}} \left(\frac{\delta}{D}\right)^{\frac{1}{3}}$

5. Conclusions and Recommendations

Annular two-phase flows in vertical pipes widely exist in many industrial applications, which have been investigated since the early years of fluid mechanics. This is also the case with annular flow in building drainage systems. Aiming to provide a general understanding of the current state of understanding on the topic, this work provides a review of the research in the most recent two decays, with a particular focus on the experimental techniques and advanced understanding of the mechanism underpinning annular flow.

Experimental techniques used in the field are summarised and their advantages and limitations are discussed in Section 2.15. These challenges, such as distortion of the transparent wall in optical measurements, strong fluctuation of the liquid film, and disturbance induced by the inserted probes, further imply the need for new or improved techniques to capture the detailed flow properties of annular two-phase flow, particular with these characteristics, high accuracy, simplicity and suitability for wide industrial applications. Lots of analytical expressions and empirical correlations of the liquid film thickness, disturbance wave property, droplet entrainment, drop size and void fraction has been proposed. on the other hand, this indicates the lack of a well-accepted model that can accurately predict the flow parameters in a wide range of flow and pipe conditions.

With the development of experimental techniques and computing capacity in the last two decays, the understanding of the complex two-phase annular flow in vertical pipes has been significantly advanced, for instance, the clear capture of the liquid film fluctuation and droplet generation. It is not surprising to note that all previous investigations are focused on steady flow conditions, including the transient flow region near the inlet of a stable annular flow, which is more frequently used, easier to measure and simpler to model. However, unsteady annular two-phase flow, which can be also seen in some cases and is always found in a building drainage system, has not been widely investigated and is not well understood. The current understanding of unstable annular flow is generally derived from the steady-state annular flow. Therefore, investigation of unsteady annular flow and transient region of a stable flow is recommended with particular focus on the wavy liquid film, gas core and entrained drops. It would therefore be reasonable to suggest that building drainage codes (particularly for tall buildings) should incorporate elements of this new understanding.

It should be also emphasized that the gas flow in the central core region is not an essential requirement for annular two-phase flow in vertical pipes. When gas velocity in the central region is set to zero, or the ventilation pipe is closed in a free-falling flow, similar flow features should be observed in a vertical pipe, such as film development, disturbance wave, and droplet entrainment. But many previous theoretical or empirical models have gas flow as one of the dominating mechanisms and non-negligible gas superficial velocity in the formulas. These models have obvious limitations in predicting annular flow with no gas flow in the central region. Both facts mentioned here indicate the need to further advance the understanding of the flow mechanism and to develop new theoretical or empirical models.

Author Contributions: Conceptualization, Y.X.; methodology, Y.X.; formal analysis, Y.X.; investigation, Y.X.; writing—original draft preparation, Y.X.; writing—review and editing, M.G., Y.X., C.S., D.K. and D.C.; project administration, M.G., D.K. and D.C.; funding acquisition, M.G., D.K. and D.C. All authors have read and agreed to the published version of the manuscript.

Funding: This research was funded by Aliaxis S.A., and the APC was funded by Aliaxis S.A.

Data Availability Statement: Not applicable.

Conflicts of Interest: The authors declare no conflict of interest.

Abbreviations

Nomenclature List

Symbols		Greek Characters	
a	Void fraction	δ	Film thickness
C	Friction factor	$\bar{\delta}$	Time-averaged film thickness
d	Diameter of the drop	ε	Entrainment rate
D	Diameter of the pipe	λ	Wavelength
E	Entrainment fraction	μ	Dynamic viscosity
f	Friction factor	ν	Kinematic viscosity
Fr	Froude number	ρ	Density
g	acceleration of gravity	σ	Surface tension coefficient
h	Disturbance wave height	τ	Shear stress
j	Superficial velocity	**Subscripts**	
k	Wave number	$*$	Friction
Ka	Kapitza number	32	Sauter diameter
L	Length	$base$	The base of the disturbance wave
$\dot{m}$	Mass flow rate	c	Gas core
N_u	Viscosity number	DW	Disturbance wave
N_{uf}	Non-dimensional viscosity number	e	Entrained
Δp	Pressure difference	m	Modified
Re	Reynolds number	max	Maximum condition
St	Strouhal number	G	Gas
u	Velocity	Gc	Critical gas state
V	Volume	L	Liquid
We	Weber number	La	Laplace length
x	Vaper quality	lf	Liquid film
X_{tt}	Lockhart-Martinelli parameter	lfc	Critical film flow
$\left(\dfrac{dp}{dz}\right)_{fric}$	Pressure gradient due to friction loss	L, ref	Liquid at reference condition (at 20 °C)
		i	Interfacial
		v	Volume mean
		w	Wall

References

1. Gormley, M.; Aspray, T.J.; Kelly, D.A. COVID-19: Mitigating transmission via wastewater plumbing systems. *Lancet Glob. Health* **2020**, *8*, e643. [CrossRef]
2. Gormley, M.; Swaffield, J.; Sleigh, P.; Noakes, C. An assessment of, and response to, potential cross-contamination routes due to defective appliance water trap seals in building drainage systems. *Build. Serv. Eng. Res. Technol.* **2012**, *33*, 203–222. [CrossRef]
3. Stewart, C.; Gormley, M.; Xue, Y.; Kelly, D.; Campbell, D. Steady-State Hydraulic Analysis of High-Rise Building Wastewater Drainage Networks: Modelling Basis. *Buildings* **2021**, *11*, 344. [CrossRef]
4. Wong, L.-T.; Mui, K.-W.; Cheng, C.-L.; Leung, P.H.-M. Time-Variant Positive Air Pressure in Drainage Stacks as a Pathogen Transmission Pathway of COVID-19. *Int. J. Environ. Res. Public Health* **2021**, *18*, 6068. [CrossRef] [PubMed]
5. Gormley, M.; Aspray, T.J.; Kelly, D.A. Aerosol and bioaerosol particle size and dynamics from defective sanitary plumbing systems. *Indoor Air* **2021**, *31*, 1427–1440. [CrossRef] [PubMed]
6. Hewitt, G.F.; Hall-Taylor, N.S. *Annular Two-Phase Flow*; Elsevier: Amsterdam, The Netherlands, 2013.
7. Azzopardi, B.J. Drops in annular two-phase flow. *Int. J. Multiph. Flow* **1997**, *23*, 1–53.
8. Berna, C.; Escrivá, A.; Muñoz-Cobo, J.L.; Herranz, L.E. Review of droplet entrainment in annular flow: Characterization of the entrained droplets. *Prog. Nucl. Energy* **2015**, *79*, 64–86. [CrossRef]

9. Berna, C.; Escrivá, A.; Muñoz-Cobo, J.L.; Herranz, L.E. Review of droplet entrainment in annular flow: Interfacial waves and onset of entrainment. *Prog. Nucl. Energy* **2014**, *74*, 14–43. [CrossRef]

10. Wu, B.; Firouzi, M.; Mitchell, T.; Rufford, T.E.; Leonardi, C.; Towler, B. A critical review of flow maps for gas-liquid flows in vertical pipes and annuli. *Chem. Eng. J.* **2017**, *326*, 350–377. [CrossRef]

11. Rahman, M.; Heidrick, T.; Fleck, B. A critical review of advanced experimental techniques to measure two-phase gas/liquid flow. *Open Fuels Energy Sci. J.* **2009**, *2*, 54–70.

12. Tibiriçá, C.B.; Nascimento, F.J.d.; Ribatski, G. Film thickness measurement techniques applied to micro-scale two-phase flow systems. *Exp. Therm. Fluid Sci.* **2010**, *34*, 463–473. [CrossRef]

13. Cherdantsev, A.V.; An, J.S.; Charogiannis, A.; Markides, C.N. Simultaneous application of two laser-induced fluorescence approaches for film thickness measurements in annular gas-liquid flows. *Int. J. Multiph. Flow* **2019**, *119*, 237–258. [CrossRef]

14. Qiao, S.; Mena, D.; Kim, S. Inlet effects on vertical-downward air–water two-phase flow. *Nucl. Eng. Des.* **2017**, *312*, 375–388. [CrossRef]

15. van Nimwegen, A.T.; Portela, L.M.; Henkes, R.A.W.M. The effect of the diameter on air-water annular and churn flow in vertical pipes with and without surfactants. *Int. J. Multiph. Flow* **2017**, *88*, 179–190. [CrossRef]

16. Van Nimwegen, A.; Portela, L.; Henkes, R. The effect of surfactants on air–water annular and churn flow in vertical pipes. Part 1: Morphology of the air–water interface. *Int. J. Multiph. Flow* **2015**, *71*, 133–145. [CrossRef]

17. Van Nimwegen, A.; Portela, L.; Henkes, R. The effect of surfactants on air–water annular and churn flow in vertical pipes. Part 2: Liquid holdup and pressure gradient dynamics. *Int. J. Multiph. Flow* **2015**, *71*, 146–158. [CrossRef]

18. Van der Meulen, G.P. Churn-Annular Gas-Liquid Flows in Large Diameter Vertical Pipes. Ph.D. Thesis, University of Nottingham, Nottingham, UK, 2012.

19. Adomeit, P.; Renz, U. Hydrodynamics of three-dimensional waves in laminar falling films. *Int. J. Multiph. Flow* **2000**, *26*, 1183–1208. [CrossRef]

20. Takamasa, T.; Hazuku, T.; Hibiki, T. Experimental Study of gas-liquid two-phase flow affected by wall surface wettability. *Int. J. Heat Fluid Flow* **2008**, *29*, 1593–1602. [CrossRef]

21. Pham, S.H.; Kawara, Z.; Yokomine, T.; Kunugi, T. Detailed observations of wavy interface behaviors of annular two-phase flow on rod bundle geometry. *Int. J. Multiph. Flow* **2014**, *59*, 135–144. [CrossRef]

22. Bhagwat, S.M.; Ghajar, A.J. Similarities and differences in the flow patterns and void fraction in vertical upward and downward two phase flow. *Exp. Therm. Fluid Sci.* **2012**, *39*, 213–227. [CrossRef]

23. Waltrich, P.J.; Falcone, G.; Barbosa, J.R. Axial development of annular, churn and slug flows in a long vertical tube. *Int. J. Multiph. Flow* **2013**, *57*, 38–48. [CrossRef]

24. Posada, C.; Waltrich, P.J. Effect of forced flow oscillations on churn and annular flow in a long vertical tube. *Exp. Therm. Fluid Sci.* **2017**, *81*, 345–357. [CrossRef]

25. Milan, M.; Borhani, N.; Thome, J.R. Adiabatic vertical downward air–water flow pattern map: Influence of inlet device, flow development length and hysteresis effects. *Int. J. Multiph. Flow* **2013**, *56*, 126–137. [CrossRef]

26. Zhang, Z.; Li, Y.; Wang, Z.; Hu, Q.; Wang, D. Experimental study on radial evolution of droplets in vertical gas-liquid two-phase annular flow. *Int. J. Multiph. Flow* **2020**, *129*, 103325. [CrossRef]

27. Raeiszadeh, F.; Hajidavalloo, E.; Behbahaninejad, M.; Hanafizadeh, P. Effect of pipe rotation on downward co-current air-water flow in a vertical pipe. *Int. J. Multiph. Flow* **2016**, *81*, 1–14. [CrossRef]

28. Vijayan, M.; Jayanti, S.; Balakrishnan, A.R. Effect of tube diameter on flooding. *Int. J. Multiph. Flow* **2001**, *27*, 797–816. [CrossRef]

29. Dasgupta, A.; Chandraker, D.K.; Kshirasagar, S.; Reddy, B.R.; Rajalakshmi, R.; Nayak, A.K.; Walker, S.P.; Vijayan, P.K.; Hewitt, G.F. Experimental investigation on dominant waves in upward air-water two-phase flow in churn and annular regime. *Exp. Therm. Fluid Sci.* **2017**, *81*, 147–163. [CrossRef]

30. Da Hlaing, N.; Sirivat, A.; Siemanond, K.; Wilkes, J.O. Vertical two-phase flow regimes and pressure gradients: Effect of viscosity. *Exp. Therm. Fluid Sci.* **2007**, *31*, 567–577. [CrossRef]

31. Kumar, A.; Bhowmik, S.; Ray, S.; Das, G. Flow pattern transition in gas-liquid downflow through narrow vertical tubes. *AIChE J.* **2017**, *63*, 792–800. [CrossRef]

32. Zhang, Z.; Wang, Z.; Liu, H.; Gao, Y.; Li, H.; Sun, B. Experimental study on entrained droplets in vertical two-phase churn and annular flows. *Int. J. Heat Mass Transf.* **2019**, *138*, 1346–1358. [CrossRef]

33. Zhang, Z.; Wang, Z.; Liu, H.; Gao, Y.; Li, H.; Sun, B. Experimental study on bubble and droplet entrainment in vertical churn and annular flows and their relationship. *Chem. Eng. Sci.* **2019**, *206*, 387–400. [CrossRef]

34. Wang, L.-S.; Liu, S.; Hou, L.; Yang, M.; Zhang, J.; Xu, J. Prediction of the liquid film reversal of annular flow in vertical and inclined pipes. *Int. J. Multiph. Flow* **2022**, *146*, 103853. [CrossRef]

35. Schmid, D.; Verlaat, B.; Petagna, P.; Revellin, R.; Schiffmann, J. Flow pattern observations and flow pattern map for adiabatic two-phase flow of carbon dioxide in vertical upward and downward direction. *Exp. Therm. Fluid Sci.* **2022**, *131*, 110526. [CrossRef]

36. Dang, Z.; Yang, Z.; Yang, X.; Ishii, M. Experimental study on void fraction, pressure drop and flow regime analysis in a large ID piping system. *Int. J. Multiph. Flow* **2019**, *111*, 31–41. [CrossRef]

37. Chen, L.; Tian, Y.S.; Karayiannis, T.G. The effect of tube diameter on vertical two-phase flow regimes in small tubes. *Int. J. Heat Mass Transf.* **2006**, *49*, 4220–4230. [CrossRef]

38. Coleman, J.W.; Garimella, S. Characterization of two-phase flow patterns in small diameter round and rectangular tubes. *Int. J. Heat Mass Transf.* **1999**, *42*, 2869–2881. [CrossRef]

39. Pan, L.M.; He, H.; Ju, P.; Hibiki, T.; Ishii, M. Experimental study and modeling of disturbance wave height of vertical annular flow. *Int. J. Heat Mass Transf.* **2015**, *89*, 165–175. [CrossRef]

40. Pan, L.M.; He, H.; Ju, P.; Hibiki, T.; Ishii, M. The influences of gas-liquid interfacial properties on interfacial shear stress for vertical annular flow. *Int. J. Heat Mass Transf.* **2015**, *89*, 1172–1183. [CrossRef]

41. Schubring, D.; Shedd, T.A.; Hurlburt, E.T. Studying disturbance waves in vertical annular flow with high-speed video. *Int. J. Multiph. Flow* **2010**, *36*, 385–396. [CrossRef]

42. Lin, R.; Wang, K.; Liu, L.; Zhang, Y.; Dong, S. Study on the characteristics of interfacial waves in annular flow by image analysis. *Chem. Eng. Sci.* **2020**, *212*, 115336. [CrossRef]

43. Lin, R.; Wang, K.; Liu, L.; Zhang, Y.; Dong, S. Application of the image analysis on the investigation of disturbance waves in vertical upward annular two-phase flow. *Exp. Therm. Fluid Sci.* **2020**, *114*, 110062. [CrossRef]

44. Moreira, T.A.; Morse, R.W.; Dressler, K.M.; Ribatski, G.; Berson, A. Liquid-film thickness and disturbance-wave characterization in a vertical, upward, two-phase annular flow of saturated R245fa inside a rectangular channel. *Int. J. Multiph. Flow* **2020**, *132*, 103412. [CrossRef]

45. Barbosa, J.R., Jr.; Govan, A.H.; Hewitt, G.F. Visualisation and modelling studies of churn flow in a vertical pipe. *Int. J. Multiph. Flow* **2001**, *27*, 2105–2127. [CrossRef]

46. Häber, T.; Gebretsadik, M.; Bockhorn, H.; Zarzalis, N. The effect of total reflection in PLIF imaging of annular thin films. *Int. J. Multiph. Flow* **2015**, *76*, 64–72. [CrossRef]

47. van Eckeveld, A.C.; Gotfredsen, E.; Westerweel, J.; Poelma, C. Annular two-phase flow in vertical smooth and corrugated pipes. *Int. J. Multiph. Flow* **2018**, *109*, 150–163. [CrossRef]

48. Zadrazil, I.; Matar, O.K.; Markides, C.N. An experimental characterization of downwards gas-liquid annular flow by laser-induced fluorescence: Flow regimes and film statistics. *Int. J. Multiph. Flow* **2014**, *60*, 87–102. [CrossRef]

49. Schubring, D.; Ashwood, A.C.; Shedd, T.A.; Hurlburt, E.T. Planar laser-induced fluorescence (PLIF) measurements of liquid film thickness in annular flow. Part I: Methods and data. *Int. J. Multiph. Flow* **2010**, *36*, 815–824. [CrossRef]

50. Schubring, D.; Shedd, T.A.; Hurlburt, E.T. Planar laser-induced fluorescence (PLIF) measurements of liquid film thickness in annular flow. Part II: Analysis and comparison to models. *Int. J. Multiph. Flow* **2010**, *36*, 825–835. [CrossRef]

51. Kokomoor, W.; Schubring, D. Improved visualization algorithms for vertical annular flow. *J. Vis.* **2014**, *17*, 77–86. [CrossRef]

52. Xue, T.; Yang, L.; Ge, P.; Qu, L. Error analysis and liquid film thickness measurement in gas–liquid annular flow. *Optik* **2015**, *126*, 2674–2678. [CrossRef]

53. Liu, J.; Xue, T. Experimental investigation of liquid entrainment in vertical upward annular flow based on fluorescence imaging. *Prog. Nucl. Energy* **2022**, *152*, 104383. [CrossRef]

54. Xue, T.; Li, Z.; Li, C.; Wu, B. Measurement of thickness of annular liquid films based on distortion correction of laser-induced fluorescence imaging. *Rev. Sci. Instrum.* **2019**, *90*, 033103. [CrossRef] [PubMed]

55. Xue, T.; Li, C.; Wu, B. Distortion correction and characteristics measurement of circumferential liquid film based on PLIF. *AIChE J.* **2019**, *65*, e16612. [CrossRef]

56. Vasques, J.; Cherdantsev, A.; Cherdantsev, M.; Isaenkov, S.; Hann, D. Comparison of disturbance wave parameters with flow orientation in vertical annular gas-liquid flows in a small pipe. *Exp. Therm. Fluid Sci.* **2018**, *97*, 484–501. [CrossRef]

57. Fan, W.; Cherdantsev, A.; Anglart, H. Experimental and numerical study of formation and development of disturbance waves in annular gas-liquid flow. *Energy* **2020**, *207*, 118309.

58. Alekseenko, S.V.; Cherdantsev, A.V.; Cherdantsev, M.V.; Isaenkov, S.V.; Markovich, D.M. Study of formation and development of disturbance waves in annular gas-liquid flow. *Int. J. Multiph. Flow* **2015**, *77*, 65–75. [CrossRef]

59. Alekseenko, S.V.; Cherdantsev, A.V.; Heinz, O.M.; Kharlamov, S.M.; Markovich, D.M. Analysis of spatial and temporal evolution of disturbance waves and ripples in annular gas-liquid flow. *Int. J. Multiph. Flow* **2014**, *67*, 122–134. [CrossRef]

60. Alekseenko, S.; Cherdantsev, A.; Cherdantsev, M.; Isaenkov, S.; Kharlamov, S.; Markovich, D. Application of a high-speed laser-induced fluorescence technique for studying the three-dimensional structure of annular gas-liquid flow. *Exp. Fluids* **2012**, *53*, 77–89. [CrossRef]

61. Alekseenko, S.; Antipin, V.; Cherdantsev, A.; Kharlamov, S.; Markovich, D. Two-wave structure of liquid film and wave interrelation in annular gas-liquid flow with and without entrainment. *Phys. Fluids* **2009**, *21*, 061701. [CrossRef]

62. Farias, P.; Martins, F.; Sampaio, L.; Serfaty, R.; Azevedo, L. Liquid film characterization in horizontal, annular, two-phase, gas–liquid flow using time-resolved laser-induced fluorescence. *Exp. Fluids* **2012**, *52*, 633–645. [CrossRef]

63. Karimi, G.; Kawaji, M. An experimental study of freely falling films in a vertical tube. *Chem. Eng. Sci.* **1998**, *53*, 3501–3512. [CrossRef]

64. Zadrazil, I.; Markides, C.N. An experimental characterization of liquid films in downwards co-current gas-liquid annular flow by particle image and tracking velocimetry. *Int. J. Multiph. Flow* **2014**, *67*, 42–53. [CrossRef]

65. Zadrazil, I.; Markides, C.; Matar, O.; Náraigh, L.; Hewitt, G. Characterisation of downwards co-current gas-liquid annular flows. In Proceedings of the Seventh International Symposium On Turbulence, Heat and Mass Transfer, Palermo, Italy, 24–27 September 2012.

66. Ashwood, A.C.; Hogen, S.J.V.; Rodarte, M.A.; Kopplin, C.R.; Rodríguez, D.J.; Hurlburt, E.T.; Shedd, T.A. A multiphase, micro-scale PIV measurement technique for liquid film velocity measurements in annular two-phase flow. *Int. J. Multiph. Flow* **2015**, *68*, 27–39. [CrossRef]

67. Charogiannis, A.; An, J.S.; Voulgaropoulos, V.; Markides, C.N. Structured planar laser-induced fluorescence (S-PLIF) for the accurate identification of interfaces in multiphase flows. *Int. J. Multiph. Flow* **2019**, *118*, 193–204. [CrossRef]

68. Hazuku, T.; Takamasa, T.; Matsumoto, Y. Experimental study on axial development of liquid film in vertical upward annular two-phase flow. *Int. J. Multiph. Flow* **2008**, *34*, 111–127. [CrossRef]

69. Okawa, T.; Goto, T.; Yamagoe, Y. Liquid film behavior in annular two-phase flow under flow oscillation conditions. *Int. J. Heat Mass Transf.* **2010**, *53*, 962–971. [CrossRef]

70. Shedd, T.A.; Newell, T. Automated optical liquid film thickness measurement method. *Rev. Sci. Instrum.* **1998**, *69*, 4205–4213. [CrossRef]

71. Hurlburt, E.; Newell, T. Optical measurement of liquid film thickness and wave velocity in liquid film flows. *Exp. Fluids* **1996**, *21*, 357–362. [CrossRef]

72. Ohba, K.; Nagae, K. Characteristics and behavior of the interfacial wave on the liquid film in a vertically upward air-water two-phase annular flow. *Nucl. Eng. Des.* **1993**, *141*, 17–25. [CrossRef]

73. Lel, V.V.; Al-Sibai, F.; Leefken, A.; Renz, U. Local thickness and wave velocity measurement of wavy films with a chromatic confocal imaging method and a fluorescence intensity technique. *Exp. Fluids* **2005**, *39*, 856–864. [CrossRef]

74. Liang, F.; Fang, Z.; Chen, J.; Sun, S. Investigating the liquid film characteristics of gas–liquid swirling flow using ultrasound Doppler velocimetry. *AIChE J.* **2017**, *63*, 2348–2357. [CrossRef]

75. Wang, M.; Zheng, D.; Xu, Y. A new method for liquid film thickness measurement based on ultrasonic echo resonance technique in gas-liquid flow. *Meas. J. Int. Meas. Confed.* **2019**, *146*, 447–457. [CrossRef]

76. Al-Aufi, Y.A.; Hewakandamby, B.N.; Dimitrakis, G.; Holmes, M.; Hasan, A.; Watson, N.J. Thin film thickness measurements in two phase annular flows using ultrasonic pulse echo techniques. *Flow Meas. Instrum.* **2019**, *66*, 67–78. [CrossRef]

77. Wang, C.; Zhao, N.; Fang, L.; Zhang, T.; Feng, Y. Void fraction measurement using NIR technology for horizontal wet-gas annular flow. *Exp. Therm. Fluid Sci.* **2016**, *76*, 98–108. [CrossRef]

78. Wang, C.; Zhao, N.; Feng, Y.; Sun, H.; Fang, L. Interfacial wave velocity of vertical gas-liquid annular flow at different system pressures. *Exp. Therm. Fluid Sci.* **2018**, *92*, 20–32. [CrossRef]

79. Li, C.; Liu, M.; Zhao, N.; Wang, F.; Zhao, Z.; Guo, S.; Fang, L.; Li, X. Void fraction measurement using modal decomposition and ensemble learning in vertical annular flow. *Chem. Eng. Sci.* **2022**, *247*, 116929. [CrossRef]

80. Zhao, Z.; Wang, B.; Wang, J.; Fang, L.; Li, X.; Wang, F.; Zhao, N. Liquid film characteristics measurement based on NIR in gas–liquid vertical annular upward flow. *Meas. Sci. Technol.* **2022**, *33*, 065014. [CrossRef]

81. Hawkes, N.J.; Lawrence, C.J.; Hewitt, G.F. Studies of wispy-annular flow using transient pressure gradient and optical measurements. *Int. J. Multiph. Flow* **2000**, *26*, 1565–1582. [CrossRef]

82. Coney, M.W.E. The theory and application of conductance probes for the measurement of liquid film thickness in two-phase flow. *J. Phys. E: Sci. Instrum.* **1973**, *6*, 903–911. [CrossRef]

83. Hewitt, G.F. Measurement of two phase flow parameters. *Nasa Sti/recon Tech. Rep. A* **1978**, *79*, 47262.

84. Damsohn, M.; Prasser, H.M. High-speed liquid film sensor with high spatial resolution. *Meas. Sci. Technol.* **2009**, *20*, 114001. [CrossRef]

85. Wang, C.; Zhao, N.; Chen, C.; Sun, H. A method for direct thickness measurement of wavy liquid film in gas-liquid two-phase annular flow using conductance probes. *Flow Meas. Instrum.* **2018**, *62*, 66–75. [CrossRef]

86. Polansky, J.; Wang, M. Vertical annular flow pattern characterisation using proper orthogonal decomposition of Electrical Impedance Tomography. *Flow Meas. Instrum.* **2018**, *62*, 281–296. [CrossRef]

87. Brown, R.C.; Andreussi, P.; Zanelli, S. The use of wire probes for the measurement of liquid film thickness in annular gas-liquid flows. *Can. J. Chem. Eng.* **1978**, *56*, 754–757. [CrossRef]

88. Zabaras, G.; Dukler, A.E.; Moalem-Maron, D. Vertical upward cocurrent gas-liquid annular flow. *AIChE J.* **1986**, *32*, 829–843. [CrossRef]

89. Karapantsios, T.D.; Paras, S.V.; Karabelas, A.J. Statistical characteristics of free falling films at high reynolds numbers. *Int. J. Multiph. Flow* **1989**, *15*, 1–21. [CrossRef]

90. Koskie, J.E.; Mudawar, I.; Tiederman, W.G. Parallel-wire probes for measurement of thick liquid films. *Int. J. Multiph. Flow* **1989**, *15*, 521–530. [CrossRef]

91. Ruder, Z.; Hanratty, T.J. A definition of gas-liquid plug flow in horizontal pipes. *Int. J. Multiph. Flow* **1990**, *16*, 233–242. [CrossRef]

92. Kumar, R.; Gottmann, M.; Sridhar, K. Film thickness and wave velocity measurements in a vertical duct. *J. Fluids Eng.* **2002**, *124*, 634–642. [CrossRef]

93. Usui, K.; Sato, K. Vertically downward two-phase flow,(I) Void distribution and average void fraction. *J. Nucl. Sci. Technol.* **1989**, *26*, 670–680. [CrossRef]

94. Fukano, T.; Furukawa, T. Prediction of the effects of liquid viscosity on interfacial shear stress and frictional pressure drop in vertical upward gas-liquid annular flow. *Int. J. Multiph. Flow* **1998**, *24*, 587–603. [CrossRef]

95. Al-Sarkhi, A.; Sarica, C.; Magrini, K. Inclination effects on wave characteristics in annular gas–liquid flows. *AIChE J.* **2012**, *58*, 1018–1029. [CrossRef]

96. Fore, L.B.; Beus, S.G.; Bauer, R.C. Interfacial friction in gas-liquid annular flow: Analogies to full and transition roughness. *Int. J. Multiph. Flow* **2000**, *26*, 1755–1769. [CrossRef]

97. Paras, S.V.; Karabelas, A.J. Properties of the liquid layer in horizontal annular flow. *Int. J. Multiph. Flow* **1991**, *17*, 439–454. [CrossRef]

98. Li, W.; Zhou, F.; Li, R.; Zhou, L. Experimental study on the characteristics of liquid layer and disturbance waves in horizontal annular flow. *J. Therm. Sci.* **1999**, *8*, 235–242. [CrossRef]

99. Alamu, M.B.; Azzopardi, B.J. Simultaneous investigation of entrained liquid fraction, liquid film thickness and pressure drop in vertical annular flow. *J. Energy Resour. Technol. Trans. ASME* **2011**, *133*, 023103. [CrossRef]

100. Alamu, M.B.; Azzopardi, B.J. Wave and drop periodicity in transient annular flow. *Nucl. Eng. Des.* **2011**, *241*, 5079–5092. [CrossRef]

101. Al-Yarubi, Q. Phase Flow Rate Measurements of Annular Flows. Ph.D. Thesis, University of Huddersfield, Huddersfield, UK, 2010.

102. Han, H.; Zhu, Z.; Gabriel, K. A study on the effect of gas flow rate on the wave characteristics in two-phase gas–liquid annular flow. *Nucl. Eng. Des.* **2006**, *236*, 2580–2588. [CrossRef]

103. Smith, T.R.; Schlegel, J.P.; Hibiki, T.; Ishii, M. Two-phase flow structure in large diameter pipes. *Int. J. Heat Fluid Flow* **2012**, *33*, 156–167. [CrossRef]

104. Posada, C.; Waltrich, P.J. Similarities and differences in churn and annular flow regimes in steady-state and oscillatory flows in a long vertical tube. *Exp. Therm. Fluid Sci.* **2018**, *93*, 272–284. [CrossRef]

105. Dao, E.K.; Balakotaiah, V. Experimental study of wave occlusion on falling films in a vertical pipe. *AIChE J.* **2000**, *46*, 1300–1306. [CrossRef]

106. Wang, G.; Zhu, Q.; Ishii, M.; Buchanan, J.R. Four-sensor droplet capable conductivity probe for measurement of churn-turbulent to annular transition flow. *Int. J. Heat Mass Transf.* **2020**, *157*, 119949. [CrossRef]

107. Zhu, Q.; Wang, G.; Schlegel, J.P.; Yan, Y.; Yang, X.; Liu, Y.; Ishii, M.; Buchanan, J.R. Experimental study of two-phase flow structure in churn-turbulent to annular flows. *Exp. Therm. Fluid Sci.* **2021**, *129*, 110397. [CrossRef]

108. Abdulkadir, M.; Azzi, A.; Zhao, D.; Lowndes, I.S.; Azzopardi, B.J. Liquid film thickness behaviour within a large diameter vertical 180° return bend. *Chem. Eng. Sci.* **2014**, *107*, 137–148. [CrossRef]

109. Wang, G.; Dang, Z.; Ishii, M. Wave structure and velocity in vertical upward annular two-phase flow. *Exp. Therm. Fluid Sci.* **2021**, *120*, 110205. [CrossRef]

110. Fukano, T. Measurement of time varying thickness of liquid film flowing with high speed gas flow by a constant electric current method (CECM). *Nucl. Eng. Des.* **1998**, *184*, 363–377. [CrossRef]

111. Wolf, A.; Jayanti, S.; Hewitt, G.F. Flow development in vertical annular flow. *Chem. Eng. Sci.* **2001**, *56*, 3221–3235. [CrossRef]

112. Sawant, P.; Ishii, M.; Hazuku, T.; Takamasa, T.; Mori, M. Properties of disturbance waves in vertical annular two-phase flow. *Nucl. Eng. Des.* **2008**, *238*, 3528–3541. [CrossRef]

113. Tiwari, R.; Damsohn, M.; Prasser, H.-M.; Wymann, D.; Gossweiler, C. Multi-range sensors for the measurement of liquid film thickness distributions based on electrical conductance. *Flow Meas. Instrum.* **2014**, *40*, 124–132. [CrossRef]

114. Setyawan, A. The effect of the fluid properties on the wave velocity and wave frequency of gas–liquid annular two-phase flow in a horizontal pipe. *Exp. Therm. Fluid Sci.* **2016**, *71*, 25–41. [CrossRef]

115. Omebere-Iyari, N.K.; Azzopardi, B.J. A Study of Flow Patterns for Gas/Liquid Flow in Small Diameter Tubes. *Chem. Eng. Res. Des.* **2007**, *85*, 180–192. [CrossRef]

116. Kaji, R.; Azzopardi, B.J. The effect of pipe diameter on the structure of gas/liquid flow in vertical pipes. *Int. J. Multiph. Flow* **2010**, *36*, 303–313. [CrossRef]

117. Wang, Z.; Gabriel, K.S.; Manz, D.L. The influences of wave height on the interfacial friction in annular gas-liquid flow under normal and microgravity conditions. *Int. J. Multiph. Flow* **2004**, *30*, 1193–1211. [CrossRef]

118. Zhao, Y.; Markides, C.N.; Matar, O.K.; Hewitt, G.F. Disturbance wave development in two-phase gas-liquid upwards vertical annular flow. *Int. J. Multiph. Flow* **2013**, *55*, 111–129. [CrossRef]

119. Azzopardi, B.J. Disturbance wave frequencies, velocities and spacing in vertical annular two-phase flow. *Nucl. Eng. Des.* **1986**, *92*, 121–133. [CrossRef]

120. Abdulkadir, M.; Zhao, D.; Azzi, A.; Lowndes, I.S.; Azzopardi, B.J. Two-phase air-water flow through a large diameter vertical 180 o return bend. *Chem. Eng. Sci.* **2012**, *79*, 138–152. [CrossRef]

121. Dang, Z.; Yang, Z.; Yang, X.; Ishii, M. Experimental study of vertical and horizontal two-phase pipe flow through double 90 degree elbows. *Int. J. Heat Mass Transf.* **2018**, *120*, 861–869. [CrossRef]

122. Alekseenko, S.V.; Aktershev, S.P.; Cherdantsev, A.V.; Kharlamov, S.M.; Markovich, D.M. Primary instabilities of liquid film flow sheared by turbulent gas stream. *Int. J. Multiph. Flow* **2009**, *35*, 617–627. [CrossRef]

123. Belt, R.J.; Westende, J.M.C.V.; Prasser, H.M.; Portela, L.M. Time and spatially resolved measurements of interfacial waves in vertical annular flow. *Int. J. Multiph. Flow* **2010**, *36*, 570–587. [CrossRef]

124. Muñoz-Cobo, J.-L.; Rivera, Y.; Berna, C.; Escrivá, A. Analysis of Conductance Probes for Two-Phase Flow and Holdup Applications. *Sensors* **2020**, *20*, 7042. [CrossRef]

125. Rivera, Y.; Muñoz-Cobo, J.L.; Cuadros, J.L.; Berna, C.; Escrivá, A. Experimental study of the effects produced by the changes of the liquid and gas superficial velocities and the surface tension on the interfacial waves and the film thickness in annular concurrent upward vertical flows. *Exp. Therm. Fluid Sci.* **2021**, *120*, 110224. [CrossRef]

126. Fershtman, A.; Robers, L.; Prasser, H.-M.; Barnea, D.; Shemer, L. Interfacial structure of upward gas–liquid annular flow in inclined pipes. *Int. J. Multiph. Flow* **2020**, *132*, 103437. [CrossRef]

127. Rivera, Y.; Berna, C.; Muñoz-Cobo, J.L.; Escrivá, A.; Córdova, Y. Experiments in free falling and downward cocurrent annular flows—Characterization of liquid films and interfacial waves. *Nucl. Eng. Des.* **2022**, *392*, 111769. [CrossRef]

128. Cuadros, J.L.; Rivera, Y.; Berna, C.; Escrivá, A.; Muñoz-Cobo, J.L.; Monrós-Andreu, G.; Chiva, S. Characterization of the gas-liquid interfacial waves in vertical upward co-current annular flows. *Nucl. Eng. Des.* **2019**, *346*, 112–130. [CrossRef]

129. Abdulkadir, M.; Mbalisigwe, U.P.; Zhao, D.; Hernandez-Perez, V.; Azzopardi, B.J.; Tahir, S. Characteristics of churn and annular flows in a large diameter vertical riser. *Int. J. Multiph. Flow* **2019**, *113*, 250–263. [CrossRef]

130. Fossa, M. Design and performance of a conductance probe for measuring the liquid fraction in two-phase gas-liquid flows. *Flow Meas. Instrum.* **1998**, *9*, 103–109. [CrossRef]

131. Devia, F.; Fossa, M. Design and optimisation of impedance probes for void fraction measurements. *Flow Meas. Instrum.* **2003**, *14*, 139–149. [CrossRef]

132. Yang, Z.; Dang, Z.; Yang, X.; Ishii, M.; Shan, J. Downward two phase flow experiment and general flow regime transition criteria for various pipe sizes. *Int. J. Heat Mass Transf.* **2018**, *125*, 179–189. [CrossRef]

133. Libert, N.; Morales, R.E.M.; da Silva, M.J. Capacitive measuring system for two-phase flow monitoring. Part 1: Hardware design and evaluation. *Flow Meas. Instrum.* **2016**, *47*, 90–99. [CrossRef]

134. Jaworek, A.; Krupa, A.; Trela, M. Capacitance sensor for void fraction measurement in water/steam flows. *Flow Meas. Instrum.* **2004**, *15*, 317–324. [CrossRef]

135. de Oliveira, P.M.; Strle, E.; Barbosa, J.R. Developing air-water flow downstream of a vertical 180° return bend. *Int. J. Multiph. Flow* **2014**, *67*, 32–41. [CrossRef]

136. Huang, Z.; Xie, D.; Zhang, H.; Li, H. Gas–oil two-phase flow measurement using an electrical capacitance tomography system and a Venturi meter. *Flow Meas. Instrum.* **2005**, *16*, 177–182. [CrossRef]

137. Zangl, H.; Fuchs, A.; Bretterklieber, T. Non-invasive measurements of fluids by means of capacitive sensors. *e & i Elektrotechnik Und Inf.* **2009**, *126*, 8–12.

138. Pawloski, J.; Ching, C.; Shoukri, M. Measurement of void fraction and pressure drop of air-oil two-phase flow in horizontal pipes. *J. Eng. Gas Turbines Power* **2004**, *126*, 107–118. [CrossRef]

139. Ahmed, H. Capacitance sensors for void-fraction measurements and flow-pattern identification in air–oil two-phase flow. *IEEE Sens. J.* **2006**, *6*, 1153–1163. [CrossRef]

140. Elkow, K.J.; Rezkallah, K.S. Void fraction measurements in gas-liquid flows using capacitance sensors. *Meas. Sci. Technol.* **1996**, *7*, 1153. [CrossRef]

141. Atkinson, C.; Huang, R. A theoretical model for capacitance measurement of liquid films in an annular flow. *Math.-Ind. Case Stud.* **2017**, *7*, 3. [CrossRef]

142. De Kerpel, K.; Ameel, B.; de Schampheleire, S.; T'Joen, C.; Canière, H.; de Paepe, M. Calibration of a capacitive void fraction sensor for small diameter tubes based on capacitive signal features. *Appl. Therm. Eng.* **2014**, *63*, 77–83. [CrossRef]

143. De Kerpel, K.; Ameel, B.; T'Joen, C.; Canière, H.; de Paepe, M. Flow regime based calibration of a capacitive void fraction sensor for small diameter tubes. *Int. J. Refrig.* **2013**, *36*, 390–401. [CrossRef]

144. Vieira, R.E.; Parsi, M.; Torres, C.F.; McLaury, B.S.; Shirazi, S.A.; Schleicher, E.; Hampel, U. Experimental characterization of vertical gas-liquid pipe flow for annular and liquid loading conditions using dual Wire-Mesh Sensor. *Exp. Therm. Fluid Sci.* **2015**, *64*, 81–93. [CrossRef]

145. Vieira, R.E.; Parsi, M.; McLaury, B.S.; Shirazi, S.A.; Torres, C.F.; Schleicher, E.; Hampel, U. Experimental characterization of vertical downward two-phase annular flows using Wire-Mesh Sensor. *Chem. Eng. Sci.* **2015**, *134*, 324–339. [CrossRef]

146. Johnson, I.D. Method and Apparatus for Measuring Water in Crude Oil. U.S. Patent 4644263A, 17 February 1987. Available online: https://patents.google.com/patent/US4644263A/en (accessed on 6 May 2020).

147. Prasser, H.M.; Böttger, A.; Zschau, J. A new electrode-mesh tomograph for gas–liquid flows. *Flow Meas. Instrum.* **1998**, *9*, 111–119. [CrossRef]

148. Almabrok, A.A.; Aliyu, A.M.; Lao, L.; Yeung, H. Gas/liquid flow behaviours in a downward section of large diameter vertical serpentine pipes. *Int. J. Multiph. Flow* **2016**, *78*, 25–43. [CrossRef]

149. Banowski, M.; Beyer, M.; Szalinski, L.; Lucas, D.; Hampel, U. Comparative study of ultrafast X-ray tomography and wire-mesh sensors for vertical gas–liquid pipe flows. *Flow Meas. Instrum.* **2017**, *53*, 95–106. [CrossRef]

150. Misawa, M.; Tiseanu, I.; Prasser, H.; Ichikawa, N.; Akai, M. Ultra-fast x-ray tomography for multi-phase flow interface dynamic studies. *Kerntechnik* **2003**, *68*, 85–90. [CrossRef]

151. Aliyu, A.M.; Lao, L.; Almabrok, A.A.; Yeung, H. Interfacial shear in adiabatic downward gas/liquid co-current annular flow in pipes. *Exp. Therm. Fluid Sci.* **2016**, *72*, 75–87. [CrossRef]

152. Aliyu, A.M.; Almabrok, A.A.; Baba, Y.D.; Archibong, A.E.; Lao, L.; Yeung, H.; Kim, K.C. Prediction of entrained droplet fraction in co-current annular gas–liquid flow in vertical pipes. *Exp. Therm. Fluid Sci.* **2017**, *85*, 287–304. [CrossRef]

153. Tompkins, C.; Prasser, H.-M.; Corradini, M. Wire-mesh sensors: A review of methods and uncertainty in multiphase flows relative to other measurement techniques. *Nucl. Eng. Des.* **2018**, *337*, 205–220. [CrossRef]
154. Velasco Peña, H.F.; Rodriguez, O.M.H. Applications of wire-mesh sensors in multiphase flows. *Flow Meas. Instrum.* **2015**, *45*, 255–273. [CrossRef]
155. Lucas, D.; Beyer, M.; Szalinski, L.; Schütz, P. A new database on the evolution of air-water flows along a large vertical pipe. *Int. J. Therm. Sci.* **2010**, *49*, 664–674. [CrossRef]
156. Da Silva, M.; Schleicher, E.; Hampel, U. Capacitance wire-mesh sensor for fast measurement of phase fraction distributions. *Meas. Sci. Technol.* **2007**, *18*, 2245. [CrossRef]
157. Zboray, R.; Prasser, H.-M. Measuring liquid film thickness in annular two-phase flows by cold neutron imaging. *Exp. Fluids* **2013**, *54*, 1596. [CrossRef]
158. Zboray, R.; Kickhofel, J.; Damsohn, M.; Prasser, H.-M. Cold-neutron tomography of annular flow and functional spacer performance in a model of a boiling water reactor fuel rod bundle. *Nucl. Eng. Des.* **2011**, *241*, 3201–3215. [CrossRef]
159. Zboray, R.; Prasser, H.-M. Optimizing the performance of cold-neutron tomography for investigating annular flows and functional spacers in fuel rod bundles. *Nucl. Eng. Des.* **2013**, *260*, 188–203. [CrossRef]
160. Stahl, P.; von Rohr, P.R. On the accuracy of void fraction measurements by single-beam gamma-densitometry for gas–liquid two-phase flows in pipes. *Exp. Therm. Fluid Sci.* **2004**, *28*, 533–544. [CrossRef]
161. Trabold, T.A.; Kumar, R.; Vassallo, P.F. Experimental study of dispersed droplets in high- pressure annular flows. *J. Heat Transf.* **1999**, *121*, 924–933. [CrossRef]
162. Omebere-Iyari, N.K.; Azzopardi, B.J.; Ladam, Y. Two-phase flow patterns in large diameter vertical pipes at high pressures. *AIChE J.* **2007**, *53*, 2493–2504. [CrossRef]
163. Adineh, M.; Nematollahi, M.; Erfaninia, A. Experimental and numerical void fraction measurement for modeled two-phase flow inside a vertical pipe. *Ann. Nucl. Energy* **2015**, *83*, 188–192. [CrossRef]
164. Zboray, R.; Guetg, M.; Kickhofel, J.; Barthel, F.; Sprewitz, U.; Hampel, U.; Prasser, H.-M. Investigating annular flows and the effect of functional spacers in an adiabatic double-subchannel model of a BWR fuel bundle by ultra-fast X-ray tomography. In Proceedings of the 14th International Topical Meeting on Nuclear Reactor Thermalhydraulics, NURETH-14, Toronto, ON, Canada, 25–30 September 2011.
165. Fore, L.B.; Ibrahim, B.B.; Beus, S.G. Visual measurements of droplet size in gas-liquid annular flow. *Int. J. Multiph. Flow* **2002**, *28*, 1895–1910. [CrossRef]
166. Simmons, M.J.H.; Hanratty, T.J. Droplet size measurements in horizontal annular gas-liquid flow. *Int. J. Multiph. Flow* **2001**, *27*, 861–883. [CrossRef]
167. Van't Westende, J.M.C. Droplets in Annular-Dispersed Gas-Liquid Pipe-Flows. Doctoral's Thesis, TU Delft, Delft, The Netherlands, 2008.
168. Van't Westende, J.M.C.; Kemp, H.K.; Belt, R.J.; Portela, L.M.; Mudde, R.F.; Oliemans, R.V.A. On the role of droplets in cocurrent annular and churn-annular pipe flow. *Int. J. Multiph. Flow* **2007**, *33*, 595–615. [CrossRef]
169. Azzopardi, B.J. Drop sizes in annular two-phase flow. *Exp. Fluids* **1985**, *3*, 53–59. [CrossRef]
170. Lopez De Bertodano, M.A.; Jan, C.S.; Beus, S.G. Annular flow entrainment rate experiment in a small vertical pipe. *Nucl. Eng. Des.* **1997**, *178*, 61–70. [CrossRef]
171. Okawa, T.; Kotani, A.; Kataoka, I. Experiments for liquid phase mass transfer rate in annular regime for a small vertical tube. *Int. J. Heat Mass Transf.* **2005**, *48*, 585–598. [CrossRef]
172. Sawant, P.; Ishii, M.; Mori, M. Prediction of amount of entrained droplets in vertical annular two-phase flow. *Int. J. Heat Fluid Flow* **2009**, *30*, 715–728. [CrossRef]
173. Sarimeseli, A. Determination of drop sizes in annular gas/liquid flows in vertical and horizontal pipes. *J. Dispers. Sci. Technol.* **2009**, *30*, 694–697. [CrossRef]
174. Azzopardi, B.J.; Teixeira, J.C.F. Detailed measurements of vertical annular two-phase flow-part I: Drop velocities and sizes. *J. Fluids Eng. Trans. ASME* **1994**, *116*, 792–795. [CrossRef]
175. Azzopardi, B.J.; Zaidi, S.H. Determination of entrained fraction in vertical annular gas/liquid flow. *J. Fluids Eng. Trans. ASME* **2000**, *122*, 146–150. [CrossRef]
176. Zaidi, S.H.; Altunbas, A.; Azzopardi, B.J. A comparative study of phase Doppler and laser diffraction techniques to investigate drop sizes in annular two-phase flow. *Chem. Eng. J.* **1998**, *71*, 135–143. [CrossRef]
177. Hay, K.J.; Liu, Z.C.; Hanratty, T.J. Relation of deposition to drop size when the rate law is nonlinear. *Int. J. Multiph. Flow* **1996**, *22*, 829–848. [CrossRef]
178. Fore, L.B.; Dukler, A.E. The distribution of drop size and velocity in gas-liquid annular flow. *Int. J. Multiph. Flow* **1995**, *21*, 137–149. [CrossRef]
179. Dodge, L.G. Calibration of the malvern particle sizer. *Appl. Opt..* **1984**, *23*, 2415–2419. [CrossRef] [PubMed]
180. Wang, Z.; Liu, H.; Zhang, Z.; Sun, B.; Zhang, J.; Lou, W. Research on the effects of liquid viscosity on droplet size in vertical gas–liquid annular flows. *Chem. Eng. Sci.* **2020**, *220*, 115621. [CrossRef]
181. Barbosa, J.R., Jr.; Hewitt, G.F.; König, G.; Richardson, S.M. Liquid entrainment, droplet concentration and pressure gradient at the onset of annular flow in a vertical pipe. *Int. J. Multiph. Flow* **2002**, *28*, 943–961. [CrossRef]

182. Oliveira, J.L.G.; Passos, J.C.; Verschaeren, R.; Geld, C.v.d. Mass flow rate measurements in gas-liquid flows by means of a venturi or orifice plate coupled to a void fraction sensor. *Exp. Therm. Fluid Sci.* **2009**, *33*, 253–260. [CrossRef]
183. Liu, W.; Lv, X.; Bai, B. Axial development of air–water annular flow with swirl in a vertical pipe. *Int. J. Multiph. Flow* **2020**, *124*, 103165. [CrossRef]
184. Kim, H.Y.; Koyama, S.; Matsumoto, W. Flow pattern and flow characteristics for counter-current two-phase flow in a vertical round tube with wire-coil inserts. *Int. J. Multiph. Flow* **2001**, *27*, 2063–2081. [CrossRef]
185. Kiran, R.; Ahmed, R.; Salehi, S. Experiments and CFD modelling for two phase flow in a vertical annulus. *Chem. Eng. Res. Des.* **2020**, *153*, 201–211. [CrossRef]
186. Spedding, P.L. Holdup prediction in vertical upwards to downwards flow. *Dev. Chem. Eng. Miner. Process.* **1997**, *5*, 43–60. [CrossRef]
187. Spedding, P.L.; Woods, G.S.; Raghunathan, R.S.; Watterson, J.K. Vertical two-phase flow. Part I: Flow regimes. *Chem. Eng. Res. Des.* **1998**, *76*, 612–619. [CrossRef]
188. Spedding, P.L.; Woods, G.S.; Raghunathan, R.S.; Watterson, J.K. Vertical two-phase flow. Part III: Pressure drop. *Chem. Eng. Res. Des.* **1998**, *76*, 628–634. [CrossRef]
189. Spedding, P.L.; Woods, G.S.; Raghunathan, R.S.; Watterson, J.K. Vertical Two-Phase Flow: Part II: Experimental Semi-Annular Flow and Hold-up. *Chem. Eng. Res. Des.* **1998**, *76*, 620–627. [CrossRef]
190. Woods, G.S.; Spedding, P.L.; Watterson, J.K.; Raghunathan, R.S. Vertical two phase flow. *Dev. Chem. Eng. Miner. Process.* **1999**, *7*, 7–16. [CrossRef]
191. Godbole, P.V.; Tang, C.C.; Ghajar, A.J. Comparison of Void Fraction Correlations for Different Flow Patterns in Upward Vertical Two-Phase Flow. *Heat Transf. Eng.* **2011**, *32*, 843–860. [CrossRef]
192. Elgaddafi, R.; Ahmed, R.; Kiran, R.; Salehi, S.; Fajemidupe, O. Experimental and modeling studies of gas-liquid flow in vertical pipes at high superficial gas velocities. *J. Nat. Gas Sci. Eng.* **2022**, *106*, 104731. [CrossRef]
193. Dong, Y.; Liao, R.; Luo, W.; Li, M. An Improved Pressure Drop Prediction Model Based on Okiszewski's Model for Low Gas-liquid Ratio Two-Phase Upward Flow in Vertical Pipe. *Int. J. Heat Technol.* **2022**, *40*, 17–22. [CrossRef]
194. Kumar, A.; Das, G.; Ray, S. Void fraction and pressure drop in gas-liquid downflow through narrow vertical conduits-experiments and analysis. *Chem. Eng. Sci.* **2017**, *171*, 117–130. [CrossRef]
195. Hurlburt, E.; Hanratty, T. Measurement of drop size in horizontal annular flow with the immersion method. *Exp. Fluids* **2002**, *32*, 692–699. [CrossRef]
196. Han, H.; Gabriel, K. Flow physics of upward cocurrent gas-liquid annular flow in a vertical small diameter tube. *Microgravity Sci. Technol.* **2006**, *18*, 27–38. [CrossRef]
197. Jayanti, S.; Hewitt, G. Hydrodynamics and heat transfer in wavy annular gas-liquid flow: A computational fluid dynamics study. *Int. J. Heat Mass Transf.* **1997**, *40*, 2445–2460. [CrossRef]
198. Han, H. A Study of Entrainment in Two-Phase upward Cocurrent Annular Flow in a Vertical Tube. Ph.D. Thesis, University of Saskatchewan, Saskatoon, SK, Canada, 2005.
199. Han, H.; Gabriel, K. A numerical study of entrainment mechanism in axisymmetric annular gas-liquid flow. *J. Fluids Eng. Mar.* **2007**, *129*, 293–301. [CrossRef]
200. Xie, Z.; Hewitt, G.F.; Pavlidis, D.; Salinas, P.; Pain, C.C.; Matar, O.K. Numerical study of three-dimensional droplet impact on a flowing liquid film in annular two-phase flow. *Chem. Eng. Sci.* **2017**, *166*, 303–312. [CrossRef]
201. Alipchenkov, V.; Nigmatulin, R.; Soloviev, S.; Stonik, O.; Zaichik, L.; Zeigarnik, Y. A three-fluid model of two-phase dispersed-annular flow. *Int. J. Heat Mass Transf.* **2004**, *47*, 5323–5338. [CrossRef]
202. Liu, Y.; Li, W.Z. Numerical simulation of droplet size distribution in vertical upward annular flow. *J. Fluids Eng. Trans. ASME* **2010**, *132*, 121402. [CrossRef]
203. Liu, Y.; Cui, J.; Li, W.Z. A two-phase, two-component model for vertical upward gas-liquid annular flow. *Int. J. Heat Fluid Flow* **2011**, *32*, 796–804. [CrossRef]
204. Saxena, A.; Prasser, H.M. A study of two-phase annular flow using unsteady numerical computations. *Int. J. Multiph. Flow* **2020**, *126*, 103037. [CrossRef]
205. Kishore, B.N.; Jayanti, S. A multidimensional model for annular gas-liquid flow. *Chem. Eng. Sci.* **2004**, *59*, 3577–3589. [CrossRef]
206. Hassani, M.; Motlagh, M.B.; Kouhikamali, R. Numerical investigation of upward air-water annular, slug and bubbly flow regimes. *J. Comput. Appl. Res. Mech. Eng.* **2020**, *9*, 331–341.
207. Shaban, H.; Tavoularis, S. On the accuracy of gas flow rate measurements in gas–liquid pipe flows by cross-correlating dual wire-mesh sensor signals. *Int. J. Multiph. Flow* **2016**, *78*, 70–74. [CrossRef]
208. Joseph, D.D.; Bannwart, A.C.; Liu, Y.J. Stability of annular flow and slugging. *Int. J. Multiph. Flow* **1996**, *22*, 1247–1254. [CrossRef]
209. Belt, R.J.; Van't Westende, J.M.C.; Portela, L.M. Prediction of the interfacial shear-stress in vertical annular flow. *Int. J. Multiph. Flow* **2009**, *35*, 689–697. [CrossRef]
210. Mudawwar, I.A.; El-Masri, M.A. Momentum and heat transfer across freely-falling turbulent liquid films. *Int. J. Multiph. Flow* **1986**, *12*, 771–790. [CrossRef]
211. Vassallo, P. Near wall structure in vertical air-water annular flows. *Int. J. Multiph. Flow* **1999**, *25*, 459–476. [CrossRef]
212. Hajiloo, M.; Chang, B.H.; Mills, A.F. Interfacial shear in downward two-phase annular co-current flow. *Int. J. Multiph. Flow* **2001**, *27*, 1095–1108. [CrossRef]

213. Klyuev, N.I.; Solov'eva, E.A. Method for the annular gas-liquid mixture flow regime in a vertical cylindrical channel. *Russ. Aeronaut.* **2009**, *52*, 68–71. [CrossRef]
214. Alves, M.V.C.; Waltrich, P.J.; Gessner, T.R.; Falcone, G.; Barbosa, J.R., Jr. Modeling transient churn-annular flows in a long vertical tube. *Int. J. Multiph. Flow* **2017**, *89*, 399–412. [CrossRef]
215. Schubring, D.; Shedd, T.A. A model for pressure loss, film thickness, and entrained fraction for gas-liquid annular flow. *Int. J. Heat Fluid Flow* **2011**, *32*, 730–739. [CrossRef]
216. Schubring, D.; Shedd, T.A. Critical friction factor modeling of horizontal annular base film thickness. *Int. J. Multiph. Flow* **2009**, *35*, 389–397. [CrossRef]
217. Ishii, M.; Grolmes, M.A. Inception criteria for droplet entrainment in two-phase concurrent film flow. *AIChE J.* **1975**, *21*, 308–318. [CrossRef]
218. Henstock, W.H.; Hanratty, T.J. The interfacial drag and the height of the wall layer in annular flows. *AIChE J.* **1976**, *22*, 990–1000. [CrossRef]
219. Tatterson, D.F.; Dallman, J.C.; Hanratty, T.J. Drop sizes in annular gas-liquid flows. *AIChE J.* **1977**, *23*, 68–76. [CrossRef]
220. Hori, K.; Nakasatomi, M.; Nishikawa, K.; Sekoguchi, N. Study of ripple region in cyclic two-phase flow: 3rd report, Effect of liquid viscosity on gas-liquid interface properties and friction coefficient. *Trans. Jpn. Soc. Mech. Eng.* **1978**, *44*, 3847–3856. [CrossRef]
221. Hori, K.; Nakazatomi, M.; Nishikawa, K.; Sekoguchi, K. On Ripple of Annular Two-Phase Flow: 3. Effect of Liquid Viscosity on Characteristics of Wave and Interfacial Friction Factor. *Bull. JSME* **1979**, *22*, 952–959. [CrossRef]
222. Ambrosini, W.; Andreussi, P.; Azzopardi, B.J. A physically based correlation for drop size in annular flow. *Int. J. Multiph. Flow* **1991**, *17*, 497–507. [CrossRef]
223. Okawa, T.; Kitahara, T.; Yoshida, K.; Matsumoto, T.; Kataoka, I. New entrainment rate correlation in annular two-phase flow applicable to wide range of flow condition. *Int. J. Heat Mass Transf.* **2002**, *45*, 87–98. [CrossRef]
224. MacGillivray, R.M. Gravity and Gas Density Effects on Annular Flow Average Film Thickness and Frictional Pressure Drop. Masters's Thesis, University of Saskatchewan, Saskatoon, SK, Canada, 2004.
225. Rahman, M.; Stevens, J.; Pardy, J.; Wheeler, D. An Improved Film Thickness Model for Annular Flow Pressure Gradient Estimation in Vertical Gas Wells. *J. Pet. Environ. Biotechnol.* **2017**, *8*, 1.
226. Ju, P.; Liu, Y.; Yang, X.; Ishii, M. Wave characteristics of vertical upward adiabatic annular flow in pipes. *Int. J. Heat Mass Transf.* **2019**, *145*, 118701. [CrossRef]
227. Ju, P.; Brooks, C.S.; Ishii, M.; Liu, Y.; Hibiki, T. Film thickness of vertical upward co-current adiabatic flow in pipes. *Int. J. Heat Mass Transf.* **2015**, *89*, 985–995. [CrossRef]
228. Ghajar, A.J.; Bhagwat, S.M. Effect of void fraction and two-phase dynamic viscosity models on prediction of hydrostatic and frictional pressure drop in vertical upward gas-liquid two-phase flow. *Heat Transf. Eng.* **2013**, *34*, 1044–1059. [CrossRef]
229. Tandon, T.N.; Varma, H.K.; Gupta, C.P. A void fraction model for annular two-phase flow. *Int. J. Heat Mass Transf.* **1985**, *28*, 191–198. [CrossRef]
230. Cioncolini, A.; Thome, J.R. Void fraction prediction in annular two-phase flow. *Int. J. Multiph. Flow* **2012**, *43*, 72–84. [CrossRef]
231. Kumar, P.; Das, A.K.; Mitra, S.K. Physical understanding of gas-liquid annular flow and its transition to dispersed droplets. *Phys. Fluids* **2016**, *28*, 072101. [CrossRef]
232. Schadel, S.A.; Leman, G.W.; Binder, J.L.; Hanratty, T.J. Rates of atomization and deposition in vertical annular flow. *Int. J. Multiph. Flow* **1990**, *16*, 363–374. [CrossRef]
233. McNeil, D.A.; Stuart, A.D. The effects of a highly viscous liquid phase on vertically upward two-phase flow in a pipe. *Int. J. Multiph. Flow* **2003**, *29*, 1523–1549. [CrossRef]
234. Kocamustafaogullari, G.; Smits, S.R.; Razi, J. Maximum and mean droplet sizes in annular two-phase flow. *Int. J. Heat Mass Transf.* **1994**, *37*, 955–965. [CrossRef]
235. Trabold, T.A.; Kumar, R. Vapor core turbulence in annular two-phase flow. *Exp. Fluids* **2000**, *28*, 187–194. [CrossRef]
236. Kataoka, I.; Ishii, M.; Mishima, K. Generation and size distribution of droplet in annular two-phase flow. *J. Fluids Eng. Trans. ASME* **1983**, *105*, 230–240. [CrossRef]
237. Azzopardi, B.J. *Gas-Liquid Flows*; Begell House: New York, NY, USA, 2006.
238. Kolev, N.I.; Kolev, N. *Multiphase Flow Dynamics*; Springer: Berlin/Heidelberg, Germany, 2005; Volume 1.
239. Kataoka, I.; Ishii, M.; Nakayama, A. Entrainment and desposition rates of droplets in annular two-phase flow. *Int. J. Heat Mass Transf.* **2000**, *43*, 1573–1589. [CrossRef]
240. Lopez de Bertodano, M.A.; Assad, A.; Beus, S.G. Experiments for entrainment rate of droplets in the annular regime. *Int. J. Multiph. Flow* **2001**, *27*, 685–699. [CrossRef]
241. Okawa, T.; Kataoka, I. Correlations for the mass transfer rate of droplets in vertical upward annular flow. *Int. J. Heat Mass Transf.* **2005**, *48*, 4766–4778. [CrossRef]
242. Ryu, S.H.; Park, G.C. A droplet entrainment model based on the force balance of an interfacial wave in two-phase annular flow. *Nucl. Eng. Des.* **2011**, *241*, 3890–3897. [CrossRef]
243. Liu, L.; Bai, B. Generalization of droplet entrainment rate correlation for annular flow considering disturbance wave properties. *Chem. Eng. Sci.* **2017**, *164*, 279–291. [CrossRef]
244. Wang, G.; Sawant, P.; Ishii, M. A new entrainment rate model for annular two-phase flow. *Int. J. Multiph. Flow* **2020**, *124*, 103185. [CrossRef]

245. Oliemans, R.V.A.; Pots, B.F.M.; Trompe, N. Modelling of annular dispersed two-phase flow in vertical pipes. *Int. J. Multiph. Flow* **1986**, *12*, 711–732. [CrossRef]
246. Ishii, M.; Mishima, K. Droplet entrainment correlation in annular two-phase flow. *Int. J. Heat Mass Transf.* **1989**, *32*, 1835–1846. [CrossRef]
247. Utsuno, H.; Kaminaga, F. Prediction of Liquid Film Dryout in Two-Phase Annular-Mist Flow in a Uniformly Heated Narrow Tube Development of Analytical Method under BWR Conditions. *J. Nucl. Sci. Technol.* **1998**, *35*, 643–653. [CrossRef]
248. Petalas, N.; Aziz, K. A Mechanistic Model for Multiphase Flow in Pipes. *J. Can. Pet. Technol.* **2000**, *39*, 000604. [CrossRef]
249. Pan, L.; Hanratty, T.J. Correlation of entrainment for annular flow in vertical pipes. *Int. J. Multiph. Flow* **2002**, *28*, 363–384. [CrossRef]
250. Sawant, P.; Ishii, M.; Mori, M. Droplet entrainment correlation in vertical upward co-current annular two-phase flow. *Nucl. Eng. Des.* **2008**, *238*, 1342–1352. [CrossRef]
251. Cioncolini, A.; Thome, J.R. Prediction of the entrained liquid fraction in vertical annular gas-liquid two-phase flow. *Int. J. Multiph. Flow* **2010**, *36*, 293–302. [CrossRef]
252. Cioncolini, A.; Thome, J.R. Entrained liquid fraction prediction in adiabatic and evaporating annular two-phase flow. *Nucl. Eng. Des.* **2012**, *243*, 200–213. [CrossRef]
253. Zhang, R.; Liu, H.; Liu, M. A probability model for fully developed annular flow in vertical pipes: Prediction of the droplet entrainment. *Int. J. Heat Mass Transf.* **2015**, *84*, 225–236. [CrossRef]

Article

Flow Boiling Heat Transfer Intensification Due to Inner Surface Modification in Circular Mini-Channel

Aleksandr V. Belyaev [1],*, Alexey V. Dedov [1], Nikita E. Sidel'nikov [1], Peixue Jiang [2], Aleksander N. Varava [1] and Ruina Xu [2]

[1]   Department of General Physics and Nuclear Fusion, The National Research University "MPEI", Krasnokazarmennaya 14, 111250 Moscow, Russia
[2]   Department of Thermal Engineering, Tsinghua University, Haidian District, Beijing 100084, China
*   Correspondence: beliayevavl@mpei.ru

Abstract: This work aimed to study the intensification of flow boiling heat transfer and critical heat flux (CHF) under conditions of highly reduced pressures due to a modification of the inner wall surface of a mini-channel. Such research is relevant to the growing need of high-tech industries in the development of compact and energy-efficient heat exchange devices. We present experimental results of the surface modification effect on hydrodynamics and flow boiling heat transfer, including data on the CHF. A description of the experimental stand and method for modifying the test mini-channel is also presented. The studies were carried out with freon R-125 in a vertical mini-channel with a diameter of 1.1 mm and a length of 50 mm, in the range of mass flow rates from $G = 200$ to 1400 kg/(m^{2}s) and reduced pressures between $p_r = p/p_{cr} = 0.43$ and 0.56. The maximum surface modification effect was achieved at a reduced pressure of $p_r = 0.43$, the heat transfer coefficient increased up to 110%, and the CHF increased up to 22%.

Keywords: flow boiling; critical heat flux; heat transfer enhancement; high reduced pressure; mini-channel

Citation: Belyaev, A.V.; Dedov, A.V.; Sidel'nikov, N.E.; Jiang, P.; Varava, A.N.; Xu, R. Flow Boiling Heat Transfer Intensification Due to Inner Surface Modification in Circular Mini-Channel. *Water* **2022**, *14*, 4054. https://doi.org/10.3390/w14244054

Academic Editors: Giuseppe Pezzinga and Bommanna Krishnappan

Received: 16 November 2022
Accepted: 7 December 2022
Published: 12 December 2022

Publisher's Note: MDPI stays neutral with regard to jurisdictional claims in published maps and institutional affiliations.

1. Introduction

One way to increase the heat flux density during flow boiling is surface modification. The increased interest in studying the methods that intensify heat transfer (primarily due to boiling) is associated with the increasing requirements for overall efficiency and reduced environmental impact in modern energy-efficient and energy-intensive installations. The requirements for heat removal in microelectronics elements increase the heat flux density, which reaches 2–5 MW/m^2. To provide the necessary heat removal in such devices, highly efficient mini-channel heat exchangers are being developed and implemented [1,2], where various dielectric liquids and freons can be used as coolants. It is also possible to use boiling as an efficient mechanism for heat removal and additional intensification by modifying the boiling surface.

Surface Modification Methods

Technological growth in areas associated with high heat flux density requires the creation of appropriate heat exchangers. To date, there are many articles that describe various methods for modifying the boiling surface [3]. The main part of the intensification methods was laid down in the twentieth century. However, the development of surface treatment technologies and methods for measuring the characteristics of non-stationary processes, as well as simulation techniques, allows modern research to be carried out at a fundamentally new level. Improved methods of mechanical surface treatment can achieve a significant increase in heat flux. One of the long-standing and well-known ways to increase the heat flux during fluid flow in a channel is by changing the relief using the method of deforming cutting (MDC) [4–6]. This method of obtaining modified surfaces is patented and is actively used in the commercial sphere and in experimental studies. For example, in the article [7], experiments were carried out on low-finned pipes and pipes of the Turbo-B and Thermoexcel-E types, with a length of 152 mm and an outer diameter of 18–19 mm, under conditions of free convection at a saturation temperature of 7 °C for the refrigerants HCFC22, HFC134a, HFC125 and HFC32. Heat fluxes varied in the range of 10–80 kW/m^2. The ranges of increase in the heat transfer coefficient during boiling on a low-finned pipe, the Turbo-B type, and the Thermoexcel-E type were 1.09–1.68; 1.77–5.41 and 1.64–8.77 times, respectively. A combined modification (simultaneous use of MDC and nanocoating) was performed in [8]. The heat exchange surface was made in the form of channels, onto which a nanocoating was applied. The channel depth and rib width for all surfaces were kept constant at 400 μm and 200 μm, respectively. In comparison with a smooth surface, the increase in CHF was 228%.

Experimental investigations of surface modifications are carried out in pool boiling. There are many works devoted to various methods of intensification [9]. Modern methods of plasma and ion sputtering, chemical processing technologies, as well as various methods for creating nanoscale coatings and their combination with microstructures, make it possible to create fundamentally new types of multiscale structured surfaces. There are many different methods for surface modification: meshed surfaces [10,11], silicon oxide nanoparticle coated copper [12], cutting with a micro milling machine [13], and the use of micro-nanostructured surfaces [14–17] (which make it possible to achieve an increase in the heat transfer coefficient (HTC) of up to 200% and in the CHF of up to 120%), the use of capillary-porous structured surfaces [18] (HTC increases from 1.5 to 4 times). There are methods of cold deposition by McNamara [19], including the creation of carbon [20–22] and silicon nanotubes [23], which increase the HTC up to 3 times and the CHF up to 65%. The methods of modification using a laser, for example, laser T-shaped groove [24] and laser surface texturing [25], can increase the CHF up to 240%. Due to a very high HTC, these laser technologies have found application in many industrial fields [26]. Another common method of surface modification is the use of mesh coatings. Pool boiling on a modified surface was studied in [27] using multilayer copper gradient grid coatings. The results indicated an increase in the CHF of 3 times and an increase in the HTC of 6.6 times.

All the above methods of surface modification have been studied for pool boiling. However, it is difficult, or impossible, to implement these methods in channels. Therefore, there are currently significantly fewer methods for surface modification in channels. Grooving technology can be used to modify the inner wall of a circular channel. This technology was studied in [28], where pipes with a diameter of 9.52 mm were used with different surface structures: a smooth copper pipe and two copper pipes with internal grooves of different cut widths. The modification technology creates grooves along the inner wall of the channel, which, as a result, increases the heat exchange surface area. This method is simple to implement and does not cause considerable technical difficulties. However, the modifications have a strong influence on the channel hydrodynamics, where the pressure drop is more than twice as high as in a smooth pipe. According to [28], the best results can be obtained in pipes with the smallest groove width, which showed an increase in the HTC of more than twice compared to smooth pipes. The maximum increase in the HTC was observed at the vapor quality $x = 0.8$.

The flow boiling heat transfer intensification in a rectangular channel measuring 65 mm long, 6 mm wide, and 1 mm high was studied in [29]. One of the walls (through which the flow was heated) was made of silicon, on which a C4F8 polymer layer was deposited using a mask. The result was a bifilar surface. FC-72 was used as the working fluid. The experiments were carried out at two values of the mass flow rate: $G = 90$ and 130 kg/m^{2}s. An increase in the HTC of 26% was obtained at heat flux density $q = 9$ W/sm^2.

There are many methods of modification for pool boiling which make it possible to create effective intensifying surfaces that significantly increase heat transfer during boiling and evaporation. Currently, there are a small number of methods for surface modification in channels, especially in mini- and micro-channels. The aim of this work is to use a relatively simple method for modifying the inner wall of a mini-channel to intensify flow boiling.

2. Experimental Methods and Setup Description

2.1. Experimental Setup

An experimental setup was used to study the processes of hydrodynamics and heat transfer in the mini-channel at highly reduced pressures. A detailed description is presented in earlier works on the experimental study of the hydrodynamics, heat transfer, and CHF for a wide range of mass flow rates, vapor qualities, and reduced pressures [30,31], as well as at the most demanded flow parameters in technology [32]. The experiments were carried out using R-125, the properties of which are presented in Table 1. The saturation temperature for the operating parameters of the performed experiments was in the range of 30–45 °C, which made it possible to reduce heat losses to the environment during boiling in the channel. Also, the heat of vaporization and the critical pressure was much lower than in water, which made it possible to achieve the required parameters with less energy.

Table 1. Physical properties of R-125.

Saturation temperature (1 atm), °C	−48.1
Critical point pressure, MPa	3.618
Liquid density (25 °C), kg/m^3	1190
Vapor density (boiling temperature), kg/m^3	6.7
Surface tension (25 °C), N/m	0.014
Liquid specific heat, kJ/(kg·K)	1.399
Latent heat of vaporization (25 °C), kJ/kg	164
Liquid thermal conductivity (25 °C), W/(m·K)	0.062
Liquid viscosity (25 °C), mPa·s	0.141

The working tube was placed vertically and made of 12 × 18H10T stainless steel with a heated length of 50 mm, an inner diameter of 1.1 mm, and an outer diameter of 1.6 mm. The tube was heated by alternating the current. The inlet and outlet pressures and pressure drops were measured using pressure sensors and a differential manometer. The inlet and outlet temperatures were measured using Chromel-Copel thermocouples with a cable diameter of 0.7 mm. The wall temperatures were measured using Chromel-Copel thermocouples (diameter 0.2 mm) on five cross-sections of the tube (T1–T5, see Table 2). The inner wall temperatures were calculated using a correction for the wall conductivity. Heat losses were taken into account when calculating the heat balance of regimes with convective heat transfer. An estimate of the measurement uncertainty is presented in Table 3.

Table 2. Coordinates of the cross-sections (mm).

T1	T2	T3	T4	T5
2.5	15.5	28.5	40	48

Table 3. Uncertainty parameters in the analysis.

Parameter	Uncertainty
Current	±0.9%
Voltage	±0.5 mV
Mass flow rate	±0.2%
Inlet and outlet temperatures	±0.1 °C
Wall temperature	±0.8%
Inlet and outlet pressure sensors	±1%
Pressure drop sensor	±0.2%
Tube diameter	±0.05 μm

2.2. Surface Modification

Modification of the inner wall was carried out using the action of a laser pulse on the outer surface of the tube. As a result, exposure formations of different heights and diameters, depending on the pulse power (p1–p14), were formed on the inner wall (Figure 1a). At some values of the pulse power, the outer surface of the tube was not damaged (Figure 1b).

Figure 1. (**a**) View of the inner surface of the wall after exposure to laser pulses of different powers. (**b**) External surface after exposure with the technological current value I = 130 A.

Figure 2 shows a photograph and a formation profile of the inner wall of the mini-channel after the outer wall was exposed to a laser pulse with a technological current value of I = 130 A. The diameter of the formation was d = 390 μm. The formations have the shape of a rounded cone with a dip in the center, located at the point of the laser pulse impact. At such a power, pores and craters with different diameters, ranging from 5 to 60 microns, form on the upper part of the cone. The size distribution of the pores is most uniform in the center of the dip where their diameter is about 15 μm.

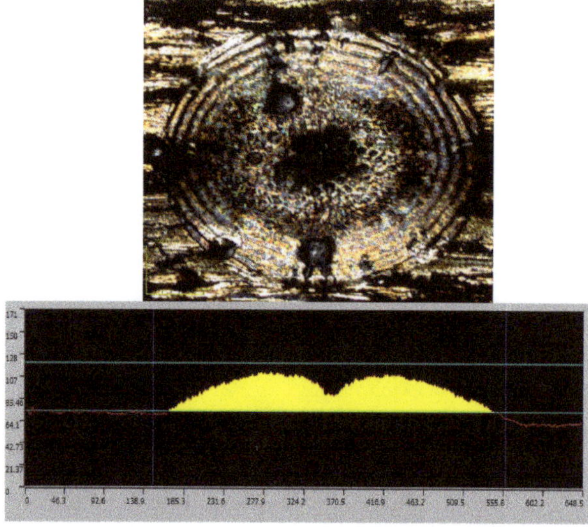

Figure 2. Photograph and view of the formation profile along the tube resulting from exposure to a laser pulse of I = 130 A, d = 390 μm.

At the pulse current $I = 110$ A, the formation diameter decreased to 240 μm, and the pattern of the crater location changed. Figure 3 shows that large pores with a diameter from 15 to 30 μm are located both in the peripheral and the central regions. The pores with the smallest diameter from 2 to 10 μm are formed in the annular region between the center and periphery. At current $I = 110$ A, the average pore diameter is smaller than at current $I = 130$ A.

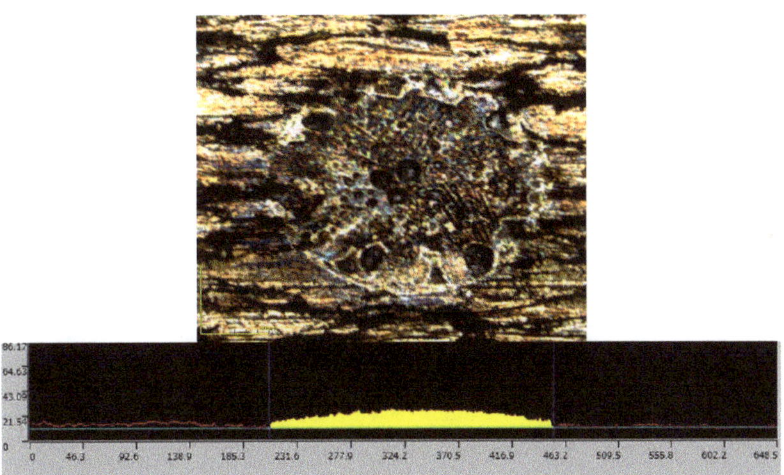

Figure 3. Photograph and view of the formation profile along the tube resulting from exposure to a laser pulse of $I = 110$ A, $d = 240$ μm.

The parameters of the laser pulse were selected based on the results of the analysis of the obtained formations. In this work, the channel modification was performed using a laser pulse with the technological current $I = 130$ A. Along the inner wall surface of the mini-channel, the laser pulse made formations with a high density of arrangement in the amount of 300 pieces (6 rows of 50 formations in each, in a checkerboard pattern, see Figure 1b).

3. Experimental Results

The experiments were carried out with R-125 in a vertical channel with a diameter of 1.1 and length of 50 mm at two values of reduced pressure: 0.43 ($T_s = 30$ °C) and 0.56 ($T_s = 40$ °C). Regimes of convective and flow boiling heat transfer were obtained in each experiment. The mass flow rate varied in the range of $G = 200–1400$ kg/m²s. The inlet flow temperature was close to room temperature ($T_{in} \approx 25$ °C). The values of inlet and outlet temperatures, the wall temperature at five sections along the tube, the inlet and outlet pressures, the pressure drop, and the mass flow rate were measured. The measurements were obtained using an automated data acquisition system after a stationary regime was established. The maximum heat flux was limited by the critical heat flux. The CHF was observed and recorded as a sharp increase in the wall temperature according to the readings of the thermocouple located near the outlet of the tube.

3.1. Effect of the Modification on the Hydrodynamics

The effects of the modification on the channel hydrodynamics were studied without heating the test tube. The total pressure drop, including the inlet and outlet pressure drops, was measured at a constant inlet temperature with an increasing mass flow rate. The prediction of the total pressure drop was carried out as follows. The pressure drop in the heated part of the mini-channel was calculated using Formula (1):

$$\Delta p_{tube} = \zeta \frac{\rho w^2}{2} \frac{L}{d}, \quad \zeta = (1.82 \lg(Re) - 1.64)^{-2}. \tag{1}$$

The test tube had unheated parts at the inlet and outlet, which passed into supply tubes with a diameter of 4 mm, forming a diffuser and confuser. Formula (1) was used to calculate the pressure drop on the unheated parts. The calculation methods of [33] were used to calculate the pressure drop in the confuser and diffuser. The resulting pressure drops at the inlet and outlet were calculated using Formulas (2) and (3):

$$\Delta p_{inlet} = 0.5 \left(1 - \frac{d^2}{D^2}\right)^{\frac{3}{4}} \frac{\rho w^2}{2} \zeta \frac{\rho w^2}{2} \frac{L_{in}}{d}, \tag{2}$$

$$\Delta p_{outlet} = \left(1 - \frac{d^2}{D^2}\right)^{2} \frac{\rho w^2}{2} + \zeta \frac{\rho w^2}{2} \frac{L_{out}}{d}, \tag{3}$$

where L_{in} and L_{out} are the lengths of the inlet and outlet unheated parts, respectively; d is the diameter of the mini-channel; and D is the diameter of the supply tubes.

In Figure 4, the filled triangles show the experimental data of the total pressure drop without modification. The line shows the sum of the calculated pressure drops in all three parts according to Formula (4):

$$\Delta p_{calc} = \Delta p_{inlet} + \Delta p_{outlet} + \Delta p_{tube}. \tag{4}$$

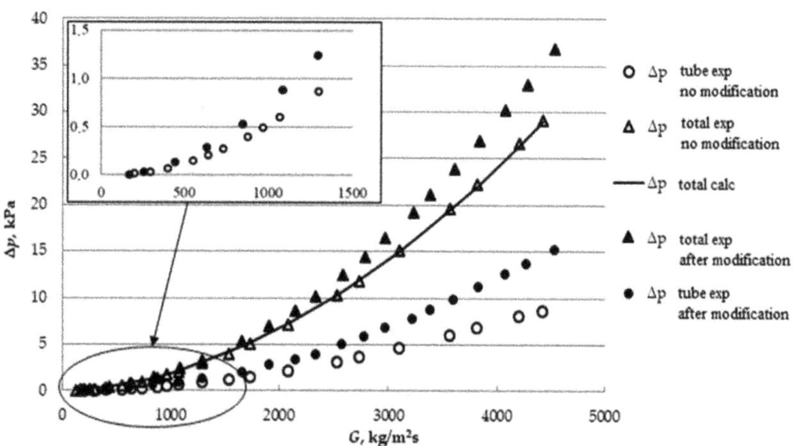

Figure 4. Pressure drop versus mass flow rate.

It is clear that the difference between the experimental data and the calculated data is insignificant, which suggests that the calculation of pressure drop at the inlet and outlet of the tube was carried out correctly. Thus, from the experimental data on the total pressure drop obtained before and after the modification, the experimental pressure drops in the heated part of the mini-channel were determined using Formula (5):

$$\Delta p_{tube.exp.} = \Delta p_{exp} - \Delta p_{inlet} - \Delta p_{outlet} \tag{5}$$

In Figure 4, the unfilled dots show the pressure drop in the tube without modification, whereas the black dots show the pressure drop after the modification. The greatest increases in the pressure drop of 50% to 90% were observed at mass flow rates of $G > 2000$ kg/m²s.

The maximum pressure drop did not exceed 40% within the range of mass flow rates that was investigated in the current work (200–1400 kg/m^{2}s).

3.2. Heat Transfer Enhancement

Primary data on the temperature at the inlet and outlet of the tube, the wall temperature at five cross sections, and the heat flux density at various mass flow rates and pressures were obtained during the experiments. In each regime, at fixed values of the mass flow rate and the inlet temperature and pressure, the tube was gradually heated until it reached the CHF. Figure 5 shows an example of the wall temperature versus the heat flux density in section T4 before and after modification at p_r = 0.56 and G = 1260 kg/(m^{2}s). The regimes of convective and boiling heat transfer are clearly visible.

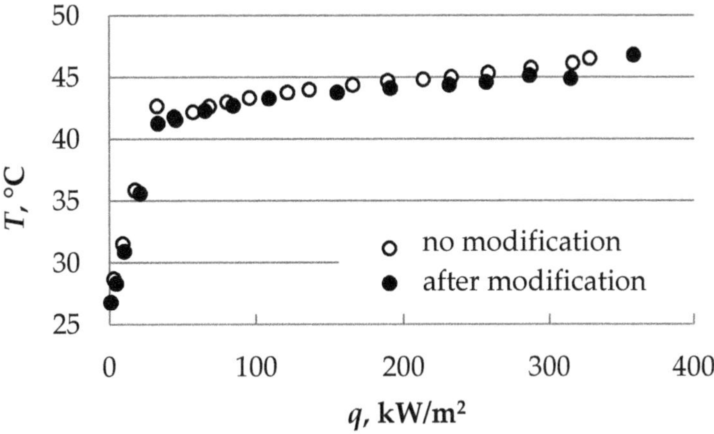

Figure 5. Wall temperature in the T4 thermocouple section versus heat flux density.

Based on the obtained data, the heat transfer coefficients for the regimes of convective and boiling heat transfer were calculated using $h_{boil} = q/(T_w - T_l)$ and $h_{con} = q/(T_w - T_s)$, where T_w, T_l, and T_s are the wall, liquid, and saturation temperatures, respectively. The HTC obtained before the tube modification was then compared to the HTC after modification.

Figures 6–8 show the results of the effect of the modification on the HTC in nucleate boiling regimes for the mass flow rates G = 420, 800, 1260 kg/(m^{2}s) obtained at reduced pressures of p_r = 0.43 and p_r = 0.56. The HTC with increasing heat flux density and mass flow rate increased as usual. A decrease in the HTC was observed upon reaching the pre-crisis values of the heat flux. As a result of the modification to the mini-channel, the nucleate boiling HTC increased. The greatest effect of the modification was observed at p_r = 0.43 for the mass flow rate G = 1260 kg/(m^{2}s) where the average increase in the HTC was 110% (Figure 8a). The average increase in the HTC for mass flow rates G = 420 kg/(m^{2}s) (Figure 6a) and G = 800 kg/(m^{2}s) (Figure 7a) were 65% and 100%, respectively.

The results obtained at the reduced pressure p_r = 0.56 are similar, but with a lower degree of intensification. The greatest effect of the modification at a highly reduced pressure was observed at G = 800 kg/(m^{2}s) where the average increase in the HTC was 23% (Figure 7b). For mass flow rates G = 420 and 1260 kg/(m^{2}s), the average increase in the HTC was 11% and 14%, respectively. With increasing pressure, the critical diameter of vapor bubbles decreases, which leads to an increase in the density of vaporization centers due to natural roughness. This method of surface modification at the highly reduced pressure p_r = 0.56 probably becomes less effective because the size of some pores is too large in relation to the critical bubble diameter to be the centers of vaporization.

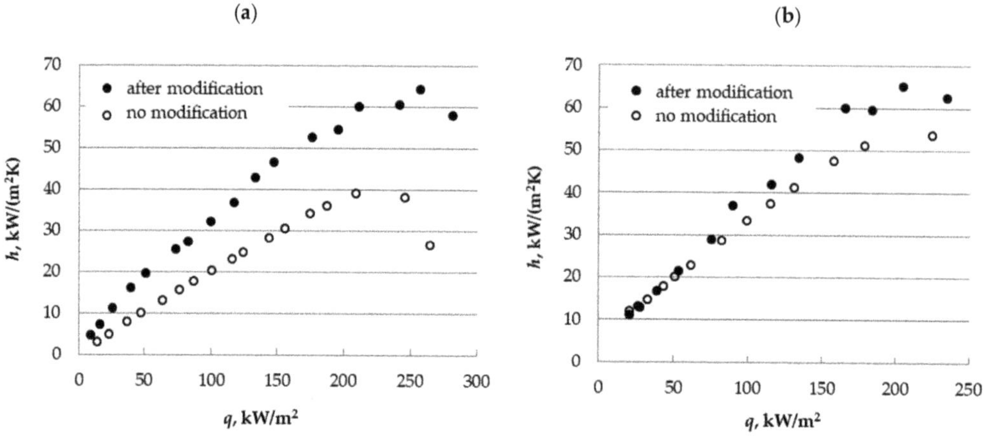

Figure 6. Effect of the modification on HTC at G = 420 kg/(m²s) in the T4 thermocouple section, (**a**) p_r = 0.43, (**b**) p_r = 0.56.

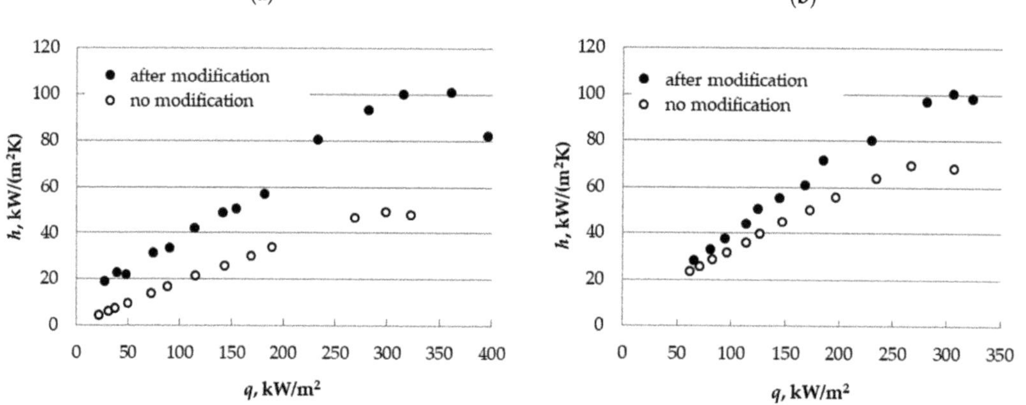

Figure 7. Effect of the modification on HTC at G = 800 kg/(m²s) in the T4 thermocouple section, (**a**) p_r = 0.43, (**b**) p_r = 0.56.

The heat transfer data were obtained at different pressures. It is clearly seen from the data analysis that an increase in the reduced pressure from p_r = 0.43 to p_r = 0.56 caused the HTC to increase significantly, on average from 25% to 100%, depending on the mass flow rate (the higher the mass flow rate, the greater the increase in the HTC). The greatest effect of the modification of the inner wall of the mini-channel on the intensity of flow boiling heat transfer was observed in the range of the mass flow rate G = 1260 kg/(m²s) at the reduced pressure p_r = 0.43. The HTC data obtained on the modified tube at the reduced pressure p_r =0.43 are comparable with the data obtained before the modification at the pressure p_r =0.56. Increasing the pressure from p_r = 0.43 to p_r = 0.56 led to an increased HTC in the tube without modification, which, on average, was comparable to the increase in the HTC due to surface modification at p_r = 0.43.

To evaluate the effect of the modification on convective heat transfer, the values of the experimental convective HTC obtained before and after the modification at the same flow parameters were compared.

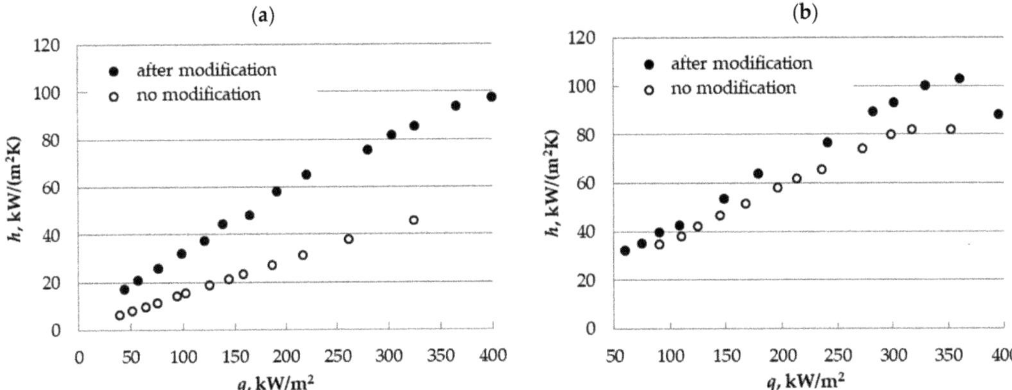

Figure 8. Effect of the modification on HTC at $G = 1260$ kg/(m^{2}s) in the T4 thermocouple section, (**a**) $p_r = 0.43$, (**b**) $p_r = 0.56$.

Figure 9 shows a comparison of the dependences of the convective HTC on the heat flux density for the non-modified and modified tubes for a maximum mass flow rate, at which the effect of convection is maximum. The data are in good agreement, indicating that there is no significant effect of the surface modification on the convective heat transfer compared to the non-modified tube. It can be concluded that the increased HTC in the nucleate boiling regimes, resulting from the modification, occurred due to the vaporization process. This conclusion is supported by the obtained boiling curves. Figure 10 shows an example of the boiling curves obtained before and after modification with the same inlet flow parameters. Comparing the values of $\Delta T_s = (T_w - T_s)$ at the constant value of the heat flux density $q \approx 250$ kW/m^2, it is clear that ΔT_s for the modified surface decreased by about 1.5 degrees, which indicated the intensification of boiling.

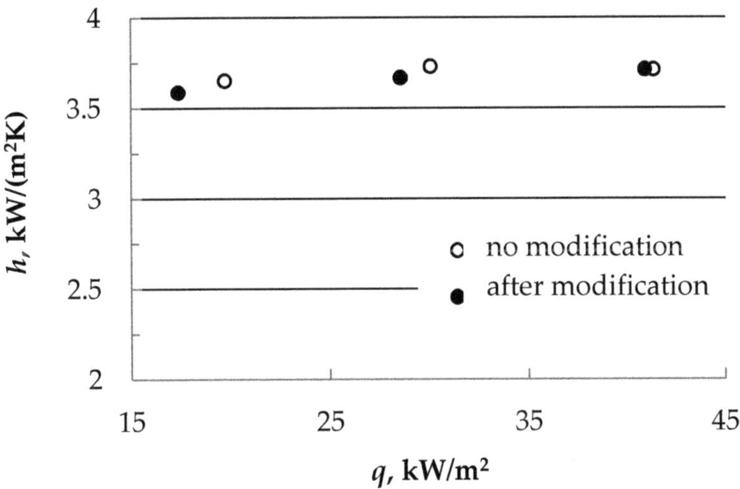

Figure 9. Convective HTC in the T4 thermocouple section versus the heat flux density at $p_r = 0.56$ and $G = 1260$ kg/(m^{2}s).

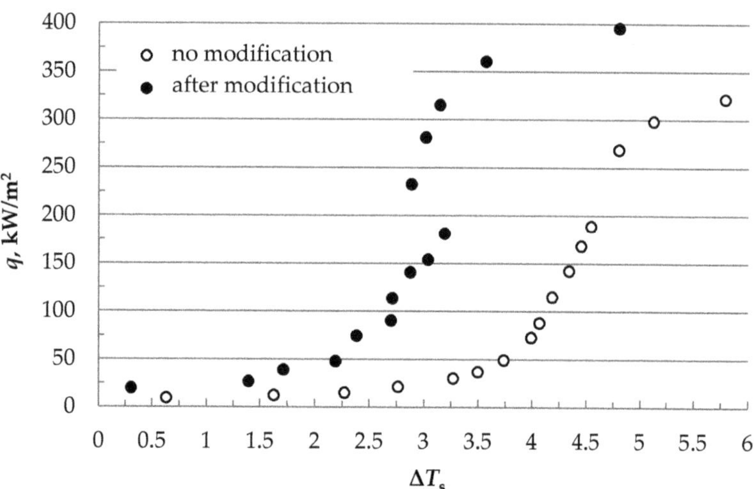

Figure 10. Boiling curves for T4 thermocouple section at p_r = 0.43 and G = 800 kg/(m²s).

3.3. Effect of Intensification on the CHF

In each regime, the tube heat flux was gradually increased at a fixed mass flow rate and inlet temperature.

Depending on the flow parameters, the heat flux increase stopped either when departure nucleate boiling (DNB) or dryout occurred. The critical heat flux values were recorded when there was a sharp increase in the wall temperature at the outlet tube part in the T5 thermocouple section.

A comparative analysis of the CHF data obtained before and after the channel modification was performed. Figure 11 shows a comparison of the results for the CHF values dependent on the mass flow rate at different reduced pressures. The analysis showed that the modification to the inner wall caused an increase in the CHF for the range of mass flow rate G > 800 kg/(m²s): the average increase was 22% at p_r = 0.43 (Figure 11a) and 7 % at p_r =0.56 (Figure 11b).

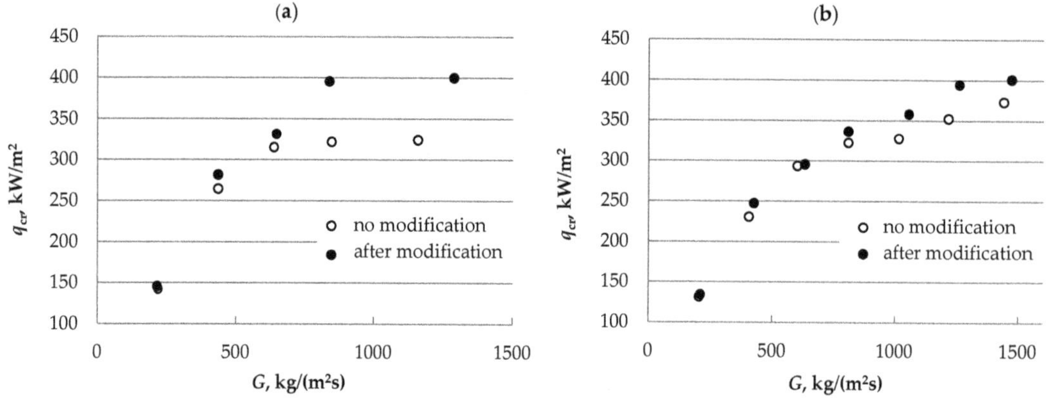

Figure 11. CHF data versus mass flow rate, (a) p_r = 0.43, (b) p_r = 0.56.

In the range G < 800 kg/(m²s), no significant increases in the CHF were observed. As can be seen from the same CHF data versus vapor quality shown in Figure 12, no increases were observed because the heat transfer crisis occurs in the range of vapor quality x > 0.7 when the mass flow rate is decreasing. Dryout conditions also occur at this point.

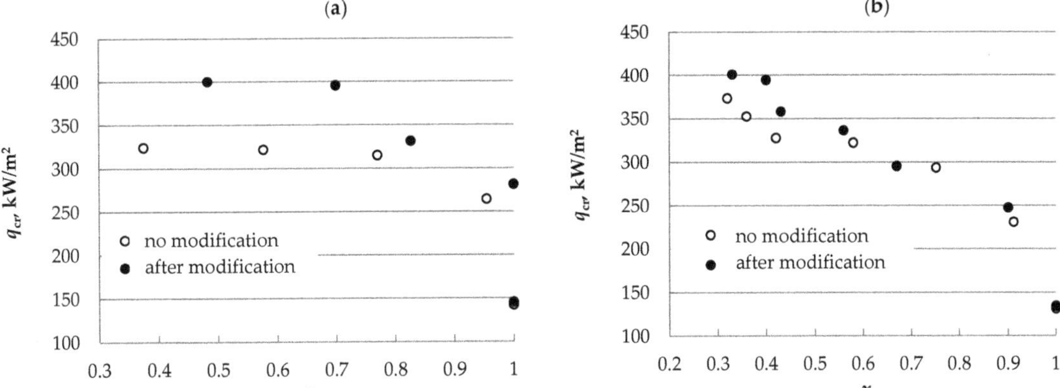

Figure 12. CHF data versus vapor quality, (**a**) $p_r = 0.43$, (**b**) $p_r = 0.56$.

Therefore, the greatest effect of the modification on the CHF was observed at the moderately reduced pressure $p_r = 0.43$ in the range of the mass flow rate $G > 800 \text{ kg}/(\text{m}^2\text{s})$ where the average increase was 22%.

4. Conclusions

A method for modifying the inner surface of a mini-channel by treatment of the outer wall with a laser pulse has been developed.

We completed experimental studies on the effect of the modification to heat transfer during flow boiling of R-125 in the vertical channel with a diameter of 1.1 mm and length of 50 mm at two values of the reduced pressure: $p_r = 0.43$ and $p_r = 0.56$. The mass flow rate was varied in the range of $G = 200$–$1200 \text{ kg}/\text{m}^2\text{s}$. The maximum heat flux was limited by the CHF.

As a result of the modification, the heat transfer coefficient and the CHF increased. The greatest effect of the surface modification on heat transfer was observed at the reduced pressure $p_r = 0.43$, where the increase in the HTC was up to 110% in the range of the mass flow rate $G = 1260 \text{ kg}/(\text{m}^2\text{s})$. The maximum increase in the CHF was 22% at the reduced pressure $p_r = 0.43$. It should be noted that the pressure increase led to a significant increase in the HTC by up to 100%. At the highly reduced pressure $p_r = 0.56$, the effect of the surface modification on heat transfer decreased, where the HTC increased from 11% to 23% depending on the mass flow rate.

These research results showed that such modifications can be used to increase the intensity of heat transfer in various heat exchange devices, where low and moderate reduced pressures can also be used. Our experiments have shown that at moderately reduced pressures of $p_r \approx 0.4$, the values of the HTC can be significantly increased to those observed at high reduced pressures of $p_r \approx 0.6$ when using the presented method of inner surface modification in circular mini-channel.

Author Contributions: Writing-review and editing, A.V.B. and R.X.; investigation, A.V.B.; Conceptualization, A.V.D. and A.N.V.; supervision, A.V.D.; project administration, A.V.D.; funding acquisition, A.V.D.; software, N.E.S.; formal analysis, N.E.S. and P.J.; data curation, P.J.; methodology, A.N.V.; data curation, R.X. All authors have read and agreed to the published version of the manuscript.

Funding: This work was supported by RSF Grant 19-19-00410.

Conflicts of Interest: The authors declare no conflict of interest.

Nomenclature

d	diameter, m
G	mass flow rate, $kg/(m^2 s)$
p	pressure, Pa
T	temperature, K
x	vapor quality
w	velocity, m/c^2
Greek symbols	
α	heat transfer coefficient, $W/m^2 \cdot K$
ξ	hydraulic friction factor
ρ	density, kg/m^3
Subscripts	
l	liquid
g	gas
boil	boiling
con	convective
calc	calculated
exp	experimental
cr	critical
in	inlet
s	saturated
r	reduced

References

1. Li, Y.; Wang, Q.; Li, M.; Ma, X.; Xiao, X.; Ji, Y. Investigation of Flow and Heat Transfer Performance of Double-Layer Pin-Fin Manifold Microchannel Heat Sinks. *Water* **2022**, *14*, 3140. [CrossRef]
2. Kumar, S.R.; Singh, S. Experimental Study on Microchannel with Addition of Microinserts Aiming Heat Transfer Performance Improvement. *Water* **2022**, *14*, 3291. [CrossRef]
3. Volodin, O.A.; Pecherkin, N.I.; Pavlenko, A.N. Heat transfer enhancement at boiling and evaporation of liquids on modified surfaces-a review. *Teplofizika Vysokikh Temperatur* **2021**, *59*, 280–312. [CrossRef]
4. Fujikake, J. 1980 Heat Transfer Tube for Use in Boiling Type Heat Exchangers and Method of Producing the Same. U.S. Patent 4216826, 12 August 1980.
5. Zubkov, N.N. Obtaining subsurface cavities by deforming cutting to intensify bubble boiling. *Vestn. Mashinostr.* **2014**, *11*, 75–79.
6. Volodin, O.A.; Pecherkin, N.; Pavlenko, A.N.; Zubkov, N.N. Heat transfer and crisis phenomena at boiling of refrigerant films falling down the surfaces obtained by deformational cutting. *Interfacial Phenom. Heat Transf.* **2017**, *5*, 215–222. [CrossRef]
7. Jung, D.; An, K.; Park, J. Nucleate Boiling Heat Transfer Coefficients of HCFC22, HFC134a, HFC125, and HFC32 on Various Enhanced Tubes. *Int. J. Refrigeration.* **2004**, *27*, 202. [CrossRef]
8. Jaikumar, A.; Kandlikar, S.G. Ultra-High Pool Boiling Performance and Effect of Channel Width with Selectively Coated Open Microchannels. *Int. J. Heat Mass Transfer.* **2016**, *95*, 795. [CrossRef]
9. Dedov, A.V. A Review of Modern Methods for Enhancing Nucleate Boiling Heat. *Therm. Eng.* **2019**, *66*, 881–915. [CrossRef]
10. Dąbek, L.; Kapjor, A.; Orman, Ł.J. Ethyl Alcohol Boiling Heat Transfer on Multilayer Meshed Surfaces AIP Conference Proceedings. *AIP Publ. LLC* **2016**, *1745*, 020005.
11. Hu, B.; Zhang, S.; Wang, D.; Chen, X.; Si, X.; Zhang, D. Experimental study of nucleate pool boiling heat transfer of self-rewetting solution by surface functionalization with TiO_2 nanostructure. *Can. J. Chem. Eng.* **2019**, *97*, 1399–1406. [CrossRef]
12. Das, S.; Kumar, D.S.; Bhaumik, S. Experimental study of nucleate pool boiling heat transfer of water on silicon oxide nanoparticle coated copper heating surface. *Appl. Therm. Eng.* **2016**, *96*, 555–567. [CrossRef]
13. Gouda, R.K.; Pathak, M.; Khan, M.K. Pool boiling heat transfer enhancement with segmented finned microchannels structured surface. *Int. J. Heat Mass Transf.* **2018**, *127*, 39–50. [CrossRef]
14. Hendricks, T.J.; Krishnan, S.; Choi, C.; Chang, C.H.; Paul, B. Enhancement of pool-boiling heat transfer using nanostructured surfaces on aluminum and copper. *Int. J. Heat Mass Transf.* **2010**, *53*, 3357–3365. [CrossRef]
15. Watanabe, Y.; Enoki, K.; Okawa, T. Nanoparticle layer detachment and its influence on the heat transfer characteristics in saturated pool boiling of nanofluids. *Int. J. Heat Mass Transf.* **2018**, *125*, 171–178. [CrossRef]
16. Prakash, C.J.; Prasanth, R. Enhanced boiling heat transfer by nano structured surfaces and nanofluids. *Renew. Sustain. Energy Rev.* **2018**, *82*, 4028–4043. [CrossRef]
17. Souza, R.R.; Manetti, L.L.; Kiyomura, I.S.; Cardoso, E.M. Liquid/surface interaction during pool boiling of DI-water on nanocoated heating surfaces. *J. Braz. Soc. Mech. Sci. Eng.* **2018**, *40*, 514. [CrossRef]

18. Surtaev, A.; Kuznetsov, D.; Serdyukov, V.; Pavlenko, A.; Kalita, V.; Komlev, D.; Ivannikov, A.; Radyuk, A. Structured capillary-porous coatings for enhancement of heat transfer at pool boiling. *Appl. Therm. Eng.* **2018**, *133*, 532–542. [CrossRef]
19. MacNamara, R.J.; Lupton, T.L.; Lupoi, R.; Robinson, A.J. Enhanced nucleate pool boiling on copper-diamond textured surfaces. *Appl. Therm. Eng.* **2019**, *162*, 114145. [CrossRef]
20. Dharmendra, M.; Suresh, S.; Kumar, C.S.; Yang, Q. Pool boiling heat transfer en- hancement using vertically aligned carbon nanotube coatings on a copper substrate. *Appl. Therm. Eng.* **2016**, *99*, 61–71. [CrossRef]
21. Xia, G.; Du, M.; Cheng, L.; Wang, W. Experimental study on the nucleate boiling heat transfer characteristics of a water-based multi-walled carbon nanotubes nanofluid in a confined space. *Int. J. Heat Mass Transf.* **2017**, *113*, 59–69. [CrossRef]
22. Kumar, G.U.; Soni, K.; Suresh, S.; Ghosh, K.; Thansekhar, M.R.; Babu, P.D. Modified surfaces using seamless graphene/carbon nanotubes-based nanostruc- tures for enhancing pool boiling heat transfer. *Exp. Therm. Fluid Sci.* **2018**, *96*, 493–506. [CrossRef]
23. Shin, S.; Choi, G.; Seok, K.B.; Cho, H.H. Flow boiling heat transfer on nanowire-coated surfaces with highly wetting liquid. *Energy* **2014**, *76*, 428–435. [CrossRef]
24. Wang, L.; Wang, Y.; Cheng, W.; Yu, H.; Xu, J. Boiling heat transfer properties of copper surface with different microstructures. *Mater. Chem. Phys.* **2021**, *267*, 124589. [CrossRef]
25. Kumar, U.G.; Suresh, S.; Kumar, C.S.; Seunghyun, B. A review on the role of laser textured surfaces on boiling heat transfer. *Int. Appl. Therm. Eng.* **2020**, *174*, 115274. [CrossRef]
26. McCarthy, M.; Gerasopoulos, K.; Maroo, S.C.; Hart, A.J. Materials, Fabrication, and Manufacturing of Micro/Nanostructured Surfaces for Phase-Change Heat Transfer Enhancement. *Nanoscale Microscale Thermophys. Eng.* **2014**, *18*, 288–310. [CrossRef]
27. Zhang, S.; Jiang, X.; Li, Y. Extraordinary Boiling Enhancement through Micro-Chimney Effects in Gradient Porous Micromeshes for High-Power Applications. *Energy Convers. Manag.* **2020**, *209*, 112665. [CrossRef]
28. Zhang, X.; Zhang, J.; Ji, H.; Zhao, D. Heat transfer enhancement and pressure drop performance for R417A flow boiling in internally grooved tubes. *Energy* **2015**, *86*, 446–454. [CrossRef]
29. Vahid, E.; Suleyman, C.; Sadaghiani, K.A.; Khellil, S.; Ali, K. Effect of Surface Biphilicity on FC-72 Flow Boiling in a Rectangular Minichannel. *Front. Mech. Eng.* **2021**, *7*, 755580.
30. Belyaev, A.V.; Varava, A.N.; Dedov, A.V.; Komov, A.T. Critical heat flux at flow boiling of refrigerants in minichannels at high reduced pressure. *Int. J. Heat Mass Transf.* **2018**, *122*, 732–739. [CrossRef]
31. Belyaev, A.V.; Varava, A.N.; Dedov, A.V.; Komov, A.T. An experimental study of flow boiling in minichannels at high reduced pressure. *Int. J. Heat Mass Transf.* **2017**, *110*, 360–373. [CrossRef]
32. Belyaev, A.V.; Dedov, A.V.; Krapivin, I.I.; Varava, A.N.; Jiang, P.; Xu, R. Study of Pressure Drops and Heat Transfer of Nonequilibrial Two-Phase Flows. *Water* **2021**, *13*, 2275. [CrossRef]
33. Idelchik, I.E. *Spravochnik po Gidravlicheskim Soprotivleniyam*, 2nd ed.; Mashinostroenie: Moskva, Russia, 1975; pp. 143–162.

Article

Intense Vortex Motion in a Two-Phase Bioreactor

Bulat R. Sharifullin [1], Sergey G. Skripkin [1], Igor V. Naumov [1,*], Zhigang Zuo [2,*], Bo Li [3] and Vladimir N. Shtern [1,4]

1.   Laboratory of Heat and Mass Transfer Problems, Kutateladze Institute of Thermophysics, Siberian Branch of the Russian Academy of Sciences, 1 Academician Lavrent'ev Avenue, 630090 Novosibirsk, Russia
2.   State Key Laboratory of Hydroscience and Engineering, Department of Energy and Power Engineering, Tsinghua University, Beijing 100084, China
3.   Beijing Key Laboratory of Information Service Engineering, Beijing Union University, Beijing 100101, China
4.   Shtern Research and Consulting, Houston, TX 77096, USA
*   Correspondence: naumov@itp.nsc.ru (I.V.N.); zhigang200@tsinghua.edu.cn (Z.Z.)

Abstract: The paper reports the results of experimental and numerical studies of vortex motion in an industrial-scale glass bioreactor (volume, 8.5 L; reactor vessel diameter D, 190 mm) filled 50–80%. The model culture medium was a 65% aqueous glycerol solution with the density $\rho_g = 1150$ kg/m^3 and kinematic viscosity $\nu_g = 15$ mm^2/s. The methods of particle image velocimetry and adaptive track visualization allow one to observe and measure the vortex motion of the culture medium. In this work, the vortex flow investigation was performed in a practical bioreactor at the operation regimes. Our research determines not only the optimal flow structure, but also the optimal activator rotation speed, which is especially important in the opaque biological culture. The main result is that, similar to the case of two rotating immiscible liquids, a strongly swirling jet is formed near the axis, and the entire flow acquires the pattern of a miniature gas–liquid tornado. The aerating gas interacts with the liquid only through the free surface, without any mixing. This intensifies the interphase mass transfer due to the high-speed motion of the aerating gas.

Keywords: vortex reactor; complex vortex; vortex flow modeling; phase boundary; immiscible liquids; free surface

Citation: Sharifullin, B.R.; Skripkin, S.G.; Naumov, I.V.; Zuo, Z.; Li, B.; Shtern, V.N. Intense Vortex Motion in a Two-Phase Bioreactor. *Water* **2023**, *15*, 94. https://doi.org/10.3390/w15010094

Academic Editors: Maksim Pakhomov and Pavel Lobanov

Received: 27 November 2022
Revised: 22 December 2022
Accepted: 23 December 2022
Published: 28 December 2022

1. Introduction

The evolution of fluid mechanics has been driven by the connection between science and the practical needs throughout history, from the hydromechanics of "the ancients" through the Newtonian era and up to the present day. It is safe to say that vortex motion is one of the fundamental states of a liquid medium [1,2]. According to the vivid expression provided in the Saffman's book [3], vortices are "sinews and muscles of fluid motion".

Although the existence of vortices of various types has been known for a long time [3], diverse metamorphoses of vortex formation and vortex transfer of mass and energy require deeper analysis and understanding [4–6]. An important and interesting fact that needs additional interpretation is the formation of vortex motion during the interaction between various liquid and gaseous media (e.g., a gas vortex in an aqueous medium [7,8] and slippage in flows containing nanoparticles [9]) or immiscible liquid media differing both in density and viscosity [10].

In technical applications, vortex mixing is a widespread method of mass transfer enhancement. In vortex combustion chambers, vortex cells stabilize the flame and reduce harmful emissions. In chemical gas and biological two-liquid reactors, the organization of vortex motion contributes to mixing of ingredients, thus increasing the yield of the net product [4]. A convenient model of a vortex reactor is a vertical cylindrical container, where movement of the filling liquid is generated through an intermediate liquid or a gaseous medium by rotating one of the end disks [4–6]. Notably, in the case of a single-component liquid, it is not that important which of the end disks is rotating; for the case of using two media or a free boundary, it is possible to compensate for the influence of gravity on the shape of the interface between the immiscible components by choosing the size of the inductor generating a vortex structure [11]. Naumov et al. [12] reported a similarity in the patterns of formation and evolution of the recirculation zones in the center and on the free end of an intensely swirled flow generated in a non-uniformly filled cylinder due to the opposite end rotation. The flow topology appears to be independent of properties of the medium (liquid or gas) at the free boundary, thus limiting fluid circulation and location of the disk generating the vortex structure [13,14].

Many vortex devices allow for interaction between a rotating working fluid and a layer of another liquid or air (e.g., when the reactor is partially filled with the working fluid). Recently, researchers have focused on two-fluid rotating flows in the context of developing vortex-aerated bioreactors. In searching for an optimal flow shape for culture growth in vortex bioreactors, an interesting and impressive structure has been discovered: it simulates strong vortices observed in the atmosphere and ocean and can be called "a two-story tornado" [15]. It has been revealed that the tornado-like flow develops in bioreactors as well.

The bioreactor is the main component in industrial-scale microbiological synthesis [16]. Microbial cells can withstand vigorous stirring and aeration. Animal and plant cells are more fragile and sensitive to mechanical action regardless of the cultivation method (in suspension or in an immobilized state). The aim of using the bioreactor is to provide optimal conditions for the growth of cultivated biological objects and biosynthesis of the target product while maintaining the sterility and cost-effectiveness of the process. During deep aerobic biosynthesis, the bioreactor contains a system of components, which are a suspension of cells (or other aforementioned cultures) in a nutrient medium enriched with dissolved oxygen. The system requires the maintenance of certain conditions—heat and mass transfer—as well as aeration of the medium.

The technological complexity of ensuring the parameters of the mixing process for different types of cells and microorganisms explains the wide variety of designs of industrial bioreactors. On the basis of the principle of operation, traditional bioreactors are included in one of the following four groups: (a) bubbling and airlift bioreactors, where mixing is realized due to the air flow supplied through the medium; (b) bioreactors with a mechanical "stirrer" in the mixed medium; (c) combinations of the first two types; and (d) highly specialized and unique ones, created for one process.

Leaving behind the conventional devices, we explain the characteristics of the gas vortex bioreactor using a fundamentally new mixing method employing a concentrated air vortex generated by an activator installed under the bioreactor lid above the surface of the culture medium. Mixing of the culture medium is carried out by creating a three-dimensional motion of the "rotating vortex ring" type: a quasi-stationary flow with an axial counterflow generated by an aerating gas vortex due to the pressure drop above the surface and the friction force of the air flow on the liquid surface.

The absence of a mechanical stirring element in the liquid ensures energy saving (0.06–0.1 W/L, compared to 1–4 W/L for bioreactors with a mechanical stirring device and airlift bioreactors), soft and efficient mixing of liquids—including viscous ones—without foam formation, cavitation (surface aeration without using a bubbler—for cell cultures), hydraulic shocks, highly turbulent or stagnant zones, and micro zones with high temperature [16].

In contrast to previous studies, here we explore the vortex flow in a real-world bioreactor. The aim of this work is to identify the regularities of vortex flow in an aerial vortex bioreactor at various activator rotation parameters. Knowing the rotation parameters, it is possible to determine the flow mode in any opaque culture medium. The culture is placed in a liquid that is activated by an upper medium (air) driven by the rotation of the activator. Therefore, it is responsible not only the optimal flow structure, but also for the optimal rotation speed, thus ensuring soft and thorough mixing of ingredients and preventing them from directly contacting the activator. The biological culture is not destroyed because it does not contact the rotating solid parts of the reactor. Meanwhile, the effective mixing and saturation of the working medium with nutrients and oxygen is ensured.

2. Materials and Methods

2.1. Experimental Setup

The patterns of vortex motion were studied in an industrial-scale GV FBR 10-I glass cylindrical bioreactor manufactured by the Center for Vortex Technologies (volume, 8.5 L; diameter of the reactor vessel D = 190 mm; and height H = 300 mm) (Figure 1a). A 65% aqueous glycerol solution was used as a model liquid (density ρ_g = 1150 kg/m^3; kinematic viscosity v_g = 15 mm^2/s). In the reactor, mixing of the cell suspension was carried out by creating a quasi-stationary rotational motion generated by a swirling gas flow. The tornado-like swirling gas flow was generated by an impeller (activator) above the surface of the culture medium (Figure 1b). The aerating gas interacted with the cell suspension only through the free surface of the latter, without mixing with it. As a result, the interfacial mass transfer was intensified due to the high velocity of the aeration gas, and the suspension was mixed evenly, without stagnant zones.

(a)

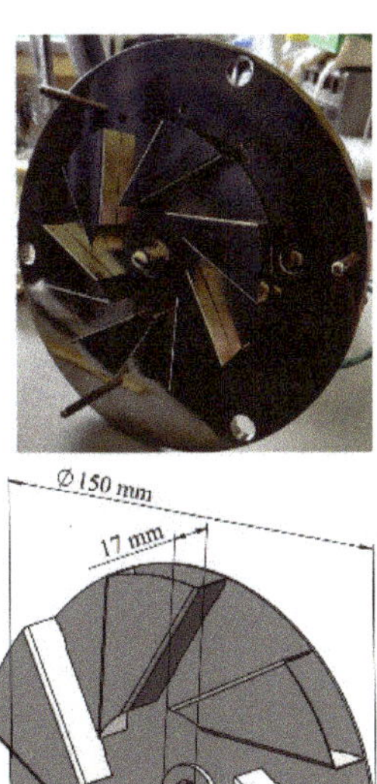

(b)

Figure 1. (**a**) A gas–liquid bioreactor: a photo of the test bench; (**b**) an impeller (activator) generating vortex motion in the bioreactor.

Due to the interface friction and pressure difference between the periphery and the center of the gas vortex, such a swirl of the aerating gas ensures the movement of the model liquid in the form of a vortex ring, rotating with respect to the tank axis with simultaneous downward liquid motion at the periphery of the tank and ascending motion in the axial zone [14,15].

Figure 2 shows the organization of vortex motion in a bioreactor and the diagram of the test bench. The gas–vortex reactor was placed in a glass container sized 300 × 300 × 400 mm to reduce optical aberrations and ensure thermal stabilization. The vortex motion of air was generated by an activator at a rotation speed Ω of up to 1000 rpm at the minimal filling of 50% (h_g = 150 mm) and, by continuously adding the working model medium, to the maximal filling of 80% (h_g = 240 mm).

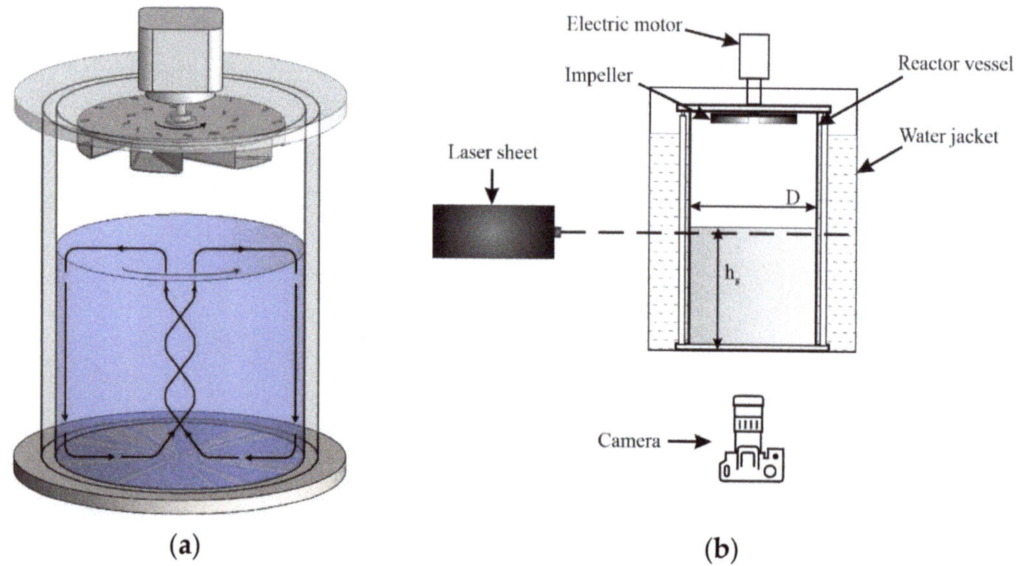

Figure 2. The experimental setup: (**a**) organization of vortex motion and (**b**) setup diagram.

The velocity fields were measured by Particle Image Velocimetry (PIV). The PIV method provided an instantaneous velocity distribution in the investigated cross-section and an instantaneous flow pattern within the two-dimensional plane of the light sheet. The PIV and adaptive track visualization methods were used to observe the vortex motion pattern [13,14,17]. Polyamide beads (density, 1030 kg/m^3; diameter, ~10 µm) were employed as seeding light-scattering particles for both flow visualization and PIV measurements. The air flow was seeded by a fog generator. Measurements were performed in a horizontal section in the vicinity of the interface, at a distance of 2 mm in air and liquid and near the bottom of the reactor vessel, as well as in a vertical section passing through the axis of the cylinder. The light cross-section was formed by a laser sheet of the PIV measuring system [14], and the image was recorded by a camera through the transparent glass bottom and sidewall of the bioreactor.

The PIV system consisted of a double-pulsed Nd:YAG Beamtech Vlite-200 laser (wavelength, 532 nm; repetition rate, 15 Hz; pulse duration, 10 ns; pulse energy, 200 mJ), an IMPERX IGV-B2020 CCD camera (8 bits per pixel; matrix resolution 2056 × 2060 pixels, 1.3″ optical format) equipped with a Nikon SIGMA 50 mm f/2.8D lens, and a synchronizing processor. We calculated the two-dimensional velocity fields using the commercial ActualFlow software, Version 1.18.8.0. The thickness of the laser light sheet formed by a cylindrical lens to illuminate tracer particles was ~0.8 mm in the measurement plane. The distance between the camera and the laser sheet was 800 mm and 720 mm for the horizontal and vertical planes, respectively.

For every set of experimental conditions, we accumulated 500 images and averaged them to increase the signal-to-noise ratio. Time delay between the two images varied from 5 to 120 ms depending on the fluid type. The velocity fields were calculated using the iterative cross-correlation algorithm with a continuous window shift and deformation and 75% overlap of the interrogation windows in order to have a relatively large dynamic range (the span between the maximal and minimal velocity values). The threshold value for concentration of seeding particles was 5–8 tracers per 64 × 64-pixel area. The sub-pixel interpolation of a cross-correlation peak was performed over three points using one-dimensional approximation by the Gaussian function. The inaccuracy of position determination did not exceed 0.1 pixel. Thus, the velocity measurement–error estimates were 1% and 4% for tracer sizes of 8 and 2 pixels, respectively. The resulting spatial resolution of the velocity fields was approximately one vector per 1.4 mm. A more detailed description of the applied methods of experimental diagnostics and the composition of measuring systems can be found in [13,14,17].

The paper considers the structure of the circulation motion of the working fluid in the mode implying the formation of a laminar soft circulating movement of the working fluid for growing cultures that are not resistant to unfavorable hydrodynamic conditions and require soft mixing. This mode of bubble-free aeration is implemented using an activator: a high-speed gas vortex is formed above the surface and swirls the culture medium, thus creating a weak circulation movement and simultaneously saturating it with oxygen. The optimal operating range of the activator rotation with neither interface fluctuations nor foaming is 180–900 rpm.

2.2. The CFD Approach

Numerical calculations were carried out in order to complete the flow structure in the upper medium (air) where PIV measurements are difficult to perform, especially in the vertical section. The main focus of the calculations was placed on the vortex motion of the air medium generated by the rotating impeller. The numerical simulations were performed using the OpenFOAM software [18,19]. The finite volume method (FVM) was used to solve the incompressible Navier–Stokes equations, and the volume of fluid (VOF) method was adopted for capturing the liquid–gas interface [18–20]. The Multiple Reference Frame (MRF) method was used to calculate the rotating domain. The MRF approach was employed to control volume with multiple zones at different rotational speeds. The fluids (gas and water) were treated as incompressible and immiscible Newtonian fluids where the heat and mass transfer are neglected. The equations of continuity and momentum were solved:

$$\frac{\partial \rho}{\partial t} + \nabla \cdot (\rho U) = 0 \tag{1}$$

$$\frac{\partial (\rho U)}{\partial t} + \nabla \cdot (\rho U U) = -\nabla p + \nabla \cdot s + \rho g + F \tag{2}$$

where $\rho = \alpha \rho_l + (1 - \alpha)\rho_a$; subscripts '$l$' and '$a$' represent the liquid and gas phases, respectively. F is the source term due to surface tension, which is modeled using the Continuum Surface Force (CSF) method. $s = \mu(\nabla U + \nabla U^T - 2/3(\nabla \cdot U))I$ is the viscous stress tensor; I is the identity tensor; and $\mu = \alpha \mu_l + (1 - \alpha)\mu_a$ is the dynamic viscosity.

The incompressible Navier–Stokes equation in the relative frame with absolute velocity could be written as [21–23]

$$\frac{\partial U_R}{\partial t} + \frac{d\Omega}{dt} \times r + \nabla \cdot (U_R \otimes U_S) + \Omega \times U_S = -\nabla \left(\frac{p}{\rho}\right) + \nu \nabla \cdot \nabla (U_S)$$
$$\nabla \cdot U_S = 0 \tag{3}$$

where U is the absolute velocity; the subscripts R and S indicate the rotary and stationary components, respectively. Ω is the rotational speed of the reference frame for a stationary observer, which is equal to the rotational speed of the rotor. p and ρ denote the pressure and density, respectively. The density is $\rho_g = 1150$ kg/m^3 for glycerol and $\rho_a = 1.2$ kg/m^3 for air. The kinematic viscosities are $\nu_g = 15$ mm^2/s for glycerol and $\nu_a = 18$ mm^2/s for air; the surface tension is $\sigma = 0.05$ N/m.

The NS equations were solved using the Direct Numerical Simulations (DNS). The time ranged from 0 to 10 s. The Courant number was set at 0.1, and the time increment was adjustable. In order to eliminate the prime error, we averaged the results during the last 5 s.

The computational domain and mesh are shown in Figure 3. For simplicity, the rotor shaft was neglected in simulations. The domain diameter was 180 mm, and the height was 250 mm. It contained 364,421 hexahedral grids. The refinement mesh was adopted in the gas domain, as shown in Figure 3. Furthermore, the boundary condition of the enclosed vessel and rotor walls was no-slip. The initial condition for the pressure over the free surface was set at $p_0 = 101{,}325$ Pa. The gravity was considered here, where $g = 9.8$ m/s^2.

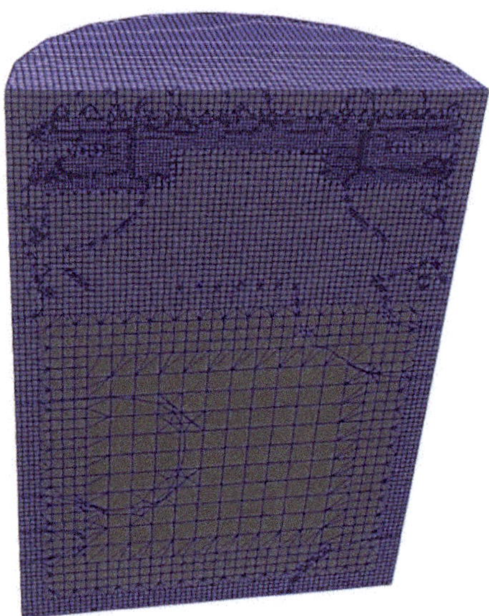

Figure 3. The computational domain and the mesh system.

3. Results

3.1. Experimental

An experimental study of the flow structure was carried out for several stationary modes at $\Omega = 90$, 180, 360, 540, 720, and 900 rpm using the PIV method and the developed hardware and software of the adaptive track visualization complex. The adaptive visualization method made it possible to qualitatively investigate the flow structure. For the motion formation, there is an analogy with the case of a two-liquid system [10,13], where the lower heavy fraction has a divergent flow under the interface and a converging flow near the bottom. The study was carried out in a horizontal section in the vicinity of the interface and bottom in the liquid at a distance of 2 mm.

Table 1 shows the maximum values of the measured tangential velocity components (V_{tgm}) depending on the velocity of the activator (Ω) at $h = 0.5$ and 0.8. These values serve to determine the characteristic Re_g. In this case, the Reynolds number for the liquid is defined as $\mathrm{Re}_g = V_{tgm}R/\nu_g$, where V_{tgm} is the maximum value of the tangential component of the velocity, R is the radius of the reactor vessel, and ν is the kinematic viscosity of the working fluid. The use of a 2.3 MP sensor Sony IMX174 with 10 Hz framing and a delay of up to 100 ms allowed us to significantly improve quality of the recorded images, and the preliminary subtraction of the background image reduced the noise level. Figure 4 shows examples of converging and diverging helical motion of the working fluid under the interface (Figure 4a) and in the vicinity of the bottom (Figure 4b), which illustrates the circulating meridional movement of the working fluid generated by an intense air vortex in the bioreactor.

Table 1. The maximum values of the tangential velocity under the interface and the Reynolds number in the liquid, depending on the rotation of the activator.

	Ω, rpm	V_{tgm}, mm/s	Re
	180	7.7	48.8
	360	18.1	114.6
$h = 0.5$	540	29.2	184.9
	720	36	228
	900	46.8	296.5
	180	9.7	61.3
	360	18.5	117.4
$h = 0.8$	540	32.4	205.5
	720	52.5	332.7
	900	77.1	488.6

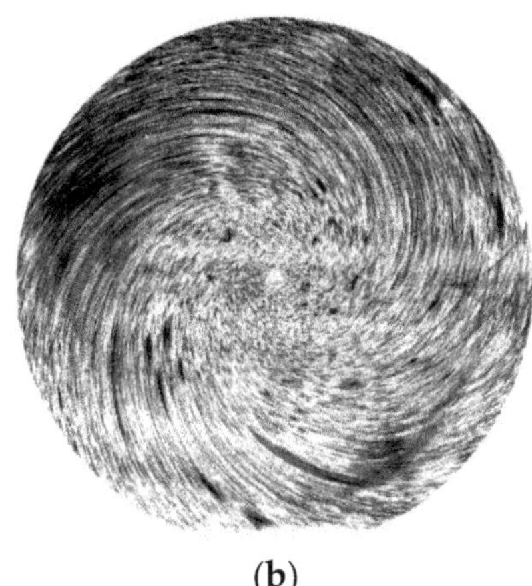

(a)   **(b)**

Figure 4. Visualization of the motion of the model fluid: (**a**) the flow diverging from the axis below the interface and (**b**) the flow converging to the axis near the bottom at $\Omega = 900$ rpm, $h_g = 0.5h$.

The PIV method was used to obtain and analyze the velocity fields and profiles of the azimuthal and radial velocity components in air and in liquid at distances of 2 mm from the interface. Figure 5 shows the streamlines reconstructed from the vector fields obtained in a horizontal section near the interface when the reactor vessel was filled by 50%.

At low Ω, there is a strongly converging flow in the air above the interface, where the centrifugal force acts on the medium adjacent to the swirler and meridional air circulation is observed, like in the case of the two-liquid system. As Ω increases, a turbulent flow is formed in the air near the reactor axis; it appears as a diverging flow near the interface (Figure 4, left column). In the liquid, the processes occur according to the classical scenario of vortex interactions in two-fluid systems [6,24,25]. At low Ω, a centrifugal cell of meridional circulation arises near the axis, thus increasing its intensity with rising Ω. This scenario of vortex development and circulating fluid motion is observed both at 50% and 80% filling of the bioreactor with a model medium.

Figures 6–8 show the profiles of the tangential V_{tg} and radial V_r velocity components in the horizontal section in the vicinity of the interface in air and the model liquid at low and high rotation speeds of the activator $\Omega = 90, 180, 360, 540,$ and 720 rpm. Further, for convenience, the heights of reactor filling with the working fluid $h_g = 210$ mm and 340 mm are presented in dimensionless quantities $h = H/h_g = 0.5$ and $h = H/h_g = 0.8$, respectively.

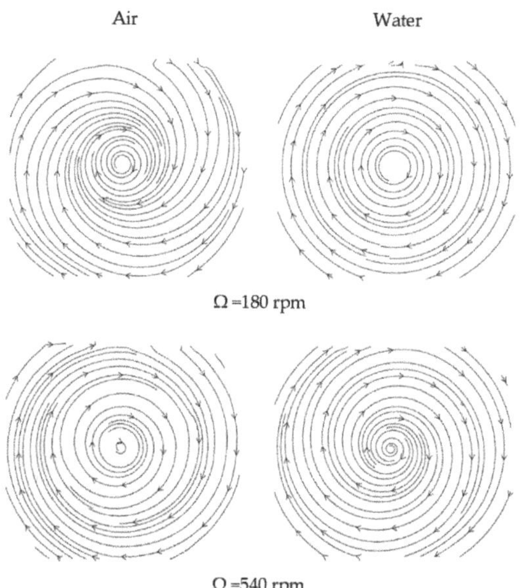

Figure 5. The streamlines reconstructed from the vector fields obtained in a horizontal section near the interface ((**left**)—in air; (**right**)—in the working model fluid).

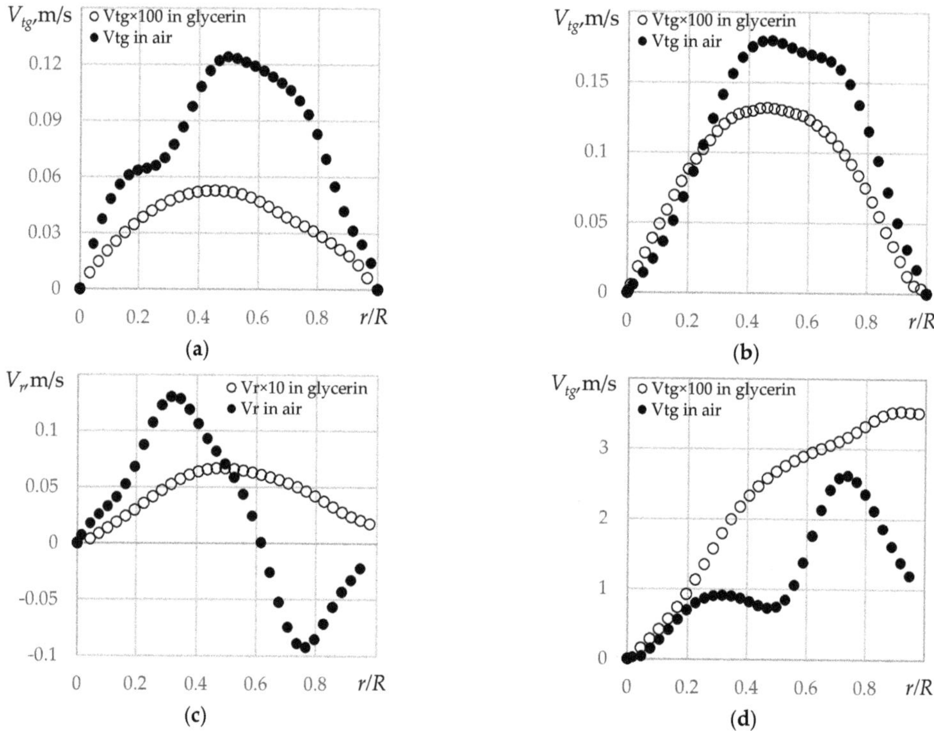

Figure 6. The tangential velocity profiles at Ω = 90 rpm: (**a**) h = 0.5; (**b**) h = 0.8; and (**c,d**) the radial and tangential velocities at Ω = 720 rpm and h = 0.5, respectively.

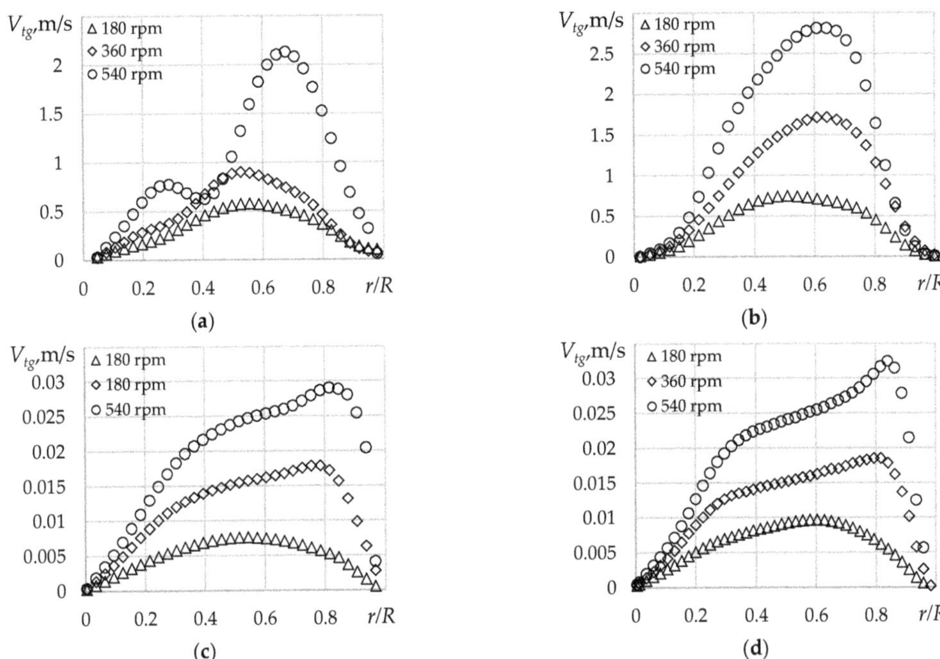

Figure 7. Tangential velocity profiles in air (**a**) at $h = 0.5$, (**b**) at $h = 0.8$, and in glycerin (**c**) at $h = 0.5$ and (**d**) at $h = 0.8$.

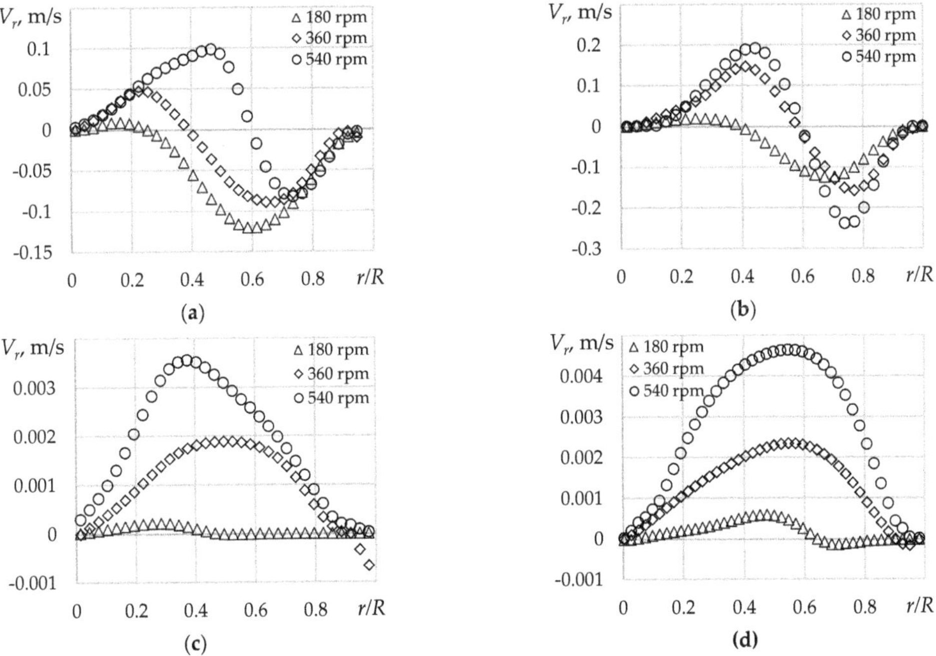

Figure 8. The radial velocity profiles in air (**a**) at $h = 0.5$, (**b**) at $h = 0.8$, and in glycerol (**c**) at $h = 0.5$ and (**d**) at $h = 0.8$.

Figure 6 compares the profiles of the tangential and radial velocity components in air and liquid at Ω = 90 and 720 rpm. It shows only the tangential velocity profiles for Ω = 90 rpm, since the radial velocity of the fluid in this mode is too low to be measured using the PIV method. Since the maximum tangential velocity in a liquid is two orders of magnitude lower than that in air, for convenience, Figure 6 shows the values of the tangential velocity of the liquid multiplied by 100; Figure 6c presents the values of the radial velocity of the liquid multiplied by 10.

Figure 6 shows that the maximum values of the profiles at Ω = 90 rpm are much smaller than those at Ω = 720 rpm; due to the considerable difference in densities of liquid and air, the velocity profiles differ by one or two orders of magnitude. In order to avoid big differences in the velocity values in the same diagram, the data were grouped by types of the medium, so Figures 7 and 8 show the profiles only for Ω = 180, 360, and 540 rpm.

Figure 7 shows the tangential component of velocity. Although the maximum velocities in air and in the liquid differ by two orders of magnitude, even with small flow swirls, at Ω = 180 rpm, the vortex motion is generated in the model liquid. As Ω increases to 540 rpm, this trend persists and becomes more pronounced despite the fact that the maximum values still differ by two orders of magnitude. Meanwhile, as it can be seen from Figure 5a, the knee of the tangential velocity component profile increases at h = 0.5. Visual observations in the vertical section indicate that a non-stationary vortex appears in the air as Ω on the axis increases, while centrifugal circulation is observed at the periphery: the air flow sinks along the wall to the bottom and rises up about half the radius before reaching the axis. A double vortex is presumably formed in the air: an internal non-stationary vortex and an external toroidal one with centrifugal circulation. This assumption is confirmed by the values of the radial velocity component (Figure 7a,b) at points where the velocity values reverse their sign from positive to negative. Hence, a flow directed towards the periphery appears in the axis region with increasing Ω.

Having compared the values of the tangential velocity component in air and the model liquid, one may note that the trend of angular momentum transfer at small and large flow swirl values, as well as different volumes of the model liquid, is fully maintained. Thus, the results obtained are consistent with the earlier data for liquid vortex systems, where a less dense liquid swirled a denser liquid [24,25]. The characteristic non-stationary processes in air, which manifest themselves, among other things, in the formation of nested vortex cells characterized by changes in circulation and velocity values at which the air–liquid interface does not undergo fluctuations, are not specific for the model liquid.

Figure 8 shows that, at Ω = 180 rpm, the air flow is directed towards the axis (negative velocity values), while divergent motion from the axis to the periphery is already beginning to form in the model fluid, and the values of the radial velocity component are becoming positive. Therefore, the converging vortex air flow forms a divergent motion in the model fluid.

The diagrams in Figure 8c,d demonstrate how the central circulation appears and grows in the model fluid. At Ω = 180 rpm, the radial velocity component in the model fluid is small, while at Ω = 540 rpm, the maximum value of the radial velocity component increases by an order of magnitude to reach values of several millimeters per second. In the meantime, the profiles of both tangential and radial velocity do not have abrupt bends, which indicates that both vortex and meridional motions of the model fluid are formed in the entire volume of the bioreactor.

As it has been shown in an experimental study focusing on the structure of a closed vortex flow of two immiscible liquids [25], there is a similarity in the development of the cellular structure in the cases of the one- and two-liquid systems, and the Reynolds number of the lower liquid is close to that for the one-liquid system. This analogy is also true for the case when the vortex motion in the working medium is formed by a gas vortex.

In order to detect an increase in the circulation cell in the model fluid, the flow structure in a vertical section passing through the cylinder axis was studied at Ω = 180, 360, 540, 720, and 900 rpm. The velocity fields were obtained and analyzed using the PIV method.

Figure 9 shows the profiles of the axial component of velocity Vax in the liquid on the reactor axis, demonstrating the enlargement of the circulation cell and formation of an upward flow along the cylinder axis at Ω = 360, 540, 720, and 900 rpm and h = 0.5 and 0.8.

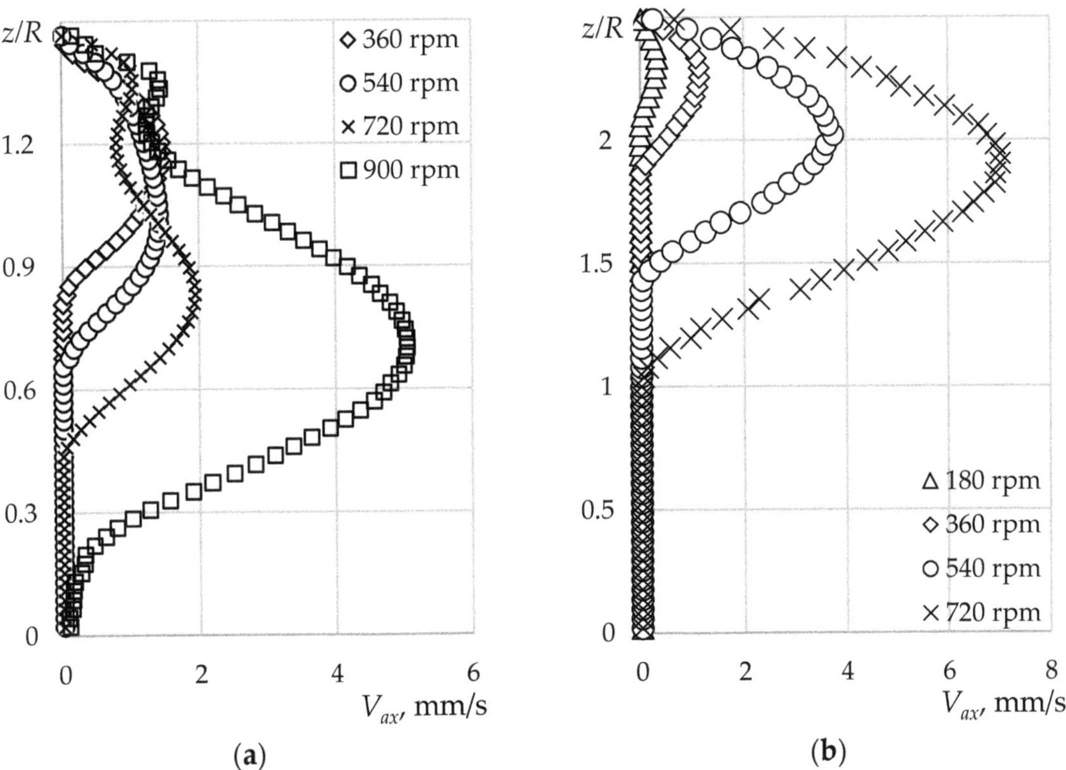

Figure 9. The profiles of the axial velocity component at (**a**) h = 0.5 and (**b**) 0.8.

The results presented in Figure 9 show that the development of vortex motion in a gas–vortex bioreactor and propagation of the vortex structure deep into the volume of the working fluid occur concordantly, like in the case of two immiscible liquids [25]. The vortex cell reaches the bottom only at the maximum values of the activator rotation Ω = 900 rpm, which must be taken into account when cultivating viscous media, at v_g > 10 cCt. When working with low-viscosity media (e.g., when cultivating microalgae in which the solution viscosity is close to that of water), full meridional circulation, according to Table 1 and the results of work [25], will already be achieved at the value of the activator rotation Ω > 180 rpm.

3.2. Numerical Data

The data on the structure of the vortex air flow were obtained using numerical simulations. The numerical calculations were verified by individual velocity profiles and showed acceptable compliance when comparing the velocity profiles over the interface (Figure 10). At high activator rotational speeds, the differences were somewhat greater; nevertheless, the numerical calculations reproduced the main features of the flow, such as the shape of the tangential velocity profile and the change in direction of the radial velocity component.

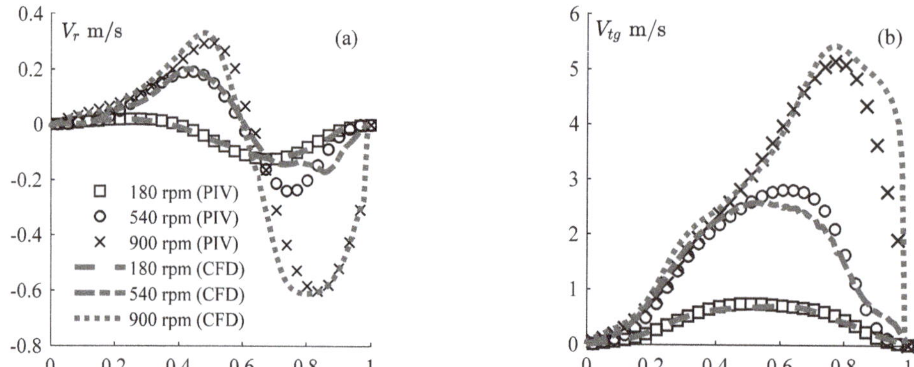

Figure 10. Comparison between the numerical and experimental (**a**) radial and (**b**) tangential profiles above the air–water interface at Ω = 180, 360, 720, and 900 rpm.

Figure 11 shows the numerical simulation data for three different impeller speeds Ω = 180, 540, and 900 rpm for h = 0.8 in the vertical cross-section. The color shows the absolute velocity value, the arrows show the direction of the flow in a vertical cross-section, the white region represent vane. The fields were obtained by time averaging during 5 s. The velocity maximum corresponds to the area near the impeller, and its value depends almost linearly on the rotational speed. The structure of the flow in the upper part of the bioreactor is turbulent and unsteady; the Reynolds numbers in the experiments and calculations lie in the range of 5000–25,000.

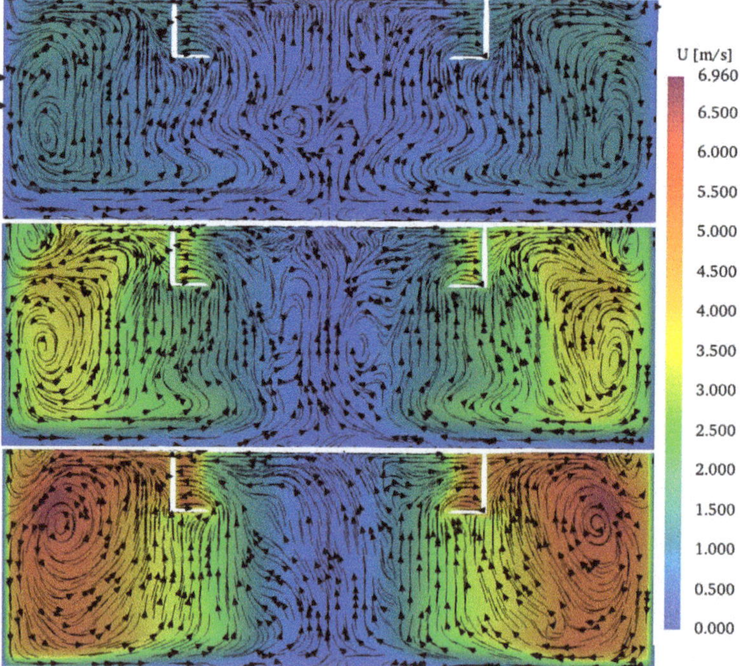

Figure 11. The velocity fields and streamlines in the vertical cross-section: Ω = 180 rpm (**top**), Ω = 540 rpm (**middle**), and Ω = 900 rpm (**bottom**).

In Figure 12, one can see the different slope angle of the streamlines. The arrows show the direction of the flow in a horizontal cross-section. For the Ω = 900 rpm flow regime, the radial velocity component is degenerated, and the central vortex area is formed; in turn, spiral motion towards the center is observed for Ω = 180 rpm, which agrees well with the current lines in Figure 5. The good agreement between the numerical data and the experimental findings allows for the numerical calculations to be used to find the optimal bioreactor filling as well as to investigate important upscaling issues.

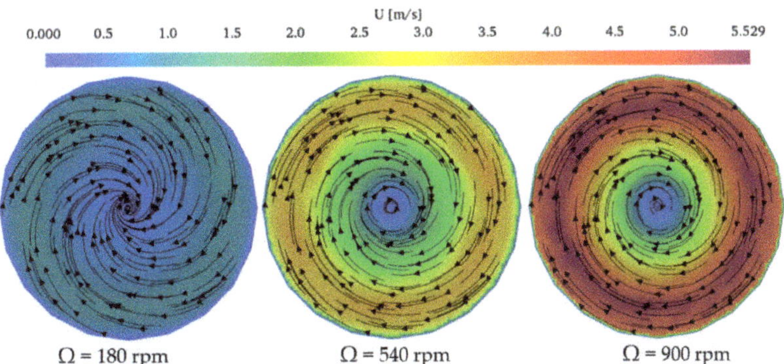

Figure 12. The velocity fields and streamlines are 2 mm above the interface in air.

Both the experimental and numerical results indicate an interesting and paradoxical fact that a spiral flow of a lighter upper media (air) converging above the interface generates a diverging spiral motion of a denser lower fluid (water) under the interface like it occurs in the two-fluid systems [26].

4. Conclusions

This paper reveals and explains the flow structure in a gas–vortex reactor at different flow swirl parameters. We show that there are common features of the vortex motion of the working medium depending on its volume, viscosity, and activator rotation intensity. The main finding is that, similar to the case of two rotating liquids [12,13], a strongly swirled jet is formed at the reactor axis, and the entire flow acquires a structure of a miniature gas–liquid tornado. When the activator rotates, a meridional and circulating flow occurs in the working liquid. Centrifugal circulation cells appear under the interface and increase into the depth of the reactor with increasing speed of the activator rotation. The centrifugal circulation of the working fluid appears similar to that in a closed vortex flow for the one-liquid system, and in the lower liquid, to that in a system of two swirling immiscible liquids [23]. A striking feature is that the radial velocity component slips at the interface, while the air flow descending to the reactor axis forms a divergent vortex motion of the model liquid medium despite the difference in densities being more than three orders of magnitude.

In this study, the vortex flow investigation was performed in a real-world bioreactor at the operation regimes. Our research determined not only the optimal flow structure, but also the optimal activator rotation speed, which is especially important for opaque biological cultures.

Author Contributions: Conceptualization, I.V.N. and V.N.S.; methodology, S.G.S., Z.Z. and B.R.S.; validation, I.V.N., Z.Z. and V.N.S.; formal analysis, I.V.N., V.N.S. and B.R.S.; investigation, B.R.S., Z.Z., B.L. and S.G.S.; writing—original draft preparation, I.V.N., V.N.S. and Z.Z.; writing—review and editing, I.V.N. and Z.Z.; visualization, B.R.S., B.L. and S.G.S.; supervision and project administration, I.V.N.; funding acquisition, I.V.N. All authors have read and agreed to the published version of the manuscript.

Funding: This research was funded by the Russian Science Foundation, grant number 19-19-00083.

Institutional Review Board Statement: Not applicable.

Informed Consent Statement: Not applicable.

Data Availability Statement: Not applicable.

Conflicts of Interest: The authors declare no conflict of interest.

Nomenclature

D	diameter of the reactor vessel
F	source term due to surface tension
g	gravitational acceleration
H	height of the reactor vessel
h_g	height of model fluid
$h = h_g/H$	dimensionless height of model fluid
I	identity tensor
p	pressure
p_0	initial pressure above the free surface
R	radius of the reactor vessel
$Re = V_{tgm}R/\nu_g$	the Reynolds number
$s = \mu(\nabla U + \nabla U^{\mathrm{T}} - 2/3(\nabla \cdot U))I$	viscous stress tensor
U	absolute velocity
V_{ax}	axial component of velocity
V_r	radial component of velocity
V_{tg}	tangential component of velocity
V_{tgm}	the maximum value of the tangential component of velocity
Subscripts	
a	air
g	aqueous glycerol solution
l	liquid
r	radial component
R	rotary
S	stationary
tg	tangential component
m	maximum value
Greek	
$\mu = \alpha\mu L + (1 - \alpha)\mu_a$	dynamic viscosity
$\rho = \alpha\rho_l + (1 - \alpha)\rho_a$	density
ρ_a	density of air
ρ_g	density of aqueous glycerol solution
ν_a	kinematic viscosity of air
ν_g	kinematic viscosity of aqueous glycerol solution
Ω	rotation speed of activator
σ	surface tension
Acronym	
CFD	Computational Fluid Dynamics modeling
CSF	Continuum Surface Force method
DNS	Direct Numerical Simulations
FVM	Finite Volume Method
MRF	Multiple Reference Frame method
NS	Navier–Stokes
PIV	Particle Image Velocimetry
VOF	Volume of Fluid method

References

1. Poincaré, H. *Théories des Tourbillons*; George Carre: Paris, France, 1893.
2. Marusic, I.; Broomhall, S. Leonardo da Vinci and Fluid Mechanics. *Annu. Rev. Fluid Mech.* **2021**, *53*, 1–25. [CrossRef]
3. Saffman, P.G. *Vortex Dynamics*; Cambridge University Press: New York, NY, USA, 1993.
4. Shtern, V. Cellular Flows. In *Topological Metamorphoses in Fluid Mechanics*; Cambridge University Press: New York, NY, USA, 2018.
5. Stepanova, E.V.; Chaplin, T.O.; Chashechkin, Y.D. Experimental investigation of oil transport in a compound vortex. *J. Appl. Mech. Tech. Phy.* **2013**, *54*, 79–86. [CrossRef]
6. Shtern, V.N.; Naumov, I.V. Swirl-decay mechanism generating counterflows and cells in vortex motion. *J. Eng. Thermophys.* **2021**, *30*, 19–39. [CrossRef]
7. Byalko, A.V. Underwater gas tornado. *Phys. Scripta* **2013**, *155*, 014030. [CrossRef]
8. Faugaret, A.; Duguet, Y.; Fraigneau, Y.; Witkowsky, L. A simple model for arbitrary pollution effects on rotating free-surface flows. *J. Fluid Mech.* **2022**, *932*, A2. [CrossRef]
9. Hejazi, H.A.; Khan, M.I.; Raza, A.; Smida, K.; Khan, S.U.; Tlili, I. Inclined surface slip flow of nanoparticles with subject to mixed convection phenomenon: Fractional calculus applications. *J. Indian Chem. Soc.* **2022**, *99*, 100564. [CrossRef]
10. Naumov, I.V.; Sharifullin, B.R.; Kravtsova, A.Y.; Shtern, V.N. Velocity jumps and the Moffatt eddy in two-fluid swirling flows. *Exp. Therm. Fluid. Sci.* **2020**, *116*, 110116. [CrossRef]
11. Naumov, I.V.; Chaplina, T.O.; Stepanova, E.V. Investigation of the influence of a vortex flow inductor on the shape of the interface between two immiscible liquids. *Process. Geoenviron.* **2020**, *1*, 599–610. (In Russian)
12. Naumov, I.V.; Kashkarova, M.V.; Okulov, V.L.; Mikkelsen, R.F. The structure of the confined swirling flow under different phase boundary conditions at the fixed end of the cylinder. *Thermophys. Aeromech.* **2020**, *27*, 93–98. [CrossRef]
13. Naumov, I.V.; Sharifullin, B.R.; Tsoy, M.A.; Shtern, V.N. Dual vortex breakdown in a two-fluid confined flow. *Phys. Fluids* **2020**, *32*, 061706. [CrossRef]
14. Naumov, I.V.; Skripkin, S.G.; Shtern, V.N. Counter flow slip in a two-fluid whirlpool. *Phys. Fluids* **2021**, *33*, 061705. [CrossRef]
15. Naumov, I.V.; Gevorgiz, R.G.; Skripkin, S.G.; Sharifullin, B.R. Experimental investigation of vortex structure formation in a gas-vortex bioreactor. *Thermophys. Aeromech.* **2022**, *29*, 719–724.
16. Mertvetsov, N.P.; Ramazanov, Y.A.; Repkov, A.P.; Dudarev, A.N.; Kislykh, V.I. Gas-vortex bioreactors "BIOK". In *Use in Modern Biotechnology*; Novosibirsk Branch of the Publishing House "Nauka": Novosibirsk, Russia, 2002. (In Russian)
17. Skripkin, S.G.; Tsoy, M.A.; Naumov, I.V. Visualization the different type of vortex breakdown in conical pipe flow with high cone angle. *J. Flow Vis. Image Process.* **2021**, *28*, 43–53. [CrossRef]
18. Wu, S.; Li, B.; Zuo, Z.; Liu, S. Dynamics of a single free-settling spherical particle driven by a laser-induced bubble near a rigid boundary. *Phys. Rev. Fluids* **2021**, *6*, 093602. [CrossRef]
19. Wang, Z.; Liu, S.; Li, B.; Zuo, Z.; Pan, Z. Large cavitation bubbles in the tube with a conical-frustum shaped closed end during a transient process. *Phys. Fluids* **2022**, *34*, 063312. [CrossRef]
20. Zeng, Q.; An, H.; Ohl, C.D. Wall shear stress from jetting cavitation bubbles: Influence of the stand-off distance and liquid viscosity. *J. Fluid. Mech.* **2022**, *932*, A14. [CrossRef]
21. Siddiqui, M.S.; Rasheed, A.; Tabib, M.; Kvamsdal, T. Numerical investigation of modeling frameworks and geometric approximations on NREL 5MW wind turbine. *Renew. Energ.* **2019**, *132*, 1058–1075. [CrossRef]
22. Mizzi, K.; Demirel, Y.K.; Banks, C.; Turan, O.; Kaklis, P.; Atlar, M. Design optimisation of Propeller Boss Cap Fins for enhanced propeller performance. *Appl. Ocean Res.* **2017**, *62*, 210–222. [CrossRef]
23. Helal, M.M.; Ahmed, T.M.; Banawan, A.A.; Kotb, M.A. Numerical prediction of sheet cavitation on marine propellers using CFD simulation with transition-sensitive turbulence model. *Alex. Eng. J.* **2018**, *57*, 3805–3815. [CrossRef]
24. Naumov, I.V.; Sharifullin, B.R.; Shtern, V.N. Vortex breakdown in the lower fluid of a two-fluid swirling flow. *Phys. Fluids* **2020**, *32*, 014101. [CrossRef]
25. Sharifullin, B.R.; Naumov, I.V. Angular momentum transfer across the interface of two immiscible liquids. *Thermophys. Aeromech.* **2021**, *28*, 67–78. [CrossRef]
26. Naumov, I.V.; Sharifullin, B.R.; Skripkin, S.G.; Tsoy, M.A.; Shtern, V.N. A Two-Story Tornado in a Lab. In *Processes in GeoMedia—Volume V*; Chaplina, T., Ed.; Springer: Cham, Germany, 2022; pp. 25–34.

 *water*

Article

Oil–Water Separation on Hydrophobic and Superhydrophobic Membranes Made of Stainless Steel Meshes with Fluoropolymer Coatings

Alexandra Melnik [1,2], Alena Bogoslovtseva [1,2], Anna Petrova [1,2], Alexey Safonov [1,*] and Christos N. Markides [1,3]

1   Kutateladze Institute of Thermophysics, Siberian Branch, Russian Academy of Science (IT SB RAS), Lavrentyev Ave. 1, Novosibirsk 630090, Russia
2   Department of Physics, Novosibirsk State University (NSU), Pirogova Str. 2, Novosibirsk 630090, Russia
3   Clean Energy Processes (CEP) Laboratory, Department of Chemical Engineering, Imperial College London, South Kensington Campus, London SW7 2AZ, UK
*   Correspondence: safonov@itp.nsc.ru

Abstract: In this work, membranes were synthesized by depositing fluoropolymer coatings onto metal meshes using the hot wire chemical vapor deposition (HW CVD) method. By changing the deposition parameters, membranes with different wetting angles were obtained, with water contact angles for different membranes over a range from $130° \pm 5°$ to $170° \pm 2°$ and a constant oil contact angle of about $80° \pm 2°$. These membranes were used for the separation of an oil–water emulsion in a simple filtration test. The main parameters affecting the separation efficiency and the optimal separation mode were determined. The results reveal the effectiveness of the use of the membranes for the separation of emulsions of water and commercial crude oil, with separation efficiency values that can reach over 99%. The membranes are most efficient when separating emulsions with a water concentration of less than 5%. The pore size of the membrane significantly affects the rate and efficiency of separation. Pore sizes in the range from 40 to 200 μm are investigated. The smaller the pore size of the membranes, the higher the separation efficiency. The work is of great economic and practical importance for improving the efficiency of the membrane separation of oil–water emulsions. It lays the foundation for future research on the use of hydrophobic membranes for the separation of various emulsions of water and oil products (diesel fuel, gasoline, kerosene, etc.).

Keywords: hydrophobic; fluoropolymer; oil–water separation; stainless steel mesh; superhydrophobic

Citation: Melnik, A.; Bogoslovtseva, A.; Petrova, A.; Safonov, A.; Markides, C.N. Oil–Water Separation on Hydrophobic and Superhydrophobic Membranes Made of Stainless Steel Meshes with Fluoropolymer Coatings. *Water* **2023**, *15*, 1346. https://doi.org/10.3390/w15071346

Academic Editors: Maksim Pakhomov and Pavel Lobanov

Received: 23 January 2023
Revised: 13 March 2023
Accepted: 24 March 2023
Published: 30 March 2023

1. Introduction

The separation of oil and water is an important task for oil production, ecology, wastewater treatment, and other applications [1–3]. Existing separation techniques have limitations due to high energy consumption, the formation of secondary pollutants, low separation efficiency, etc. [4]. The use of membranes for the separation of oil–water emulsions was proposed as early as the meddle of the 20th century [5]. Furthermore, attempts were made to modify the membrane surface by imparting hydrophobic or hydrophilic properties [6]. Due to this, the separation efficiency of membranes increased significantly, and the method has begun to compete with others [7]. The rapid development of methods for the manufacture of separating membranes with different wettability started. Three types of materials are usually used as the basis of the membrane: metal meshes [8–11], textiles [12,13], and polymer meshes [14,15]. The modifier is applied to the base in various ways. The membrane surface modifiers are materials: polymers (fluoropolymer [16,17], polystyrene [18], polydimethylsiloxane [19,20], polybenzoxazine [21], etc.); minerals (diatomite coating [22]; silicon dioxide [23]; graphene oxide [24,25]); and metal oxides [26–28], metal hydroxides [29], etc. Modifiers are usually applied to the substrate by using the following methods: dip coating [30], spray coating [31], spin-coating [32], electrodeposition [33], acid–alkaline treatment [34], heat treatment [35], plasma deposition [36], ion beam irradiation [37], chemical etching [38], chemical vapor deposition (CVD) [39], and others [40].

Many materials with different wetting properties are used in the manufacture of membranes, yet there are no strong theories concerning how wetting affects separation efficiency and what wetting range is necessary for effective separation. One parameter of wettability is the water contact angle (WCA). The data presented in various works on the effect of WCA on the separation process are ambiguous and have a wide range of results. For example, Hou and Cao [41] obtained a separation efficiency of 98–99% on fluoropolymer-coated membranes with a WCA of $150° \pm 2°$, while Dong et al. [42] obtained a stable separation efficiency of 95% on a membrane with a laser-structured copper surface with a WCA of $151° \pm 1°$. Yin [29] obtained a separation efficiency of over 99% and excellent stability even after 10 uses on a superhydrophobic-copper-hydroxide-coated membrane with a WCA of $154° \pm 2°$. Undoubtedly, in addition to the WCA, it is also necessary to compare other parameters of the membrane (modifier material, pore size, etc. [4]).

The present work considers the possibility of creating separation membranes by depositing hydrophobic and superhydrophobic fluoropolymer coatings onto the surface of metal meshes by applying the hot wire CVD method. A feature of this method is the possibility of obtaining fluoropolymer coatings with different wetting properties. Depending on the deposition parameters, it is possible to obtain coatings with a contact angle from $120°$ to $170°$ [43]. Thus, this approach makes it possible to study the effect of wettability on the efficiency and separation rate within the same material. Fluoropolymers were chosen due to a combination of their unique properties: high hydrophobicity, chemical inertness, heat resistance, etc. The main purpose of the work is to study the influence of the wetting properties of hydrophobic fluoropolymer coatings on metal meshes during the membrane separation process of water–oil emulsions. In addition to the effect of membrane wettability, this work examines the influence of the membrane pore size on the separation process and the stability of the properties of the resulting coatings. Studies make it possible to determine the main parameters of the membrane (WCA, mesh weaving) that affect the rate and efficiency of separation in order to establish the optimal characteristics of separation membranes. The advantage of this work is that commercial crude oil was used in the studies, while many works describe the use of oil simulators. This approach makes it possible to accurately simulate the use of developed membranes in oil and oil refinery industries.

2. Experimental Methods

2.1. Materials and Reagents

AISI 304 stainless steel meshes (Mesh 100, Mesh 180, Mesh 300, and Mesh 400) were purchased from Sunshinelantian Store (Anping, China). The precursor gas was a mixture consisting of hexafluoropropylene oxide (C3F6O) and 1.3% argon produced by LLC "Polymer Kirov-Chepetsk Chemical Plant" (TU standard 95-783-80) (Kirov-Chepetsk, Russia). The used nichrome filament (diameter 0.5 mm) consisted of 77%—Ni; 20%—Cr; and 1.5%—Fe, and the remaining 1.5% included Ti, Al, Si, C, Mn, P, and S (GOST 12766.1-90) produced by ZAO "Sverdlovsk Metallurgical Plant" (Ekaterinburg, Russia). The commercial crude oil (GOST R 51858-2002) was purchased from Rikom LLC (Novosibirsk, Russia).

2.2. Membrane Fabrication

Separation membranes were made by depositing a hydrophobic fluoropolymer coating on variously woven stainless steel meshes with a diameter of 40 mm. The meshes (Mesh 100, Mesh 180, Mesh 300, and Mesh 400) had a flow area of 40, 65, 90, and 200 μm, respectively. The surface of the samples was subjected to preliminary cleaning and processing to remove persistent organic (grease, oil, etc.) and non-organic (metal dust) compounds. Grease, oil, and dust were removed by the preliminary cleaning of meshes in an Ultrasonic Cleaner JP-010S (Skymen Cleaning Equipment Shenzhen Co., Shenzhen, China) with surfactants (sodium lauryl sulfate, 0.1 mol/L) at 90 °C for 30 min. After that, the samples were rinsed with distilled water and then alcohol and dried with a stream of dry argon.

Coatings were deposited using the hot wire chemical vapor deposition (HW CVD) method [43–48]. The main idea of the method is to activate the precursor gas flow with a resistively heated catalytic metal filament, that is, an activator. The experimental setup is described in detail in study [43]. This work used a nichrome filament (Ni80/Cr20) as an activator and hexafluoropropylene oxide (C_3F_6O) as a precursor gas. As a result of activation, radicals formed. These radicals then reach the surface, where they form a coating via polymerization. The structure of the resulting coating depends on the temperature of the activator filament T_f, the precursor gas pressure P, the distance between the activator and the substrate R, the deposition time t, and the temperature of the substrate holder T_s. The influence of these parameters on the structure of the deposited coating is described in detail in [43]. The wettability of membranes depends on the structure of the fluoropolymer coating. According to the Wenzel and Cassie–Baxter law, roughness is important for this [49,50]. During the formation of a fluoropolymer coating, depending on the deposition parameters, nano- and microroughnesses can form. It is their size and shape that determine the wetting properties of the surface in this case.

Three types of fluoropolymer coatings were chosen, which were previously obtained by the authors for other applications [43–45]. Depending on morphology, the coatings have different stable wetting properties when WCA = 130°, 150°, and 170°. Thus, each coating is hydrophobic but differs significantly in WCA value from others. The coatings had the same thickness of about 1 μm. The deposition parameters are presented in Table 1.

Table 1. Parameters of the coating deposition process in the manufacture of separation membranes.

	WCA, °	T_f, °C	P, Pa	R, mm	T_s, °C	t, min
Type 1	130	640	67	50	30	180
Type 2	150	580	67	50	30	90
Type 3	170	680	133	50	100	60

2.3. Membrane Characterization

The morphology of the resulting fluoropolymer coatings on the surface of a metal mesh was studied using a JEOL JSM6700F scanning electron microscope (Tokyo, Japan). The morphology of the obtained samples was studied at a magnification of ×140 (general view of the membrane network), ×20,000, and ×50,000 (morphology of the fluoropolymer coating). Imaging parameters were selected individually depending on the electrical properties of the sample, namely, the accelerating voltage was 5–15 keV, and the operating focal length was 3.0–15.1 mm.

Figure 1 shows micrographs of a metal mesh after the deposition of hydrophobic fluoropolymer coatings and their water- and oil-wetting properties. The morphology of the fluoropolymer coating is different for each type of coating. However, the morphology is homogeneous over the entire mesh surface for each type of coating.

The wetting properties of the resulting membranes were determined by measuring the water contact angle (WCA) or oil contact angle (OCA) using a DSA-E100 drop-shape analyzer (KRUSS, Hamburg, Germany). A drop of a characteristic volume of 3 μL was placed on the membrane surface. The image of the drop was captured using the shadow technique. The program recognizes the contour, and then the geometric model of the contour is selected. The baseline (contact line) was set in the program to more accurately determine the contact angle. The measured angle between the tangent to the drop contour and the baseline is the contact angle. Two methods were used to analyze the drop's contour: the Young–Laplace method and the conic section method. Both methods evaluate the full contour of a lying drop. To accurately determine the wettability of the membranes, measurements were carried out in five different places, three times for each. The OCA of the obtained coatings was found to be similar at ~80° ± 2°, while the WCA was found to depend strongly on the morphology of the coating, varying in the range from ~130° ± 5° to 170° ± 2°.

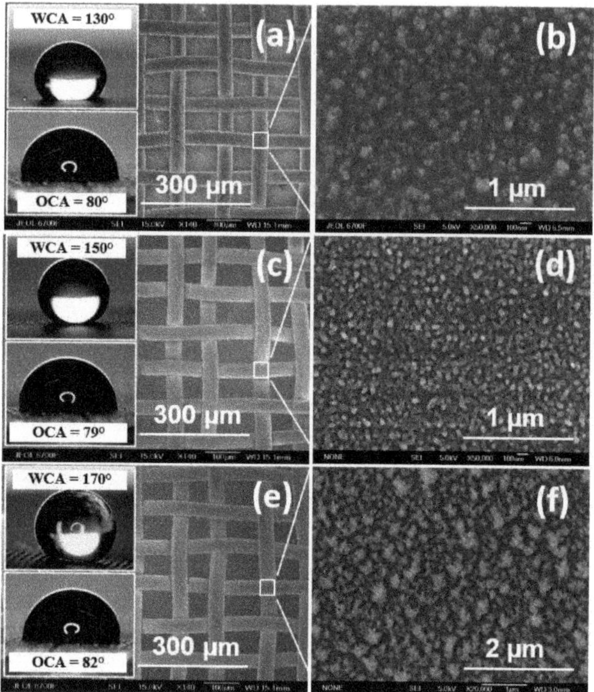

Figure 1. Micrograph of the surface of membranes with various fluoropolymer coatings: (**a**,**b**) Type 1, (**c**,**d**) Type 2, and (**e**,**f**) Type 3. Insets include WCA values and photographs of water and oil droplets on the membrane surface.

2.4. Emulsion Preparation

The emulsion for experiments was prepared by mixing commercial crude oil and distilled water in various proportions. The use of commercial crude oil made it possible to accurately simulate the use of membranes in real conditions because oil contains a large number of various low- and high-molecular fractions, salts, and other impurities. All these components can interact with the membrane surface in different ways. For experiments, emulsions were prepared with different volume concentrations of water and oil. The mixture was stirred with a Braun MQ-5237 BK kitchen blender (Bucharest, Romania) for two minutes. The initial water content in commercial oil is less than 0.5% (GOST R 51858-2002), and it is neglected. The concentration was determined from:

$$\varphi_w = \frac{V_w}{\Sigma V_e} \times 100\%, \tag{1}$$

where φ_w is the concentration of water in the emulsion, V_w is the volume of water mixed into the oil, and ΣV_e is the total volume of the prepared emulsion. In the experiments, the values of concentration φ_w were as follows: 5%, 10%, 25%, 50%, and 90%. For each separation cycle, an emulsion was prepared in a volume of ΣV_e = 50 mL.

2.5. Emulsion Separation Arrangement and Mechanism

The experimental arrangement used for the separation tests is shown in Figure 2a. A membrane filter is fixed on a laboratory stand. This filter consists of two tightly connected glass tubes, between which the obtained membranes were installed. The membrane's diameter was 33 mm.

Figure 2b shows a scheme of the separation mechanism. It is assumed that the separation on the membrane is caused by the fact that the capillary pressure of oil and water in the pores is different.

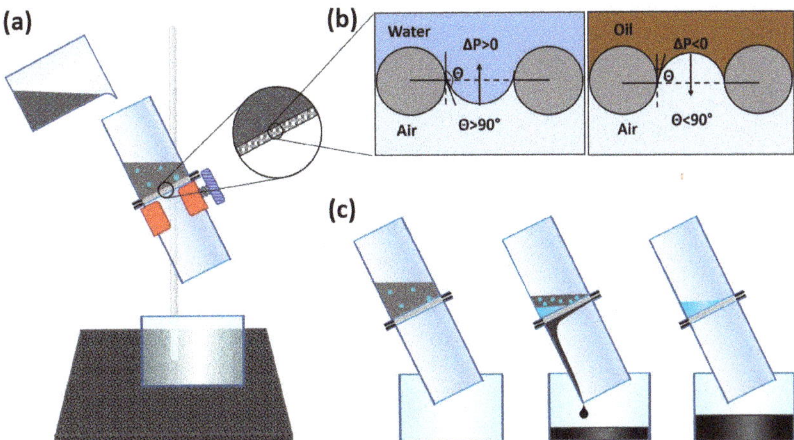

Figure 2. (a) Experimental stand for separation. (b) Schematic showing the oil–water separation mechanism on the superhydrophobic mesh. (c) Schematic of the separation process.

The pressure can be calculated by using the Young–Laplace Equation (2) [51]:

$$\Delta P = \frac{2\,\sigma}{R} = \frac{4\,\sigma\,sin\,(\theta_A - 90°)}{D} = -\frac{l\,\sigma\,cos\theta_A}{A},\tag{2}$$

where σ is the surface tension of the liquid–air interface, R is the meniscus radius, D is the center of the cross-section of adjacent, parallel stainless-steel wires, l is the mesh thickness, A is the cross-sectional area of the pore, and θ_A is the wetting angle of the liquid at the surface. It follows from this equation that in the absence of external pressure on the emulsion, WCA $\theta_A > 90°$, water will not be able to spontaneously penetrate the pores because there is a pressure drop when $\Delta P > 0$, which prevents it. However, for oil, the value is $\theta_A < 90°$, so the pressure drop becomes $\Delta P < 0$, and the oil will spontaneously penetrate the pores; that is, it will pass through the membrane. The process of the separation of water–oil emulsion will take place.

For the obtained membranes with various fluoropolymer coatings, OCA remains almost unchanged, while the WCA changes significantly. According to the equation, with an increase in hydrophobic properties, the pressure drop and separation efficiency should increase. That is, the higher the WCA value, the better and faster the separation of water and oil.

Figure 2c schematically shows the principle of the operation of the membrane filter; namely, the prepared emulsion is poured into the receiving container of the membrane filter, and the stopwatch is simultaneously started. Since the coating is hydrophobic, during the separation process, water remains above the surface of the membrane, while oil flows into the container under the filter. After separation, the volume of water is measured, recorded, and removed from the filter. The separation cycles continue as long as the water is released during the separation process. A new membrane was used each time to separate the new emulsion.

3. Results and Discussion

3.1. Effect of Wettability on Separation Efficiency and Rate

The separation efficiency was determined from:

$$\eta = \frac{V_s}{V_w} \times 100\%,\tag{3}$$

where η is the separation efficiency, V_s is the volume of separated water from the emulsion, and V_w is the volume of water mixed into the emulsion. On average, the experiments were carried out 3–5 times. Considering the statistics, the error bars are indicated in the figures. The points on the figures and the results in the tables correspond to the average value of the results obtained.

Figure 3 shows the results of the separation of emulsions with different water concentrations on membranes with different wettability. The separation of crude oil showed an efficiency close to 100%. The separation of distilled water showed an efficiency equal to η = 0%; that is, the entire volume of water remained on the membrane's surface. Next, the separation of emulsions with different concentrations from 5% to 90% was carried out. The separation process is carried out in several cycles n, on average, 5–6. The graphs show the results of all separation cycles. The maximum separation efficiency of emulsions is achieved for membranes with a WCA = 170° and is about 99%. For superhydrophobic membranes, these values remain in a wide range of water concentrations in the emulsion. For membranes with hydrophobic coatings (WCA = 130°–150°) at low water concentrations (φ_w = 5%), the separation efficiency is also high and is about 98–99%. As the water concentration φ_w increases from 5 to 90% in the emulsion, the value of the separation efficiency decreases monotonically from η = 98% to 93% for membranes with a WCA = 150°. Similarly, for membranes with a WCA = 130°, the efficiency decreases to 75%. Thus, the wetting properties of the membrane significantly affect the separation efficiency. That is, superhydrophobic membranes can be used to separate emulsions with high water concentrations, while hydrophobic membranes are effective with emulsions with low water concentrations (φ_w = 5–25%).

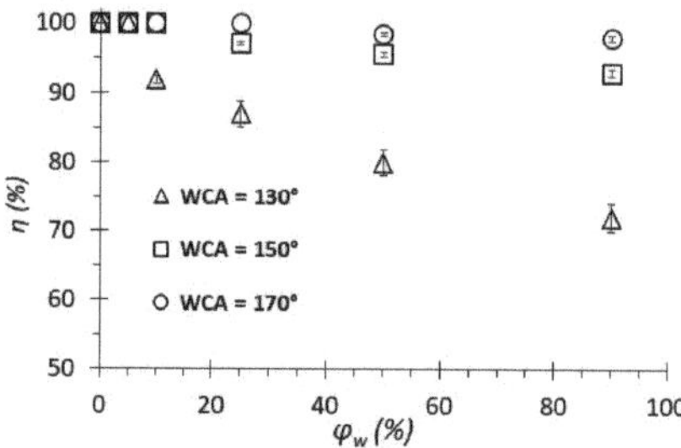

Figure 3. Dependence of the separation efficiency on the water concentration in the emulsion for coatings with different wetting contact angles.

Figure 4 shows data on the volumes of water released in each cycle during the separation of an emulsion with a water concentration φ_w = 25%. It can be noted that the main volume of water is released during separation in the first 5–6 cycles. For membranes with WCAs between 150° and 170°, the maximum volume of water is released in the first two separation cycles, and for membranes with a WCA of 130°, the maximum volume of water is in the second, third, and fourth cycles. There is a clear dependence of the separation efficiency on the WCA of the membrane. The higher the WCA of the membrane, the faster the release of water from the emulsion.

In this case, the separation time increases with increasing water concentrations (Figure 5). Firstly, when a mixture of oil and water flows through the membrane filter, oil freely penetrates through the membrane with a hydrophobic coating, while water is repelled from the mesh due to a large negative capillary effect and remains on the surface, blocking the pores and thereby increasing the separation time. Secondly, the higher the concentration of water in the emulsion, the higher its viscosity, which prevents the flow of "fresh" portions of the emulsion to the membrane's pores.

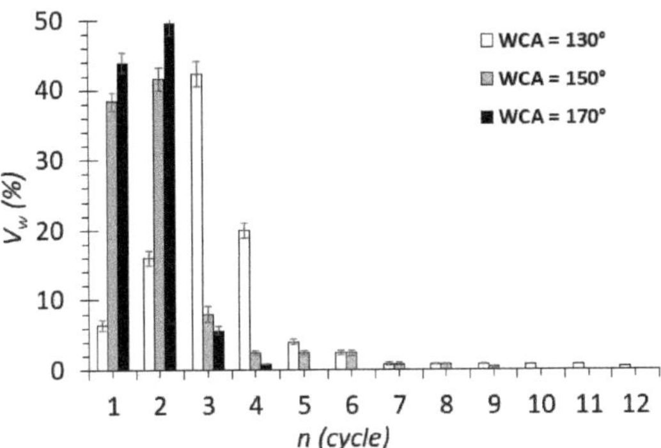

Figure 4. Dependence of the volume of released water on the cycle number for different wetting contact angles.

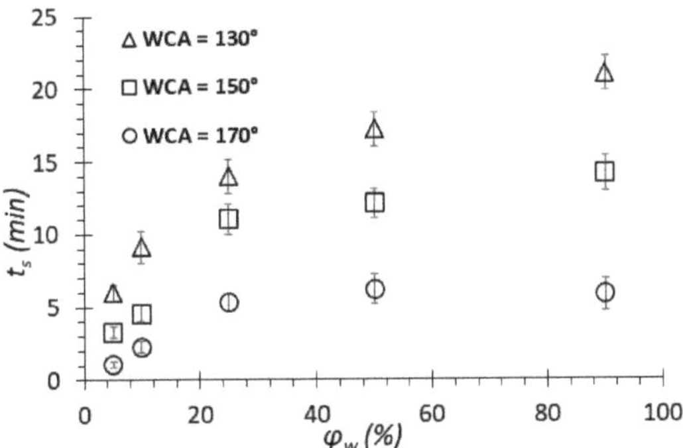

Figure 5. Dependence of the separation time on the concentration of water in the emulsion for coatings with different wetting contact angles.

The total separation time on meshes with a WCA of 170° is higher than on meshes with a WCA between 150° and 130°. This is due to the fact that capillary pressure increases with an increasing contact angle; therefore, water is more efficiently and quickly separated from oil and remains on the surface, slowing down the flow of oil.

3.2. Influence of Mesh Pore Size on Separation Efficiency

To study the dependence of the separation efficiency of an oil–water emulsion on the pore size of hydrophobic membranes, stainless steel metal meshes with various weaves were used. Pore sizes were 40, 65, 90, and 200 microns. Fluoropolymer coatings with a wetting contact angle of 130° (Type 1) were deposited on the surfaces of the meshes. In the experiments, an emulsion with a water concentration of $\varphi_w = 25\%$ was used.

Figure 6a shows the dependence of the separation efficiency of the oil–water emulsion on the pore size of the membranes D. It can be noted that with a decrease in the pore size, the separation efficiency increases. Specifically, a membrane with a pore size of 200 μm separates with an efficiency of 87% and when using a pore size of 40 μm an efficiency increases to 99%. This is due to the fact that a membrane with small pores can retain water drops (micelles) with a smaller diameter on the surface, while water drops can penetrate through larger pores of the membrane, thereby reducing the separation efficiency. However, as the membrane's pore size decreases, the separation time increases (Figure 6b). The time increases from 11 to 16 min for membranes with a pore size of 200 μm and 40 μm, respectively. This can be easily explained by an increase in hydrodynamic resistance caused by a decrease in the flow area of the membrane pores.

Apparently, the reason for the discrepancy between WCA and separation efficiency with respect to different authors is precisely the different pore sizes in the membranes.

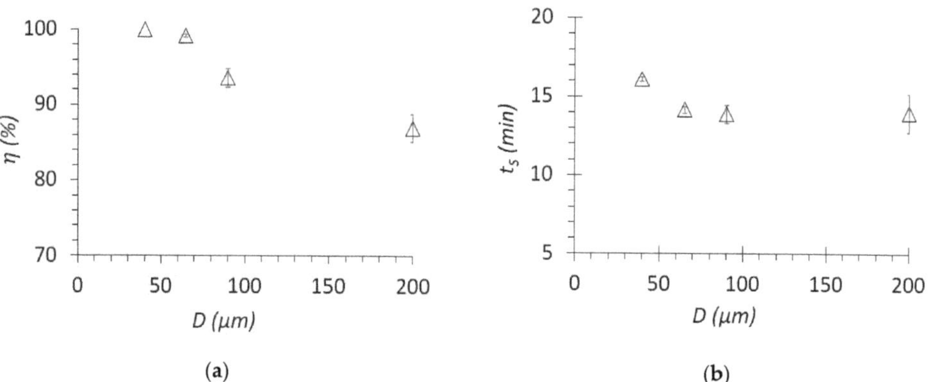

Figure 6. (**a**) Dependence of separation efficiency on the pore size of a metal mesh and (**b**) dependence of separation time on the pore size of a metal mesh.

3.3. Evaluation of Separation Efficiency on Membranes

The conducted studies have shown that it is necessary to introduce a parameter that would take into account all the above measurements and would make it possible to determine the overall efficiency of the separation process. This parameter is the coefficient of separation efficiency (C_{SE}), which is determined by the following formula:

$$C_{SE} = \frac{\eta}{t_s \cdot n},\tag{4}$$

where η is the separation efficiency, t_s is the separation time, and n is the number of separation cycles. The higher the C_{SE} value, the higher the overall separation efficiency.

Using this formula, the coefficients for each membrane were determined. The results are presented in Table 2. Analyzing the data obtained, it can be noted that it is most effective to use membranes with a superhydrophobic coating for small concentrations of water in an oil emulsion.

Separation coefficients for membranes with different pore sizes were determined in a similar way (Table 3). In this case, superhydrophobic membranes with a pore size of 40 μm are more effective. However, the influence of the membrane pore size on efficiency is not as significant as the influence of the wetting properties of the membrane.

Table 2. Efficiency coefficient for the separation of emulsions with different water concentrations for different contact angles of a hydrophobic membrane.

φ_w, %	WCA, °	t_s, min	n, Times	η, %	C_{SE}
	130 ± 5	6.0 ± 0.5	5	100	3.3
5	150 ± 3	3.3 ± 0.4	6	100	5.1
	170 ± 2	1.1 ± 0.2	4	100	23.1
	130 ± 5	9.2 ± 1.1	4	92.0 ± 0.6	2.5
10	150 ± 3	4.6 ± 0.5	6	100	3.6
	170 ± 2	2.3 ± 0.4	5	100	8.8

Table 2. *Cont.*

φ_w, %	WCA, °	t_s, min	n, Times	η, %	C_{SE}
	130 ± 5	14.0 ± 1.2	7	87.0 ± 1.8	0.9
25	150 ± 3	11.0 ± 1.1	6	97.0 ± 0.2	1.5
	170 ± 2	5.3 ± 1.0	4	100	4.7
	130 ± 5	17.2 ± 1.2	6	80.0 ± 1.9	0.8
50	150 ± 3	12.1 ± 1.0	5	96.0 ± 0.3	1.6
	170 ± 2	6.2 ± 1.0	5	99.0 ± 0.2	3.2
	130 ± 5	21.0 ± 1.2	5	72.0 ± 2.1	0.7
90	150 ± 3	14.2 ± 1.2	4	93.0 ± 0.5	0.9
	170 ± 2	6.2 ± 1.0	4	98.0 ± 0.4	4.1

Table 3. Efficiency coefficient for the separation of emulsions with a water concentration of 25% for different pore sizes of a hydrophobic membrane.

WCA, °	φ_w, %	D, µm	t_s, min	n, Times	η, %	C_{SE}
		40	16.1 ± 0.2	3	100	2.1
130 ± 5	25	65	14.2 ± 0.3	4	99.0 ± 0.2	1.8
		130	13.5 ± 1.0	4	94.0 ± 1.3	1.7
		200	14.0 ± 1.2	7	87.0 ± 1.8	0.9

The results show that the highest efficiency is achieved when separating membranes with a superhydrophobic coating (WCA = 170°) and a minimum pore size (40 µm).

3.4. Resource Tests

Resource tests were performed using the membranes obtained as described earlier in this paper. The fluoropolymer coating with a WCA = 130° had the highest wear resistance of the coatings in the study [46]. Therefore, these coatings were selected for resource testing. On a membrane with a pore size of 200 µm and a WCA = 130°, 30 successive cycles of separation of the emulsion with a water concentration φ_w = 25% were carried out. After each cycle, the efficiency and separation time were recorded. The results are shown in Figure 7. The results showed a decrease in separation efficiency after about six cycles. At the same time, for 18–20 cycles, the membrane separates the emulsion with a slight decrease in efficiency, and then, by the 30th cycle, there is a significant decrease in separation efficiency up to 70%. A similar pattern is observed with separation time. Namely, the separation time is almost unchanged up to the 6th cycle and is about 11 min, but by the 30th cycle, it increases to 40 min. The examination of the membrane with an optical microscope showed that with each new cycle, a "deposit" is formed on the membrane surface (Figure 7b). This "deposit" blocks the flow area of the pores, reducing the flow rate of the emulsion and the wetting properties of the membrane. Therefore, the efficiency decreases, and the separation time increases. It is likely that this "deposit" comprises high molecular weight hydrocarbons (paraffin, resin, etc.) contained in crude oil since it can be removed from the membrane surface by washing it for 5 min in diesel fuel. Using a microscope, small remnants of the "deposit" can be observed on the walls of the membrane pores (Figure 7b). However, this simple method of cleaning the membrane surface proved to be effective. After washing, the separation efficiency once again increased to ~92%, and the separation time decreased to 15 min.

The obtained results showed comparable or better separation efficiency in comparison with the results presented in [25,27,28].

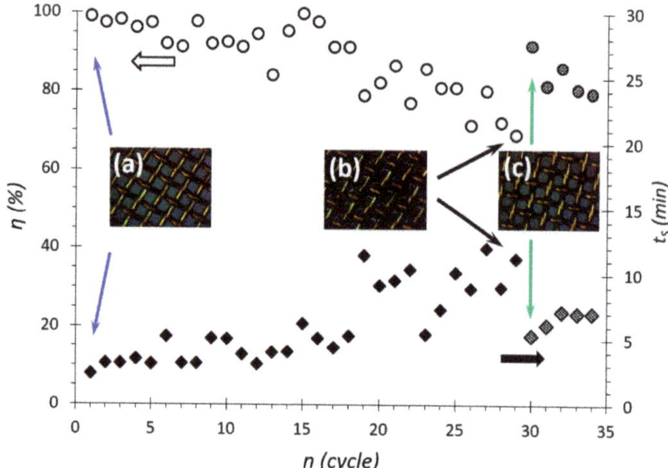

Figure 7. Dependences of efficiency and separation time on the number of separation cycles. In the inserts, the surface of (**a**) clean, (**b**) used, and (**c**) washed membranes. The arrows indicate the points where clean (blue arrows), used (black arrows) and washed (green arrows) were used.

4. Conclusions

(1) The HW CVD method can be applied to the fabrication of highly efficient hydrophobic separation membranes by depositing fluoropolymer coatings onto the surfaces of metal meshes. Depending on the deposition parameters, it is possible to obtain membranes with different surface-wetting properties; specifically, in this work, the WCA ranges from 130 to 170°, while the OCA remains constant and is about $80° \pm 2°$.

(2) Studies have shown the effectiveness of the use of the obtained membranes for the separation of emulsions of water and commercial crude oil, with separation efficiency values that can reach over 99%. The membrane-wetting properties affect the rate and efficiency of separation. The higher the WCA value of the membrane surface, the more efficient the separation. It has been established that emulsions with a lower water concentration (5%) are most effectively separated.

(3) The pore size of the membrane significantly affects the rate and efficiency of separation. The smaller the pore size of the membranes, the higher the separation efficiency, but the lower its rate.

(4) The use of the proposed coefficient of separation efficiency made it possible to determine the optimal parameters for the use of membranes for separating emulsions. The highest efficiency is achieved when separating membranes with a superhydrophobic coating (WCA = 170°) and a minimum pore size (40 μm).

(5) The experiments were performed to explore whether hydrophobic coated membranes produced by the HW CVD method can be used for several separation cycles. The used membranes can be easily washed and reused without significant reduction in separation efficiency.

(6) The work is of great economic and practical importance for improving the efficiency of the membrane separation of oil–water emulsions. It lays the foundation for future research on the use of hydrophobic membranes for the separation of various emulsions of water and oil products (diesel fuel, gasoline, kerosene, etc.).

Author Contributions: Conceptualization, A.S.; methodology, A.S.; validation, A.S.; investigation, A.M. and A.B.; resources, A.S. and A.B.; data curation, A.M. and A.S.; writing—original draft preparation, A.M. and A.S.; writing—review and editing, A.S., A.M., A.P. and C.N.M.; visualization,

A.M.; supervision, C.N.M.; project administration, A.P.; funding acquisition, C.N.M. All authors have read and agreed to the published version of the manuscript.

Funding: This work was executed from the state contract with IT SB RAS, project number 121031800218–5. The deposition of unique fluoropolymer coatings for this work was funded by the Russian Science Foundation, project number 18-79-10119. The experimental equipment was provided by the Government of the Russian Federation, under Megagrant project number 075-15-2022-1043.

Institutional Review Board Statement: Not applicable.

Informed Consent Statement: Not applicable.

Data Availability Statement: Not applicable.

Acknowledgments: The authors acknowledge shared research facilities VTAN at NSU.

Conflicts of Interest: The authors declare no conflict of interest.

References

1. Joye, S.B. Deepwater Horizon, 5 Years On. *Science* **2015**, *349*, 592–593. [CrossRef] [PubMed]
2. Schrope, M. Oil Spill: Deep Wounds. *Nature* **2011**, *472*, 152–154. [CrossRef] [PubMed]
3. Spaulding, M.L. State of the Art Review and Future Directions in Oil Spill Modeling. *Mar. Pollut. Bull.* **2017**, *115*, 7–19. [CrossRef] [PubMed]
4. Sobolčiak, P.; Popelka, A.; Tanvir, A.; Al-Maadeed, M.A.; Adham, S.; Krupa, I. Some Theoretical Aspects of Tertiary Treatment of Water/Oil Emulsions by Adsorption and Coalescence Mechanisms: A Review. *Water* **2021**, *13*, 652. [CrossRef]
5. Paul, D.; Friedrich, E.; Schwarz, H.H.; Bartsch, D.; Hicke, H.G.; Neustadt, W.; Morgenstern, S.; Sawatzki, P.; Tietze, R. Membrane filtration for oil-water emulsion separation. *Chem. Tech.* **1979**, *31*, 24–26.
6. Srijaroonrat, P.; Julien, E.; Aurelle, Y. Unstable Secondary Oil/Water Emulsion Treatment Using Ultrafiltration: Fouling Control by Backflushing. *J. Membr. Sci.* **1999**, *159*, 11–20. [CrossRef]
7. Pervez, M.N.; Mishu, M.R.; Stylios, G.K.; Hasan, S.W.; Zhao, Y.; Cai, Y.; Zarra, T.; Belgiorno, V.; Naddeo, V. Sustainable Treatment of Food Industry Wastewater Using Membrane Technology: A Short Review. *Water* **2021**, *13*, 3450. [CrossRef]
8. Tian, L.; Li, W.; Ye, H.; Zhu, L.; Chen, H.; Liu, H. Environmentally Benign Development of Superhydrophilic and Underwater Superoleophobic Mesh for Effective Oil/Water Separation. *Surf. Coat. Technol.* **2019**, *377*, 124892. [CrossRef]
9. Tudu, B.K.; Kumar, A. Robust and Durable Superhydrophobic Steel and Copper Meshes for Separation of Oil-Water Emulsions. *Prog. Org. Coat.* **2019**, *133*, 316–324. [CrossRef]
10. Zhang, X.; Pan, Y.; Gao, Q.; Zhao, J.; Wang, Y.; Liu, C.; Shen, C.; Liu, X. Facile Fabrication of Durable Superhydrophobic Mesh via Candle Soot for Oil-Water Separation. *Prog. Org. Coat.* **2019**, *136*, 105253. [CrossRef]
11. Shi, T.; Liang, J.; Li, X.; Zhang, C.; Yang, H. Improving the Corrosion Resistance of Aluminum Alloy by Creating a Superhydrophobic Surface Structure through a Two-Step Process of Etching Followed by Polymer Modification. *Polymers* **2022**, *14*, 4509. [CrossRef]
12. Shang, Y.; Si, Y.; Raza, A.; Yang, L.; Mao, X.; Ding, B.; Yu, J. An in Situ Polymerization Approach for the Synthesis of Superhydrophobic and Superoleophilic Nanofibrous Membranes for Oil–Water Separation. *Nanoscale* **2012**, *4*, 7847. [CrossRef]
13. Wang, J.; Chen, Y. Oil-Water Separation Capability of Superhydrophobic Fabrics Fabricated via Combining Polydopamine Adhesion with Lotus-Leaf-like Structure. *J. Appl. Polym. Sci.* **2015**, *132*, e42614. [CrossRef]
14. Tan, X.; Rodrigue, D. A Review on Porous Polymeric Membrane Preparation. Part II: Production Techniques with Polyethylene, Polydimethylsiloxane, Polypropylene, Polyimide, and Polytetrafluoroethylene. *Polymers* **2019**, *11*, 1310. [CrossRef]
15. Xiang, Y.; Liu, F.; Xue, L. Under Seawater Superoleophobic PVDF Membrane Inspired by Polydopamine for Efficient Oil/Seawater Separation. *J. Membr. Sci.* **2015**, *476*, 321–329. [CrossRef]
16. Feng, L.; Zhang, Z.; Mai, Z.; Ma, Y.; Liu, B.; Jiang, L.; Zhu, D. A Super-Hydrophobic and Super-Oleophilic Coating Mesh Film for the Separation of Oil and Water. *Angew. Chem.* **2004**, *116*, 2046–2048. [CrossRef]
17. Zhang, J.; Seeger, S. Polyester Materials with Superwetting Silicone Nanofilaments for Oil/Water Separation and Selective Oil Absorption. *Adv. Funct. Mater.* **2011**, *21*, 4699–4704. [CrossRef]
18. Zhang, J.; Huang, W.; Han, Y. A Composite Polymer Film with Both Superhydrophobicity and Superoleophilicity. *Macromol. Rapid Commun.* **2006**, *27*, 804–808. [CrossRef]
19. Halake, K.; Bae, S.; Lee, J.; Cho, Y.; Jo, H.; Heo, J.; Park, K.; Kim, H.; Ju, H.; Kim, Y.; et al. Strategies for Fabrication of Hydrophobic Porous Materials Based on Polydimethylsiloxane for Oil-Water Separation. *Macromol. Res.* **2019**, *27*, 109–114. [CrossRef]
20. Xiong, X.; Xie, F.; Meng, J. Preparation of Superhydrophobic Porous Coating Film with the Matrix Covered with Polydimethylsiloxane for Oil/Water Separation. *Prog. Org. Coat.* **2018**, *125*, 365–371. [CrossRef]
21. Caldona, E.B.; De Leon, A.C.C.; Thomas, P.G.; Naylor, D.F.; Pajarito, B.B.; Advincula, R.C. Superhydrophobic Rubber-Modified Polybenzoxazine/SiO$_2$ Nanocomposite Coating with Anticorrosion, Anti-Ice, and Superoleophilicity Properties. *Ind. Eng. Chem. Res.* **2017**, *56*, 1485–1497. [CrossRef]

22. Li, J.; Cui, M.; Tian, H.; Wu, Y.; Zha, F.; Feng, H.; Tang, X. Facile Fabrication of Anti-Corrosive Superhydrophobic Diatomite Coatings for Removal Oil from Harsh Environments. *Sep. Purif. Technol.* **2017**, *189*, 335–340. [CrossRef]

23. Xue, C.-H.; Ji, P.-T.; Zhang, P.; Li, Y.-R.; Jia, S.-T. Fabrication of Superhydrophobic and Superoleophilic Textiles for Oil–Water Separation. *Appl. Surf. Sci.* **2013**, *284*, 464–471. [CrossRef]

24. Chen, J.; Li, K.; Zhang, H.; Liu, J.; Wu, S.; Fan, Q.; Xue, H. Highly Efficient and Robust Oil/Water Separation Materials Based on Wire Mesh Coated by Reduced Graphene Oxide. *Langmuir* **2017**, *33*, 9590–9597. [CrossRef] [PubMed]

25. Abdullahi, B.O.; Ahmed, E.; Al Abdulgader, H.; Alghunaimi, F.; Saleh, T.A. Facile Fabrication of Hydrophobic Alkylamine Intercalated Graphene Oxide as Absorbent for Highly Effective Oil-Water Separation. *J. Mol. Liq.* **2021**, *325*, 115057. [CrossRef]

26. Cao, Q.; Zheng, S.; Wong, C.-P.; Liu, S.; Peng, Q. Massively Engineering the Wettability of Titanium by Tuning Nanostructures and Roughness via Laser Ablation. *J. Phys. Chem. C* **2019**, *123*, 30382–30388. [CrossRef]

27. Baig, U.; Dastageer, M.A.; Gondal, M.A. Facile Fabrication of Super-Wettable Mesh Membrane Using Locally-Synthesized Cobalt Oxide Nanoparticles and Their Application in Efficient Gravity Driven Oil/Water Separation. *Colloids Surf. A Physicochem. Eng. Asp.* **2023**, *660*, 130793. [CrossRef]

28. Said, A.; Al Abdulgader, H.; Alsaeed, D.; Drmosh, Q.A.; Baroud, T.N.; Saleh, T.A. Hydrophobic Tungsten Oxide-Based Mesh Modified with Hexadecanoic Branches for Efficient Oil/Water Separation. *J. Water Process Eng.* **2022**, *49*, 102931. [CrossRef]

29. Yin, X.; Wang, Z.; Shen, Y.; Mu, P.; Zhu, G.; Li, J. Facile Fabrication of Superhydrophobic Copper Hydroxide Coated Mesh for Effective Separation of Water-in-Oil Emulsions. *Sep. Purif. Technol.* **2020**, *230*, 115856. [CrossRef]

30. Huang, X.; Wen, X.; Cheng, J.; Yang, Z. Sticky Superhydrophobic Filter Paper Developed by Dip-Coating of Fluorinated Waterborne Epoxy Emulsion. *Appl. Surf. Sci.* **2012**, *258*, 8739–8746. [CrossRef]

31. Li, J.; Yan, L.; Li, H.; Li, J.; Zha, F.; Lei, Z. A Facile One-Step Spray-Coating Process for the Fabrication of a Superhydrophobic Attapulgite Coated Mesh for Use in Oil/Water Separation. *RSC Adv.* **2015**, *5*, 53802–53808. [CrossRef]

32. Long, M.; Peng, S.; Deng, W.; Yang, X.; Miao, K.; Wen, N.; Miao, X.; Deng, W. Robust and Thermal-Healing Superhydrophobic Surfaces by Spin-Coating of Polydimethylsiloxane. *J. Colloid Interface Sci.* **2017**, *508*, 18–27. [CrossRef]

33. Xu, Z.; Jiang, D.; Wei, Z.; Chen, J.; Jing, J. Fabrication of Superhydrophobic Nano-Aluminum Films on Stainless Steel Meshes by Electrophoretic Deposition for Oil-Water Separation. *Appl. Surf. Sci.* **2018**, *427*, 253–261. [CrossRef]

34. Cheng, C.; Wang, F.; Zhao, B.; Ning, Y.; Lai, Y.; Wang, L. Acid/Base Treatment of Monolithic Activated Carbon for Coating Silver with Tunable Morphology. *J. Wuhan Univ. Technol. Mater. Sci. Ed.* **2017**, *32*, 760–765. [CrossRef]

35. Tang, K.; Yu, J.; Zhao, Y.; Liu, Y.; Wang, X.; Xu, R. Fabrication of Super-Hydrophobic and Super-Oleophilic Boehmite Membranes from Anodic Alumina Oxide Film via a Two-Phase Thermal Approach. *J. Mater. Chem.* **2006**, *16*, 1741. [CrossRef]

36. Salapare, H.S.; Suarez, B.A.T.; Cosiñero, H.S.O.; Bacaoco, M.Y.; Ramos, H.J. Irradiation of Poly(Tetrafluoroethylene) Surfaces by CF4 Plasma to Achieve Robust Superhydrophobic and Enhanced Oleophilic Properties for Biological Applications. *Mater. Sci. Eng. C* **2015**, *46*, 270–275. [CrossRef]

37. Kim, D.-H.; Lee, D.-H. Effect of Irradiation on the Surface Morphology of Nanostructured Superhydrophobic Surfaces Fabricated by Ion Beam Irradiation. *Appl. Surf. Sci.* **2019**, *477*, 154–158. [CrossRef]

38. Saleh, T.A.; Baig, N. Efficient Chemical Etching Procedure for the Generation of Superhydrophobic Surfaces for Separation of Oil from Water. *Prog. Org. Coat.* **2019**, *133*, 27–32. [CrossRef]

39. Padaki, M.; Isloor, A.M.; Nagaraja, K.K.; Nagaraja, H.S.; Pattabi, M. Conversion of Microfiltration Membrane into Nanofiltration Membrane by Vapour Phase Deposition of Aluminium for Desalination Application. *Desalination* **2011**, *274*, 177–181. [CrossRef]

40. Liravi, M.; Pakzad, H.; Moosavi, A.; Nouri-Borujerdi, A. A Comprehensive Review on Recent Advances in Superhydrophobic Surfaces and Their Applications for Drag Reduction. *Prog. Org. Coat.* **2020**, *140*, 105537. [CrossRef]

41. Hou, C.; Cao, C. Superhydrophobic Cotton Fabric Membrane Prepared by Fluoropolymers and Modified Nano-SiO_2 Used for Oil/Water Separation. *RSC Adv.* **2021**, *11*, 31675–31687. [CrossRef] [PubMed]

42. Dong, Z.; Sun, X.; Kong, D.; Chu, D.; Hu, Y.; Duan, J.-A. Spatial Light Modulated Femtosecond Laser Ablated Durable Superhydrophobic Copper Mesh for Oil-Water Separation and Self-Cleaning. *Surf. Coat. Technol.* **2020**, *402*, 126254. [CrossRef]

43. Safonov, A.I.; Sulyaeva, V.S.; Gatapova, E.Y.; Starinskiy, S.V.; Timoshenko, N.I.; Kabov, O.A. Deposition Features and Wettability Behavior of Fluoropolymer Coatings from Hexafluoropropylene Oxide Activated by NiCr Wire. *Thin Solid Films* **2018**, *653*, 165–172. [CrossRef]

44. Safonov, A.I.; Bogoslovtseva, A.L.; Sulyaeva, V.S.; Kiseleva, M.S.; Zhidkov, I.S.; Starinskiy, S.V. Effect of Annealing on the Structure and Properties of Thin Fluoropolymer Coatings Prepared by HW CVD. *J. Struct. Chem.* **2021**, *62*, 1441–1446. [CrossRef]

45. Starinskiy, S.V.; Bulgakov, A.V.; Gatapova, E.Y.; Shukhov, Y.G.; Sulyaeva, V.S.; Timoshenko, N.I.; Safonov, A.I. Transition from Superhydrophilic to Superhydrophobic of Silicon Wafer by a Combination of Laser Treatment and Fluoropolymer Deposition. *J. Phys. D Appl. Phys.* **2018**, *51*, 255307. [CrossRef]

46. Martin, T.P.; Lau, K.K.S.; Chan, K.; Mao, Y.; Gupta, M.; Shannan O'Shaughnessy, W.; Gleason, K.K. Initiated Chemical Vapor Deposition (ICVD) of Polymeric Nanocoatings. *Surf. Coat. Technol.* **2007**, *201*, 9400–9405. [CrossRef]

47. Yasuoka, H.; Yoshida, M.; Sugita, K.; Ohdaira, K.; Murata, H.; Matsumura, H. Fabrication of PTFE Thin Films by Dual Catalytic Chemical Vapor Deposition Method. *Thin Solid Films* **2008**, *516*, 687–690. [CrossRef]

48. Safonov, A.; Sulyaeva, V.; Timoshenko, N.; Gatapova, E.; Kabov, O.; Kirichenko, E.; Semenov, A. Deposition and Investigation of Hydrophobic Coatings. *MATEC Web Conf.* **2015**, *37*, 01047. [CrossRef]

49. Boinovich, L.B.; Emelyanenko, A.M. Hydrophobic Materials and Coatings: Principles of Design, Properties and Applications. *Russ. Chem. Rev.* **2008**, *77*, 583–600. [CrossRef]
50. Drelich, J.; Chibowski, E.; Meng, D.D.; Terpilowski, K. Hydrophilic and Superhydrophilic Surfaces and Materials. *Soft Matter* **2011**, *7*, 9804–9828. [CrossRef]
51. Cao, H.; Gu, W.; Fu, J.; Liu, Y.; Chen, S. Preparation of Superhydrophobic/Oleophilic Copper Mesh for Oil-Water Separation. *Appl. Surf. Sci.* **2017**, *412*, 599–605. [CrossRef]

Article

Effect of Wall Proximity and Surface Tension on a Single Bubble Rising near a Vertical Wall

Raghav Mundhra [1], Rajaram Lakkaraju [1], Prasanta Kumar Das [1,*], Maksim A. Pakhomov [2]
and Pavel D. Lobanov [3]

1. Mechanical Engineering Department, Indian Institute of Technology, Kharagpur 721302, India
2. Laboratory of Thermal and Gas Dynamics, Kutateladze Institute of Thermophysics SB RAS, 630090 Novosibirsk, Russia
3. Laboratory of Physical Hydrodynamics, Kutateladze Institute of Thermophysics SB RAS, 630090 Novosibirsk, Russia
* Correspondence: pkd@mech.iitkgp.ac.in

Abstract: Path instability of a rising bubble is a complex phenomenon. In many industrial applications, bubbles encounter walls, and the interactions between the bubbles and the wall have a significant impact on flow physics. A single bubble rising near a vertical wall was experimentally observed to follow a bouncing trajectory. To investigate the near-wall dynamics of rising bubbles, 3D numerical simulations were performed based on the volume of fluid (VOF) method using the open source solver OpenFOAM. The effect of wall proximity and surface tension on the bubble trajectory was investigated. Previous studies have focused on the near-wall rising dynamics of bubbles for higher Eotvos numbers (Eo) and varied the Galilei number (Ga). The physical properties of the flow were chosen such that the free-rising bubble lies in the rectilinear regime. The Ga number was fixed and the Eo number was varied to analyze its effect on the bubble's rising trajectory. It was found that the presence of the wall increases the drag experienced by the bubble and induces an early transition from rectilinear to a planar zigzagging regime. We identify the maximum wall distance and the critical Eo number for the bubble to follow a bouncing trajectory. The amplitude, frequency and wavelength of the bouncing motion are independent of the initial wall distance, but they decrease with decreasing surface tension.

Keywords: rising bubbles; path instability; wall effect; bouncing bubbles

Citation: Mundhra, R.; Lakkaraju, R.; Das, P.K.; Pakhomov, M.A.; Lobanov, P.D. Effect of Wall Proximity and Surface Tension on a Single Bubble Rising near a Vertical Wall. *Water* **2023**, *15*, 1567. https://doi.org/10.3390/w15081567

Academic Editor: Giuseppe Pezzinga

Received: 1 March 2023
Revised: 2 April 2023
Accepted: 14 April 2023
Published: 17 April 2023

1. Introduction

Rising bubbles have been a topic of active research due to their rich physics and various applications in different domains of science and engineering. Bubble behavior has a significant impact on the flow properties and heat and mass transfer effects in gas–liquid two-phase flows encountered in industry and many natural phenomena. The application domain of bubbles is very wide, ranging from heat exchangers and bubble column reactors to targeted drug delivery and laser histotripsy.

The non-linearity and the number of parameters involved make the dynamics of rising bubbles quite complex. The dynamics of a single bubble rising due to buoyancy in an infinite liquid pool has been the focus of many experimental [1–6] and numerical studies [7–14] in the past. A bubble can show different shapes and rising trajectories depending on the interplay between surface tension, viscosity and inertia forces. The contamination of the bubble surface can alter its dynamics in a significant manner [4].

In clean water, small bubbles follow a straight vertical path maintaining a spherical shape. There are no vortices behind the bubble making the flow axisymmetric. As the bubble size increases, it rises following a plane zigzag or a three-dimensional helical path. The fascinating phenomenon of path instability of a rising bubble has been the focus of many experimental, numerical [8] and analytical studies [15]. The transition from the rectilinear to the oscillatory regime takes place when the streamwise vorticity accumulated on the bubble surface exceeds a critical value [5,16], resulting in an asymmetry in the vortical structures in the wake. The wake structures and vortical activity for the different path regimes have been extensively researched and documented [3,8,17]. Moreover, the bubbles also show significant shape deformations resulting in ellipsoidal, oblate ellipsoidal or cap shapes [4]. The shape deformations and wake instability have been considered responsible for the bubble path instability [12,18].

Gumulya et al. [19] and Senapati et al. [20] performed numerical simulations to investigate the effect of the bubble wake on a trailing bubble. They found that the leading bubble is unaffected by the proximity of the trailing bubble. The trailing bubble interacts with the wake of the leading bubble and accelerates. The path of the trailing bubble is also affected by the vorticity in the wake and the path deviation depends on the Reynolds and Eotvos numbers. Ghosh et al. [11] numerically modeled the shape distortion of one bubble due to the influence of the other and their merging.

In many industrial applications, bubbles encounter walls, and the interactions between the bubbles and the wall have a significant impact on flow physics. A single bubble rising near a vertical wall has been experimentally observed to follow a "bouncing" trajectory. Tsao and Koch [21] made observations for a single bubble rising very close to a wall at different inclination angles. They were the first to report that the bubble shows a steady bouncing behavior as it rises along the wall. After a certain critical angle, the bubble just slides along the wall. They linked this behavior to the bubble deformation dynamics. De Vries et al. [3] conducted experiments for a bubble rising very close to a vertical wall in still water. They observed that the bubble showed different behaviors such as sliding away, bouncing with constant amplitude or bouncing with increasing amplitude as the bubble size increases. Other researchers have explained this phenomenon based on the wall-induced lift force [22,23]. Podvin et al. [24] attempted to mathematically model the phenomenon using lubrication theory to model the wall-induced force. However, their model could not capture the bouncing process. Jeong and Park [25] carried out experimental studies for a high Re bubble rising at different distances from a vertical wall made of different materials. Lee and Park [26] suggested that the presence of the wall surface induces vortex shedding. Barbosa et al. [27] conducted experiments for a wide range of Reynolds and Weber numbers for bubbles rising near inclined walls. Using a force balance approach correlated with their experimental data, they formulated a condition for the critical angle for the transition from bouncing to sliding behavior.

The physics of this phenomenon is quite complex, and the mechanism of bouncing depends on several parameters. The problem of a rising bubble depends on four non-dimensional parameters:

Eotvos number-ratio of gravitational force to surface tension force $Eo = (\rho_l\, g d^2)/\sigma$
Galilei number-ratio of inertial force to viscous force $Ga = (\rho_l\, g^{1/2} d^{3/2})/\mu_l$
Density ratio $= \rho_l/\rho_g$
Viscosity ratio $= \mu_l/\mu_g$

Here, ρ, μ, σ, d and g represent the density, dynamic viscosity, surface tension coefficient, bubble diameter and gravitational acceleration, respectively. The subscripts l and g denote the liquid and gas properties, respectively.

In an experimental setting, it is tough to independently vary the different parameters governing the phenomenon. Numerical simulations come in handy to explore physics incorporating a wider parameter space. Sugioka and Tsukada [28] investigated the drag and lift forces acting on an inviscid bubble moving near a plane wall using 3D direct numerical simulations. They found that, due to the presence of the wall, the drag force increases, and the direction of the lift force depends on the Reynolds number and the bubble–wall distance. Zhang et al. [29] simulated bubbles in the presence of a vertical wall and found that the amplitude and frequency of oscillatory bubble motion decrease with decreasing Galilei number. They also noted that the wall-normal dimensionless force, amplitude and frequency are almost independent of the wall-bubble initial distance. They predicted a critical Galilei number for the transition from a steady to an oscillating regime, and it is found to be less than that for the unbounded condition. However, due to high Eo (=16), the bubble does not experience any attractive motion, and thus, bouncing motion and collision are not observed. Zhang et al. [30] performed 2D simulations for large deformable bubbles rising at different distances from the wall. They found that the bubble rises in a Z-shape and as the distance from the wall increases, the swing amplitude of the bubble gradually decreases, while the swing frequency of the bubble increases. Yan et al. [31] carried out VOF-based 3D simulations to investigate the bubble rising behavior near a vertical wall and observed the bubble bouncing motion. They found that as the Galileo number increases, the asymmetrical shedding of vortex structures is exacerbated by the presence of the wall. They predicted a critical Galilei number for the transition from rectilinear to spiral motion. The closer the bubble is to the wall, the earlier the transition occurs. Moreover, at low Galileo numbers, the bubble terminal velocity decreases due to the presence of the wall. Hasan and Hasan [32] studied the migration dynamics of an initially spherical bubble near a corner formed by two vertical walls and identified five regimes based on the bubble's migrating trajectory. Recently, Khodadadi et al. [33] performed numerical simulations over a wide range of Bond and Morton numbers for a single bubble rising near inclined walls. They identified three different bubble–wall interaction regimes—sliding, intermittent and non-contact—and plotted the regime map over the range of parameters considered.

The goal of this research is to explore the physics of bubble–wall interactions and gain insight into the bubble bouncing behavior as well as the parameters governing this phenomenon. The presence of a wall acts as a perturbation in the flow domain which can strongly affect the trajectory of the rising bubble and trigger the path instability of the bubble. For gas–liquid two-phase systems, a precise knowledge of the bubble trajectory is essential to estimate the gas-phase residence time through which the contact time for interfacial transport can be calculated. The bubble trajectory can be indirectly linked to the efficient design of gas–liquid two-phase systems. Moreover, knowledge of the parameters controlling the bubble trajectory can help us control the flow physics and heat transfer characteristics of many types of industrial equipment. Thus, a detailed investigation of the wall effect on the bubble dynamics is very much needed. This work focuses on the canonical case of a single bubble rising near a vertical wall which serves as the first step towards understanding multi-bubble systems encountered in industry.

Except for a few recent computational studies [28,30,31], the phenomenon of bouncing bubbles due to wall proximity has been studied in the past only through experiments. Experimental studies have their own constraints, and to delve deeper into the physics of any natural phenomenon, we need to look from the perspective of a wider parameter space. There is a need for a robust and well-established numerical model to capture this phenomenon. Moreover, there are still a lot of unanswered questions on bouncing bubbles, such as (1) which force plays the dominant role in governing the dynamics of bouncing; (2) how far into the liquid domain can the wall affect the bubble; (3) what is the role of the transport properties of both the fluids; (4) what are the kinematic and dynamic characteristics of the bouncing motion of the bubble, when observed in a wider parameter space; (5) a force balance model which can predict such dynamics.

To answer some of these questions, we use numerical simulations to analyze the effect of wall proximity and surface tension on the motion of a single bubble rising close to a vertical wall. In Section 2.2 we compare the parameter ranges covered in this work with the recent literature. Previous studies [25,29–31] have explored the effect of wall proximity on the dynamics of large bubbles with a higher Eotvos number. In this work, we study the effect of the initial bubble–wall distance on the dynamics of smaller bubbles with Eo and Ga corresponding to the air–water system. The bubble size is chosen such that the bubble lies in the rectilinear regime.

Tagawa et al. [34] explored the effect of surfactant concentration on the path instability of a bubble rising in a helical trajectory. For the case of bubbles rising near a vertical wall, previous works [29,31] have studied the effect of the Galilei number keeping the

Eotvos number fixed at values greater than 1. They predicted the critical *Ga* number for the transition from one regime to another. However, the authors have not come across any previous work studying the effect of the Eotvos number on bubbles rising near a vertical wall. Most of the previous research has focused on only discrete values of the *Eo*. In this work, by varying the surface tension, we systematically vary the Eotvos number over a large range (0.1–10) and find its critical value for the bubbles to start bouncing on the wall.

2. Computational Model

To study the wall effect on the rising motion of an initially spherical bubble, we initialize an air bubble of diameter *d* in a water tank of dimensions ($10d \times 60d \times 8d$) (see Figure 1). The dimensions of the computational domain have been chosen as the bubble bouncing phenomenon while keeping the computational costs and time as low as possible.

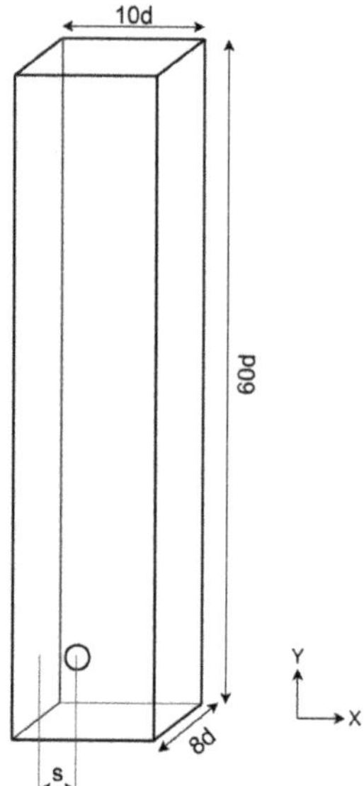

Figure 1. Schematic of the problem.

The height of the domain was taken as $60d$ so as to capture at least three bubble bounce events. Previous numerical studies [35] have shown that if the walls are at least $6d$ away from the bubble, then the computational domain can be treated as an unbounded fluid medium. The length and breadth have been taken as $10d$ and $8d$, respectively, so that there is no effect of the right wall, the front and back walls on the bubble motion. Since, the parameter range is chosen such that the bubble falls either in the rectilinear regime or the planar zigzagging regime and the bubble is going to bounce on the left wall (i.e., along the length of the domain) with negligible motion along the breadth of the domain, the length is kept greater than the breadth.

To avoid the influence of the right, bottom, back and front walls the initial co-ordinates of the bubble are set to (s, $4d$, $4d$), where s is the normalized initial distance of the bubble centroid from the left wall. Thus, only the left wall would affect the dynamics of the bubble [35,36]. The initial bubble–wall distance has been normalized with respect to the bubble diameter.

The 3D computational simulation is performed with the interFoam solver in CFD code OpenFOAM, where the VOF model, is used to reconstruct the gas–liquid interface. OpenFOAM is a very robust open source CFD library and has been validated and used for the study of bubble dynamics by many researchers [37–39]. The VOF method has been proven to be quite accurate to capture the physics of bubble–wall interactions [40,41].

2.1. Governing Equations and Numerical Model

The VOF model uses the phase fraction α to locate the phase interface. The grid cell is filled with the liquid phase when $\alpha = 1$, and when $\alpha = 0$, it is filled with the gas phase. The interface can be located as the region where α lies between 0 and 1.

The conventional VOF method involves the simultaneous solution of the continuity and momentum equations along with the solution of the transport equation for the phase fraction.

The incompressible continuity equation can be written as:

$$\nabla \cdot u = 0 \tag{1}$$

The two phases share a common set of momentum equations, which can be represented as:

$$\frac{\partial \rho u}{\partial t} + \nabla \cdot (\rho uu) = -\nabla p + \nabla \cdot \left[\mu \left(\nabla u + \nabla u^T \right) \right] + F_\sigma + \rho g \tag{2}$$

where u represents the velocity vector shared by the two fluids throughout the flow domain. t and p denote the time and static pressure, respectively. F_σ represents the surface tension force. In a single pressure system, the normal component of the pressure gradient at a stationary non-vertical solid wall with a no-slip condition must be different for each phase due to the hydrostatic component ρg when the phases are separated at the wall. In order to simplify the definition of boundary conditions, a modified pressure is defined as:

$$p_d = p - \rho g.x \tag{3}$$

where x represents the position vector of the fluid element.

Using Equation (3) in (2) yields the modified momentum equation as follows:

$$\frac{\partial \rho u}{\partial t} + \nabla \cdot (\rho uu) = -\nabla p_d - g.x\nabla \rho + \nabla \cdot \left[\mu \left(\nabla u + \nabla u^T \right) \right] + F_\sigma \tag{4}$$

Body forces due to pressure gradient and gravity are implicitly accounted for by the first two terms on the right-hand side of Equation (4).

In addition, the density ρ and viscosity μ are weighted by the phase fraction parameter as follows:

$$\rho = \alpha \rho_l + (1 - \alpha)\rho_g \tag{5}$$

$$\mu = \alpha \mu_l + (1 - \alpha)\mu_g \tag{6}$$

where α indicates the liquid phase fraction parameter. Subscripts l and g denote the liquid and gas phases, respectively.

The continuum surface force (CSF) model is used for surface tension. The surface tension force is calculated as follows:

$$F_\sigma = \sigma \kappa \nabla \alpha \tag{7}$$

where σ represents the surface tension coefficient and κ represents the local curvature at the interface which can be approximated as follows:

$$\kappa = -\nabla \left(\frac{\nabla \alpha}{|\nabla \alpha|} \right) \tag{8}$$

In order to know where the interface between the two fluids is, an additional equation for α has to be solved:

$$\frac{\partial \alpha}{\partial t} + \nabla \cdot (\alpha u) = 0 \tag{9}$$

In the interFoam solver, a two-fluid Eulerian formulation is used in which the phase fraction equation is solved separately for each individual phase [37].

$$\frac{\partial \alpha}{\partial t} + \nabla \cdot (\alpha u_l) = 0 \tag{10}$$

$$\frac{\partial (1-\alpha)}{\partial t} + \nabla \cdot ((1-\alpha)u_g) = 0 \tag{11}$$

The velocity is modeled in terms of the weighted average of the corresponding liquid and gas velocities as shown below:

$$u = \alpha u_l + (1-\alpha)u_g \tag{12}$$

Equations (10)–(12) can be combined [37] to obtain:

$$\frac{\partial \alpha}{\partial t} + \nabla \cdot (\alpha u) + \nabla \cdot [\alpha(1-\alpha)u_C] = 0 \tag{13}$$

With this formulation, a sharper interface resolution is achieved. The phase fraction transport Equation (13) contains an extra convective term that is referred to as the compression term, which is used to keep the sharpness of the phase interface. In Equation (13), u_c is called the compression velocity, which is, in essence, the relative velocity between the two phases. Since only a single velocity for both fluids is considered in the whole domain, the compression velocity needs to be modeled [37,38] as:

$$u_C = c|u|\left(\frac{\nabla \alpha}{|\nabla \alpha|}\right) \tag{14}$$

where c is the compression factor with the value of 1. Equation (14) closes the system of governing equations.

OpenFOAM solves the coupled system of Equations (1), (4) and (13) based on the finite volume method. The details of the used discretization schemes are given in Table 1. A predicted velocity field is constructed and then corrected repeatedly by using the pressure implicit with splitting of operators (PISO) iteration procedure. Details of the solution procedure followed by the solver can be found in this paper [38].

Table 1. Discretization schemes used for the different terms of the governing equations.

Term	Discretization Scheme	OpenFOAM Terminology
$\frac{\partial}{\partial t}(\rho u)$	Euler implicit time scheme	Euler
$\nabla \cdot (\rho u u)$	Total Variation Diminishing	Limited linearV 1
$\nabla \cdot (\alpha u)$	Total Variation Diminishing	Gauss vanLeer
$\nabla \cdot [\alpha(1-\alpha)u_C]$	Bounded limited scheme	InterfaceCompression
$\nabla \cdot u$	Central Differencing Scheme	Linear
$\nabla \alpha$	Central Differencing Scheme	Linear
$\nabla \cdot [\mu(\nabla u + \nabla u^T)]$	Central Differencing Scheme	Linear Corrected

All the walls were assigned a no-slip boundary condition for velocity and a zero gradient boundary condition for pressure. A static contact angle boundary [40] condition was imposed on the left wall. Both the velocity and pressure fields were initialized as a uniform field with a value of zero.

The convergence tolerance was set to 1×10^{-6} for all the governing equations. The adjustable time step setting was turned on to ensure time step convergence.

The computations were performed on a supercomputing cluster using at least 128 cores in Intel Xeon SKL G-6148 processors running in parallel. Each solution needed at least 5 days of runtime.

2.2. Parameter Space

The density and viscosity ratios have been chosen to represent an air–water gas–liquid system. For the air–water system, taking $\rho = 1000$ kg m^{-3}, $\mu = 0.001$ Pa s, g = 9.81 m s^{-2} and d = 0.001 m, we obtain Ga = 99. The case of Ga = 99 and Eo = 0.14 represent a 1 mm air bubble rising near a wall in pure water. Fixing the Ga and bubble diameter and varying the Eo from 0.1 to 10 allows us to study the effect of surface tension on the near-wall bubble behavior. In practical situations, we can control the surface tension of the air–water interface by using surfactants. These Eo values range can be considered as the value of Eo for a 1 mm air bubble rising in water having different surfactant concentrations. The knowledge of the effect of surface tension on the bubble behavior in water can be leveraged in practical situations to control the flow properties by using surfactants.

Table 2 shows a comparison between the parameters considered in previous research and this work. It is quite clear that we are exploring a parameter space that has been unexplored as of yet.

Table 2. Comparison of the range of physical parameters.

Researchers	Eo	Ga	s
Zhang et al. [29]	16	0.57, 63.36, 90.51	0.75, 1, 2
Yan et al. [31]	2	8.8, 51, 95, 133	0.75, 1, 2
Our work	0.1, 0.14, 1, 2, 10	99	0.75, 1, 1.2, 1.5

3. Results

To investigate the effect of a nearby vertical wall on the hydrodynamic behavior of a bubble, we divide this work into three parts. First, we compare the dynamics of a wall-bounded bubble with a free-rising bubble, then, we investigate the bubble dynamics at different initial wall distances. Finally, we study the effect of surface tension on the bubble behavior.

3.1. Grid Independence and Validation

We performed our simulations on a uniform mesh so as to obtain the same level of accuracy throughout the domain. To validate our model, we first chose a very coarse mesh with only 10 cells (see Figure 2a) across the bubble diameter. We then refined the mesh to obtain different meshes with 15 cells (see Figure 2b), 20 cells (see Figure 2c) and 25 cells across the bubble diameter. Simulations for an air bubble (d = 1 mm) rising in pure water were performed using these meshes and the trend of rising velocity achieved by each mesh has been plotted in Figure 3. The results for terminal velocity for each mesh have been tabulated in Table 3. Figure 2d shows the mesh in the region between the bubble and the wall.

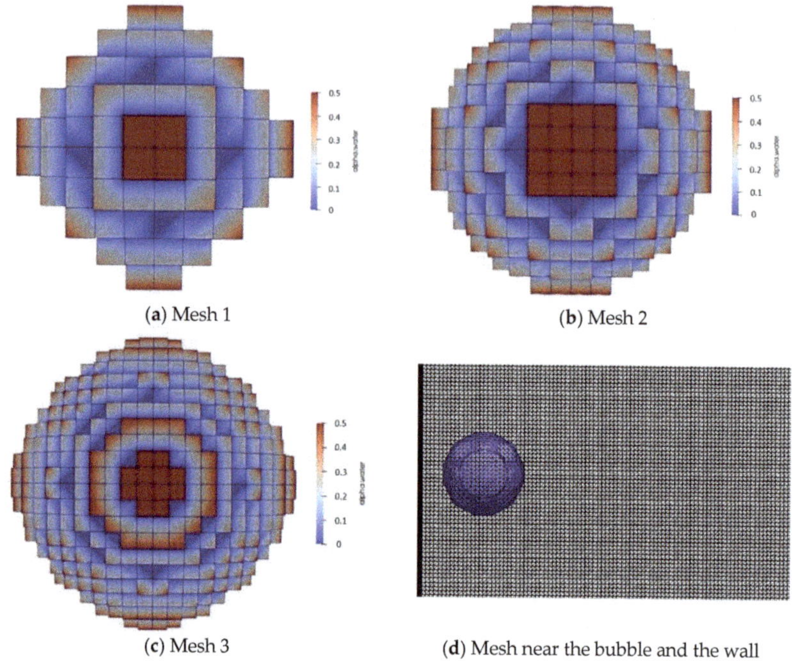

(**a**) Mesh 1

(**b**) Mesh 2

(**c**) Mesh 3

(**d**) Mesh near the bubble and the wall

Figure 2. Bubble shape resolution for different meshes.

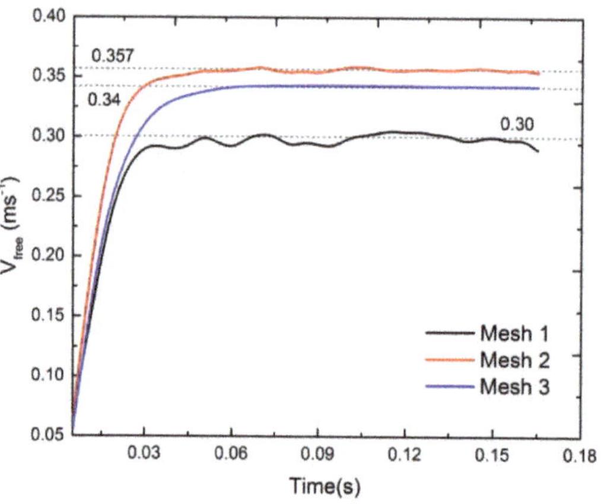

Figure 3. Bubble rise velocity for different meshes.

Table 3. Mesh refinement study.

Refinement Level	Number of Cells per Bubble Diameter	Terminal Velocity (m/s)
1	10	0.3
2	15	0.357
3	20	0.34

For simulating the phenomenon, a good resolution of the bubble surface is essential. It is quite clear from Figure 2a that Mesh 1 cannot resolve the bubble surface.

From Figure 3, we can see that the bubble velocity converges to a stable terminal velocity only for Mesh 3.

Mesh 4 (25 cells per bubble diameter) was also tested and there was not much difference in the results from Mesh 3. However, there was a considerable increase in the simulation time. Mesh 3 was thus the optimum choice for carrying out the simulations and the results from Mesh 3 can be considered to be grid independent.

The bubble terminal velocity for the same bubble size and fluid properties was calculated using the correlation of Jamialahmadi [42]. As can be seen from Table 4, the numerical results are in excellent agreement with the experimental correlation. The drag coefficient of the bubble is evaluated by balancing the buoyancy force and drag force at terminal conditions. The experimental drag coefficient for the same conditions has been obtained using Moore's correlation [4]. Both the numerical and experimental drag coefficients are shown in Table 4 below. The error between the experimental correlation and the numerical result is around 8%.

Table 4. Validation with experimental results.

	Numerical (Mesh 3)	Experimental	% Error
Terminal Velocity	0.34	0.344	1.16
Drag Coefficient	0.113	0.123	8.13

Thus, we can say that results from Mesh 3 are in agreement with experimental findings validating our model.

3.2. Near-Wall Rising Behaviour

For bubbles rising due to buoyancy, the near-wall bubble motion is very distinct from the motion of an unbounded bubble. Figure 4 shows the trajectory of a bubble ($d = 1$ mm, $Eo = 0.14$, $Ga = 99$) released at a distance of $s = 1, 4$ from the left wall. The bubble released at $s = 4$ shows a straight rising trajectory (Figure 4b) while the one released at $s = 1$ shows an almost two-dimensional zigzagging trajectory (Figure 4a). Since at $s = 4$ the rising trajectory is not affected by the wall, it can be considered an unbounded bubble.

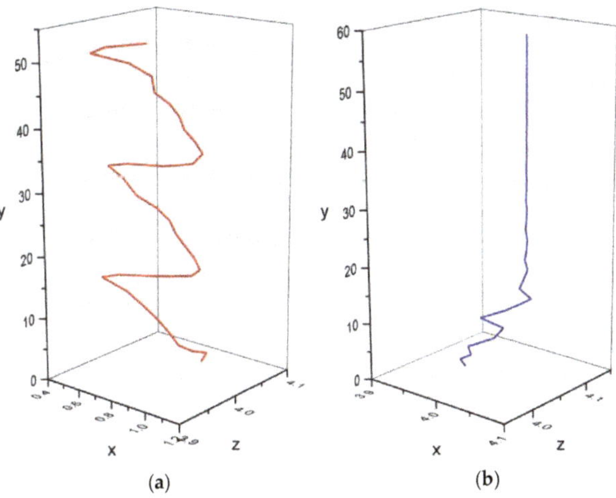

Figure 4. Comparison of the rising trajectory of a bubble released near a wall with that of an unbounded bubble: (**a**) 3D trajectory of the bubble released close to the wall; (**b**) 3D trajectory of the bubble released away from the wall. $d = 1$ mm, $Eo = 0.14$, $Ga = 99$.

It can be observed from Figure 5a that the unbounded bubble has no motion along the x-direction. However, the wall-bounded bubble shows a sinusoidal variation of its x-position with time, referred to as bouncing motion in the literature [3,21,25]. Figure 5b shows the variation of the wall-normal velocity for the wall-bounded and unbounded bubbles. After an initial transition state, the wall-normal velocity for the unbounded bubble converges to a steady value of zero, whereas the wall-normal velocity for the near-wall bubble shows a sinusoidal variation.

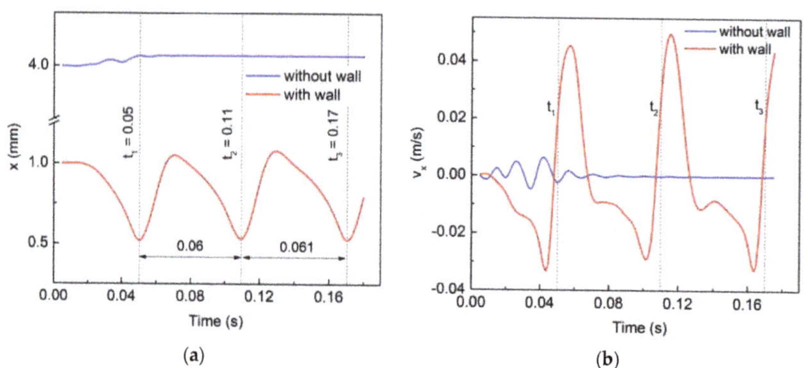

Figure 5. Comparison of x-motion of the bounded and unbounded bubble: (**a**) Time variation of x position of bubble centroid; (**b**) Time variation of x velocity of bubble centroid d = 1 mm, Eo = 0.14, Ga = 99.

The effect of the wall on the rising motion of the bubble is shown in Figure 6. The y-position of the bubble is monotonically increasing for both cases, but the slope for the bounded bubble is lesser than that for the unbounded bubble. The rise velocity of the wall-bounded bubble is fluctuating, and its terminal velocity is lower than that of the unbounded bubble.

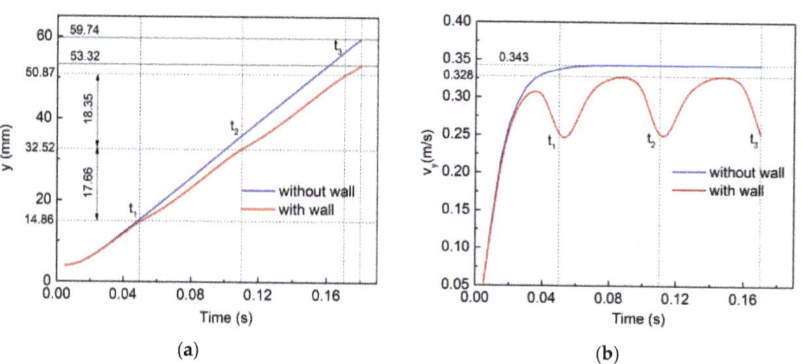

Figure 6. Comparison of y-motion of the bounded and unbounded bubble: (**a**) Time variation of y-position of bubble cen troid; (**b**) Time variation of y-velocity of bubble cen troid. d = 1 mm, Eo = 0.14, Ga = 99.

Now, if we look at the motion of the bubble in the spanwise direction (shown in Figure 7a,b) it is quite clear that the variations are of negligible magnitude compared to the variations in the other directions. For the unbounded bubble, after an initial transition state, the motion in the z-direction ceases and the z-position of the bubble gets fixed at z = 4.05 mm and z-velocity becomes zero. For the bounded bubble, random fluctuations in both the position and velocity in the z-direction are observed. However, the magnitude of these changes is very small.

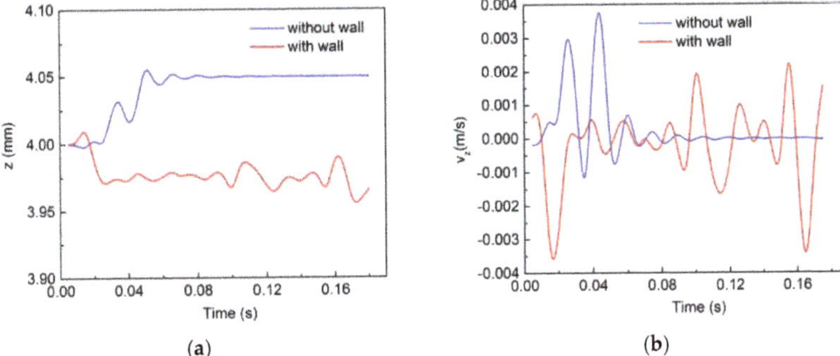

Figure 7. Comparison of *z*-motion of the bounded and unbounded bubble: (**a**) Time variation of *z*-position of bubble centroid; (**b**) Time variation of *z*-velocity of bubble centroid. *d* = 1 mm, *Eo* = 0.14, *Ga* = 99.

Figures 8 and 9 show the X and Y velocity contours, respectively, during a bounce event. It can be observed that just after the bubble–wall collision, both the x-velocity and the y-velocity in the region surrounding the bubble drop to significantly low values, thus breaking the continuity of the liquid flow around the bubble. This can be considered as the separation of the flow behind the bubble. It appears that the flow separation occurs after the bubble has bounced. This might have an effect on the wake region behind the bubble.

Figures 10 and 11 show the velocity vector plots around the wall-bounded and free bubble, respectively. It is clear that the wall provides a symmetry-breaking perturbation to the vortices in the flow field. As per Kelvin's circulation theorem, vortices in the flow domain can generate lateral forces. From Figure 10, we can see that in the YZ plane, there are two symmetrical vortices in the wake of the bubble. The lateral force generated due to the two vortices on the bubble cancels out, and thus, the resultant force on the bubble is zero, resulting in no motion in the spanwise direction.

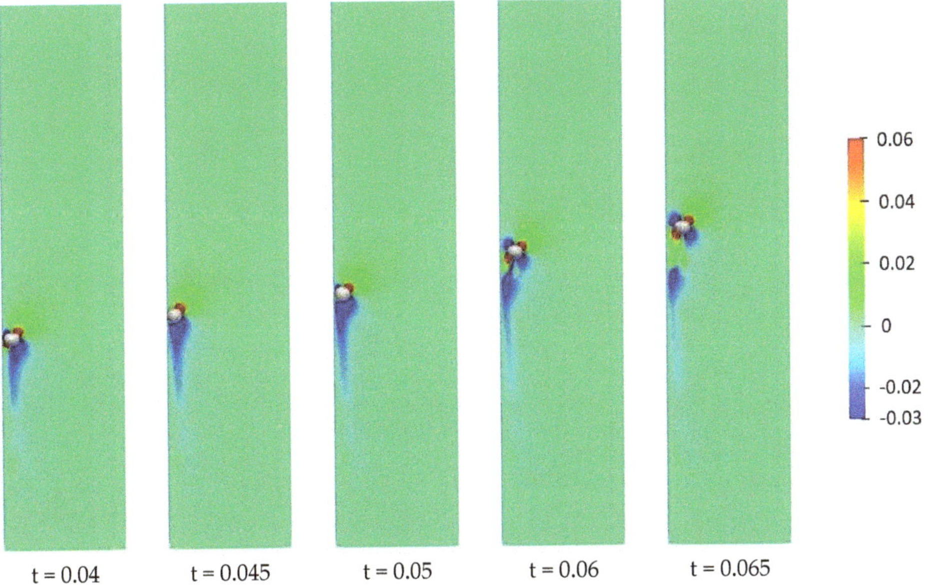

Figure 8. X-velocity contours during a bounce event (*s* = 1).

However, due to the presence of the wall, the flow structures in the XY plane become unsymmetrical. The vortex structures in the wake of the bubble are asymmetrical resulting in a net lateral force that attracts the bubble towards the wall. Thus, the asymmetry in the flow structures provides a qualitative explanation for the attractive lift force experienced by the bubble due to the presence of the wall. The bouncing phenomenon is just a consequence of this lift force. As the bubble rises near the wall, it is attracted towards the wall due to the lift force. When it becomes sufficiently close to the wall, the pressure in the thin liquid film between the bubble and the wall exerts a force that repels the bubble away from the wall. The bubble now moves away from the wall as it rises upwards. The lift force again attracts the bubble towards the wall while the pressure in the liquid film repels it. The amplitude of the bouncing motion, i.e., the maximum wall-normal distance reached by the bubble thus depends on the balance of these two forces. Moreover, the wavelength of the bouncing motion depends on the bubble rise velocity and the response time required for these two forces to balance out.

Now, for the case of a bubble rising away from the wall, the wake structures are symmetrical in both planes (Figure 11) and thus there is no net lateral force acting on the bubble and the bubble rises straight upwards following a rectilinear trajectory.

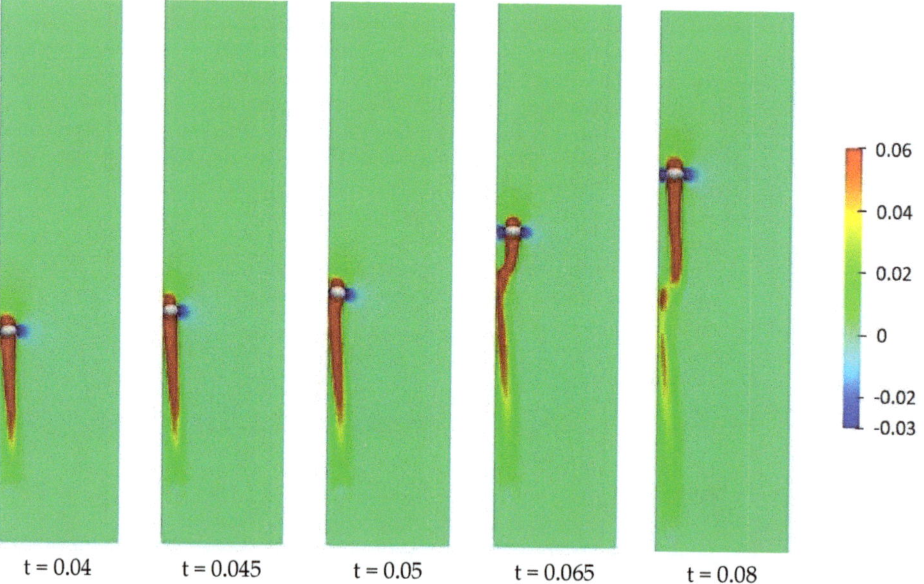

Figure 9. Y-velocity contours during a bounce event ($s = 1$).

3.3. Effect of Initial Wall Distance

To study the effect of the bubble–wall distance on the bubble rising behavior, we release the bubble at different distances from the wall. Keeping all other parameters fixed, we vary s as $s = 0.75, 1, 1.2, 1.5$. It is quite intuitive to expect the effect of the wall to decrease as the initial distance increases. This study will help us to analyze how far into the flow field the effect of the wall creeps in.

From Figure 12, we can see that as the bubble–wall initial distance increases, the bubble trajectory changes from bouncing to straight rising. The bubble shows a bouncing trajectory for s of less than 1.5.

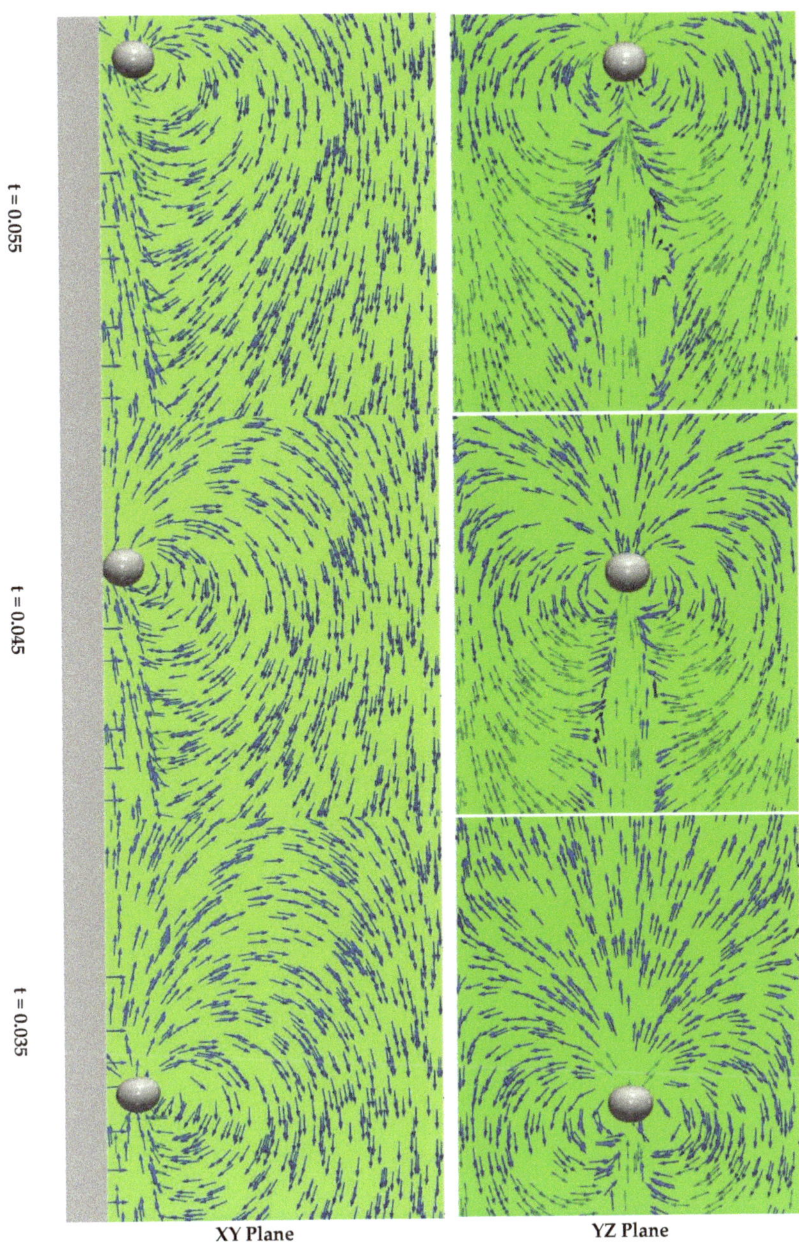

Figure 10. Velocity vector plots around a bubble rising close to the wall ($s = 1$).

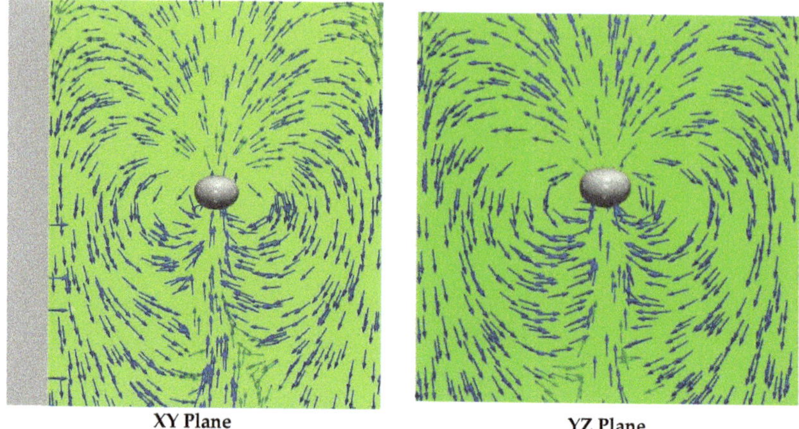

Figure 11. Velocity vector plot around a bubble rising away from the wall.

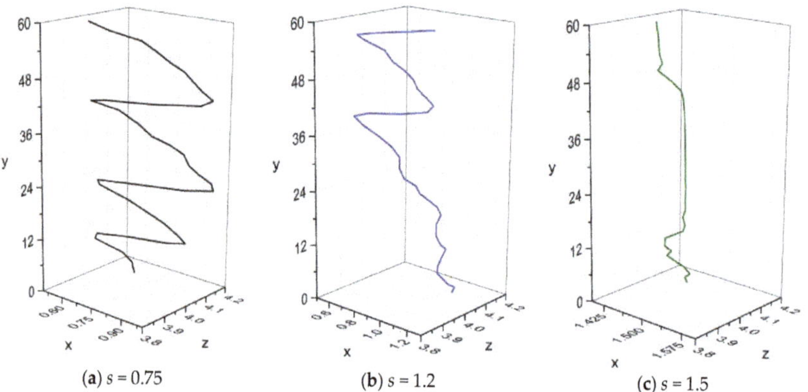

Figure 12. Comparison of the bubble trajectory when released at different distances from the wall ($Eo = 0.14$, $Ga = 99$).

Figure 13a,b shows the effect of the wall distance on the wall-normal motion of the bubble. The characteristics of the bouncing motion for different s are listed in Table 5. It is noted that for the s values for which the bubble shows a bouncing motion, the bouncing trajectory has almost similar characteristics. This implies that the initial wall distance does not impact the bouncing motion; however, there is just a phase difference of the bouncing trajectories (Figure 13a) for different s. In other words, as the distance from the wall increases, the time of onset of these instabilities also increases as we can observe for the cases of $s = 0.75$, 1, 1.2. After $s = 1.2$, the bubble does not show bouncing in the considered domain.

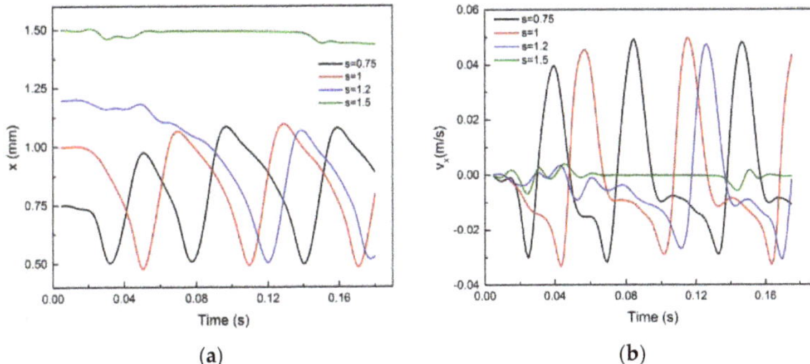

Figure 13. Comparison of *x*-motion of the bubble released at different distances from the wall: (**a**) Time variation of *x*-position of bubble centroid; (**b**) Time variation of *x*-velocity of bubble centroid. $d = 1$ mm, $Eo = 0.14$, $Ga = 99$.

Table 5. Characteristics of bouncing motion at different *s* ($Eo = 0.14$).

s	Amplitude (A) (mm)	Wavelength (λ) (mm)	Time Period (T) (s)
0.75	0.56	18.5	0.06
1	0.6	18	0.06

Figure 14a,b show the effect of the wall distance on the rising motion of the bubble. The rising velocity shows a fluctuating trend for *s* less than 1.5 and the terminal rise velocity decreases as *s* decreases.

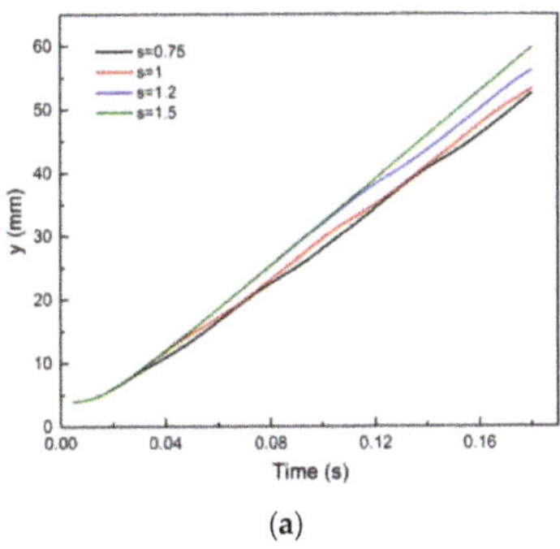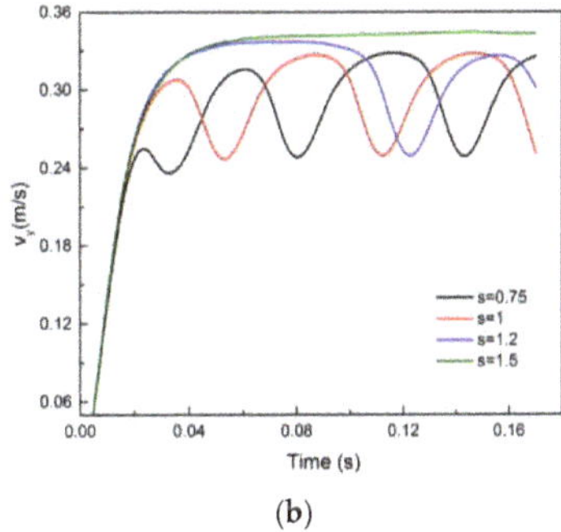

Figure 14. Comparison of *y*-motion of the bubble released at different distances from the wall: (**a**) Time variation of *y*-position of bubble centroid; (**b**) Time variation of *y*-velocity of bubble centroid. $d = 1$ mm, $Eo = 0.14$, $Ga = 99$.

Figure 15 below shows the variation of the average rise velocity of the bubble with the initial bubble–wall distance. As the wall proximity increases, the drag force on the bubble increases, and thus, the average rise velocity decreases. In terms of the drag force, s = 1.5 can be considered as the critical distance beyond which the wall has a negligible effect on the bubble motion.

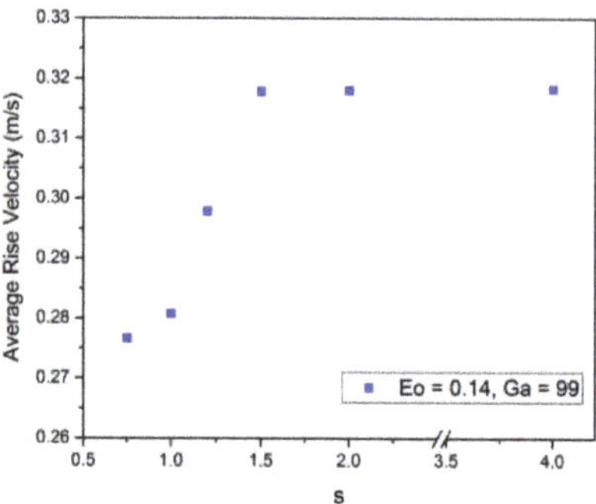

Figure 15. The variation of the average rise velocity of the bubble with initial wall distance.

As can be seen from Figure 16, the spanwise motion of the bubble is quite randomized.

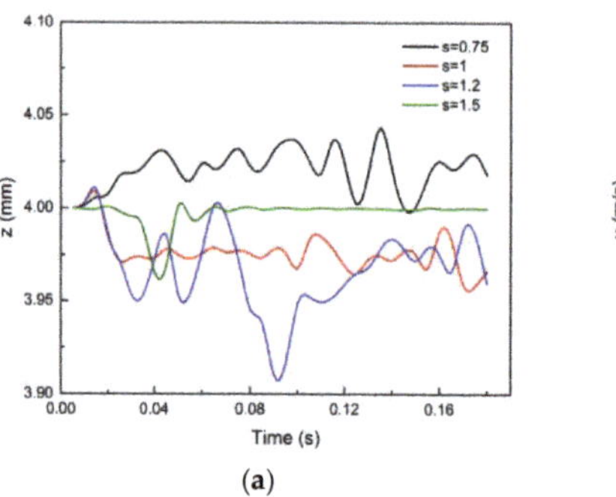

(**a**)

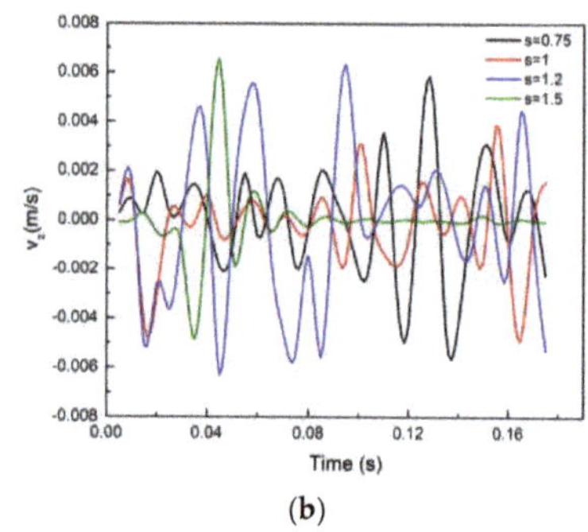

(**b**)

Figure 16. Comparison of z-motion of the bubble released at different distances from the wall: (**a**) Time variation of z-position of bubble centroid; (**b**) Time variation of z-velocity of bubble centroid. $d = 1$ mm, $Eo = 0.14$, $Ga = 99$.

However, the fluctuations in the z-component of velocity decrease as the distance from the wall increases. Moreover, the magnitude of the position variation and the velocity in spanwise direction is very small compared to the other two directions.

3.4. Effect of Surface Tension

Keeping the bubble size ($d = 1$ mm) and distance from the wall ($s = 1$) constant, we now vary the surface tension of the liquid–air interface. The different values of σ and the corresponding Eo are shown below (see Table 6):

Table 6. The Eotvos numbers vs. various surface tension coefficients.

σ	0.1	0.07	0.01	0.005	0.001
Eo	0.0981	0.14	0.981	1.96	9.81

Figure 17 shows how the bubble rising trajectory changes at different values of surface tension. After $Eo = 1$, the bubble shows a zigzagging trajectory while moving away from the wall.

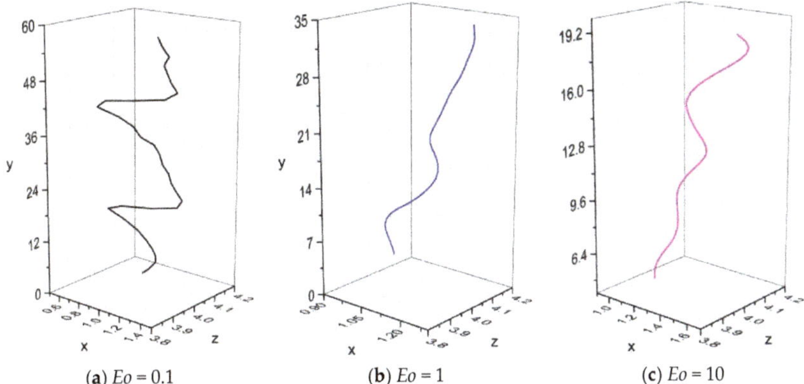

Figure 17. Comparison of the bubble trajectory at different *Eo* ($s = 1$, $Ga = 99$).

The effect of surface tension on the wall-normal motion, rising motion and spanwise motion is shown in Figures 18–21, respectively. It is evident from these figures that the wall effect on the bubble dynamics is dependent on the surface tension value. A detailed discussion of these results is performed in the following section.

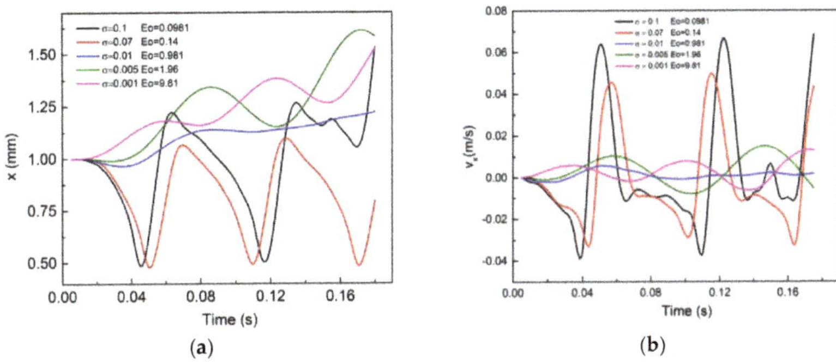

Figure 18. Comparison of x-motion of the bubble for different *Eo*: (**a**) Time variation of x-position of bubble centroid; (**b**) Time variation of x-velocity of bubble centroid. $d = 1$ mm, $Ga = 99$, $s = 1$.

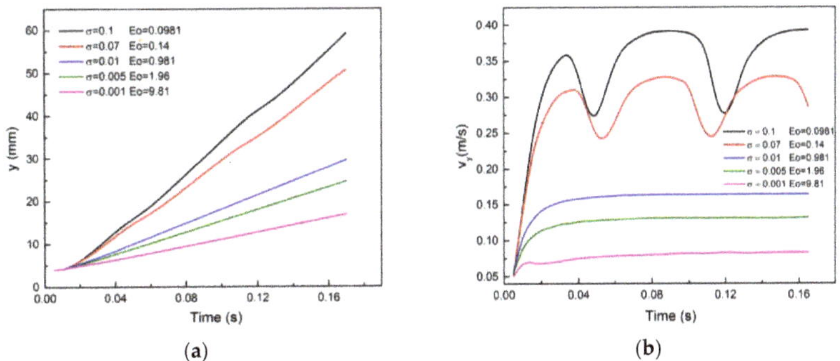

Figure 19. Comparison of y-motion of the bubble for different *Eo*: (**a**) Time variation of y-position of bubble centroid; (**b**) Time variation of y-velocity of bubble centroid. $d = 1$ mm, $Ga = 99$, $s = 1$.

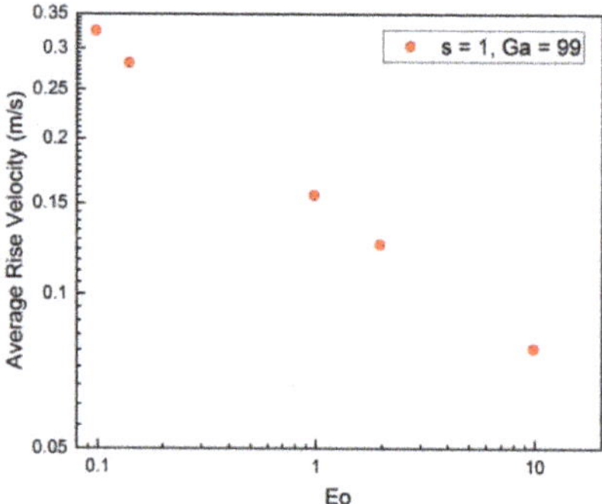

Figure 20. The variation of the average rise velocity of the bubble with Eotvos number.

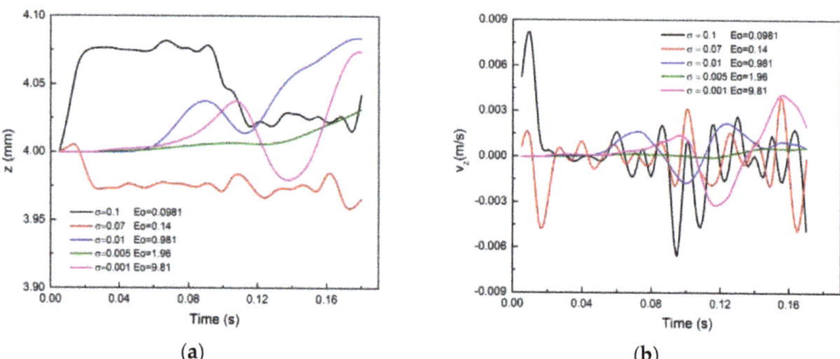

(a) (b)

Figure 21. Comparison of z-motion of the bubble for different Eo: (**a**) Time variation of z-position of bubble centroid; (**b**) Time variation of z-velocity of bubble centroid. $d = 1$ mm, $Ga = 99$, $s = 1$.

From Table 7, we can see that for the values of surface tension for which the bubble shows a bouncing trajectory (Figure 18a), as the Eotvos number increases, the amplitude, time period and wavelength of the bouncing motion decreases.

Table 7. Characteristics of bouncing motion at different Eo ($s = 1$).

Eo	Amplitude (A) (mm)	Wavelength (λ) (mm)	Time Period (T) (s)
0.1	0.73	25.1	0.07
0.14	0.6	18	0.06

Figure 20 given below shows the variation of the average rise velocity of the bubble with the Eotvos number plotted on the log scale. It is evident from the plot that with increasing Eo, the rise velocity decreases. This implies that the decrease in surface tension indirectly increases the drag force.

From the above figure, we can observe that at lower surface tension values the bubble assumes a flat shape. Figure 22a,b shows the deformation in the bubble shape at the time of bouncing.

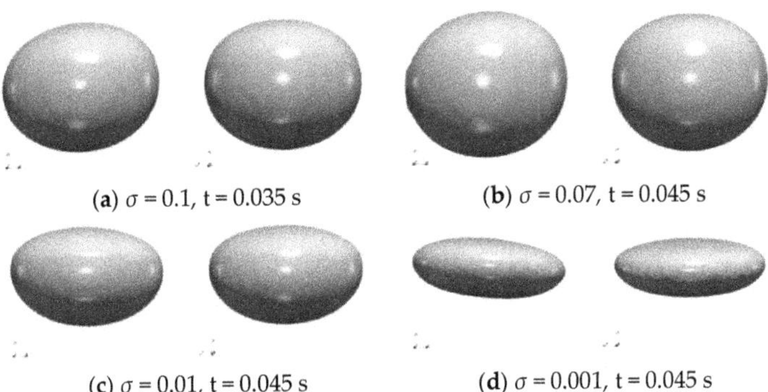

(a) $\sigma = 0.1$, t = 0.035 s   **(b)** $\sigma = 0.07$, t = 0.045 s

(c) $\sigma = 0.01$, t = 0.045 s   **(d)** $\sigma = 0.001$, t = 0.045 s

Figure 22. The bubble shape as viewed from the XY (**left**) and YZ (**right**) planes for different values of surface tension.

4. Discussion

According to phase diagram of Cano-Lozano et al. [13], a rising bubble with $Ga = 99$ and $Eo = 0.14$ lies in the rectilinear regime. From Figure 4b, it is seen that the trajectory of the bubble far from the wall is straight rising. The wall proximity induces an early transition from the rectilinear to the planar zigzagging regime as shown in Figure 4a. This zigzagging motion on the wall is referred to as bouncing motion in the literature [21,25] as the bubble keeps on colliding with the wall, and thus, seems to bounce off the wall repeatedly.

From Figure 5, we can observe that the amplitude of this bouncing motion is constant. The bubble collides with the wall at t_1, t_2 and t_3. The time between two collisions is highlighted in the figure and is almost constant, i.e., the bouncing frequency is constant. The rising motion of the bounded and unbounded bubbles have the same nature (Figure 6). Figure 6 shows the vertical distance traveled by the wall-bounded bubble between two collisions (t_1, t_2) and (t_2, t_3) and it is almost constant. Thus, the bouncing motion is very uniform with constant amplitude, frequency and rise distance in every bouncing cycle.

On analyzing the x-component of the bubble velocity (Figure 5b), it is seen that the x- motion of the unbounded bubble has an initial transition state in which the x-velocity oscillates about the mean value of zero. After reaching the steady state, the x-motion ceases, i.e., x-velocity becomes zero. For the bounded bubble, the x-velocity shows a fluctuating trend. As expected, the velocity starts to rise at the time of the bubble–wall collision. From Figure 6a, it is evident that due to the presence of the wall, the drag force increases, and thus, the average rise velocity of the bounded bubble is lesser than that for the unbounded case.

As expected, the unbounded bubble reaches a terminal state with $v_y = 0.34$ m/s. However, the y-motion of the bounded bubble cannot achieve a steady state terminal velocity. After the buoyancy and drag forces are balanced, the bubble reaches a maximum $v_y = 0.32$ m/s. From Figures 5b and 6b, we can observe that at the time of the collision, the x-velocity starts increasing, and consequently, the y-velocity drops from its maximum value. As the bubble completes its bouncing cycle, the x-velocity decreases and the y-velocity starts rising and it again reaches its maximum value and stays there for some time after which the bubble is again attracted to the wall and the rising velocity drops again.

The spanwise motion (Figure 7a,b) is very negligible compared to those in the wall-normal and wall-parallel directions. However, compared to the unbounded bubble, there is some randomized motion in the spanwise direction for the bubble rising near the wall.

4.1. Effect of Wall Proximity

From Figures 12–15, it is evident that as s increases, the effect of the wall on the bubble decreases. For $s = 0.75$ and 1, the frequency and amplitude of bouncing are almost equal and constant (Table 5). However, there is a phase lag between them. For $s = 0.75$, the bubble is attracted towards the wall quite early as expected. As the distance from the wall increases,

119

it takes more time for the bubble to be attracted towards the wall. For $s = 1.2$, initially, the bubble rises straight upwards with negligible motion in the wall-normal direction. After $t = 0.15$, the bubble is attracted to the wall, and then after colliding with the wall, it shows a bouncing trajectory with almost the same amplitude and frequency as the previous cases. For $s = 1.5$, the bubble experiences negligible attraction towards the wall and its x-position decreases very slowly with time. At this distance, we can say that there is no effect of the wall, and hence, there is no bouncing motion. The wall-normal velocity shows a sinusoidal variation of constant amplitude and frequency, as expected for the bouncing motion. For $s = 1.2$ and 1.5, the velocity shows small variations about zero, this is quite obvious as the bubble has no considerable motion in the x-direction.

The y-position of the bubble increases monotonically with an almost constant slope for all s. The slope of the line increases as s increases as observed in Figure 14a. As we can see from Figure 14b, the maximum y-velocity reached by the bubble increases as s increases. As the distance from the wall (s) increases, the drag force decreases, and thus, the average rise velocity increases. As explained in the previous section, the y-velocity magnitude shows a fluctuating pattern due to the bouncing motion. With increasing s, the bouncing motion dies out and so do the fluctuations in the y-velocity. As s increases, the maximum y-velocity achieved by the bubble increases. This is expected because, with increasing distance from the wall, the effect of the wall decreases, and thus, the effect of the wall on the rising velocity decreases. For $s = 1.5$, there are no fluctuations, and the rising velocity reaches a terminal state with $v_y \approx 1$. This implies that the terminal velocity is slightly lesser compared to the terminal velocity of the free-rising bubble. At $s = 1.5$, the wall effect is not strong enough to bring an effect in the wall-normal direction and induce a bouncing motion but it increases the drag, thus decreasing the y-velocity of the bubble.

4.2. Effect of Eotvos Number

Let us now analyze the effect of the variation of Eo on the bubble behavior. As the surface tension decreases and the Eo increases, the drag experienced by the bubble increases. This can be deduced from Figure 19. The y-position and the maximum rising velocity achieved decrease with increasing Eo number. This increase in drag can be attributed to the flattening of the bubble surface with decreasing surface tension.

At $Ga = 99$, the transition from rectilinear rising regime to planar zigzagging regime occurs at $Eo \gtrsim 1.5$ [13]. In the presence of the wall, the bubble shows zigzagging motion (18a) for the entire range of Eo considered in this study, i.e., from $Eo = 0.098$–9.8, except for $Eo = 1$. It is observed that for $Eo < 1$, the bubble bounces on the wall. Meanwhile, for $Eo > 1$, the bubble slowly moves away from the wall following a zigzagging trajectory. At $Eo = 1$, after an initial transition state, the bubble gradually slides away from the wall. Thus, at $Ga = 99$, $Eo = 1$ represents the critical value of Eo for which the wall force transitions from attractive to repulsive, thus changing the bubble motion from bouncing to sliding away.

For $Eo < 1$, with increasing surface tension and decreasing Eo, both the amplitude and the frequency of the bouncing motion increase (Figure 18a). The maximum wall-normal velocity achieved after the bubble–wall collision is also higher for lower Eo (Figure 18b). With regards to the motion in the spanwise direction, for the bouncing regime ($Eo < 1$) the oscillations of velocity in the z-direction increase with decreasing Eo, whereas for the sliding away regime ($Eo > 1$), the z-velocity magnitude is highest for the highest Eo. However, no particular trend can be fitted for the z-motion for any value of Eo and the order of magnitude of the velocity and displacement is negligible compared to those in the other two spatial directions.

5. Conclusions

In this work, VOF-based numerical simulations were performed to assess the impact of wall proximity and surface tension on the motion of a single air bubble rising in still water. For the parameters considered in this work, the bubble lies in the rectilinear regime. The following conclusions can be drawn from the obtained results:

- The presence of the wall near the bubble provides a significant perturbation to the flow structures in the fluid domain, which induces an early transition of the bubble trajectory from the rectilinear to the planar zigzagging regime. At $Ga = 99$, the critical Eo for this transition decreases from 1.5 to 0.14.
- A 1 mm diameter air bubble rising near a vertical wall in pure water ($Ga = 99$, $Eo = 0.14$) follows a bouncing trajectory as observed in previous experimental works. The bouncing motion is characterized by constant wavelength, amplitude and frequency. There is no significant motion of the bubble along the spanwise direction. The bouncing motion is thus two-dimensional.
- Due to the presence of the wall, the drag experienced by the bubble increases. The average rise velocity of the bubble rising near the wall is less compared to the unbounded bubble.
- As the bubble–wall initial distance increases, the bubble rising trajectory changes from bouncing (planar zigzagging) to straight rising. Moreover, as the distance from the wall increases, the maximum velocity and the average rise velocity also increases, i.e., the drag decreases.
- The bubble shows bouncing motion for $s < 1.5$. For the bouncing regime (i.e., $s < 1.5$), the change in wall proximity only changes the time of onset of bouncing. It does not impact the characteristics of the bouncing trajectory. The closer the bubble is to the wall, the earlier it triggers these path instabilities.
- For $Ga = 99$, $s = 1$, the bubble shows a bouncing motion for $Eo < 1$.

To have a complete picture of the effect of surface tension on the bubble bouncing behavior, the effect of wall proximity at different values of surface tension needs to be explored through more simulations in the future. Moreover, to further explore this phenomenon, the effect of bubble size on the bouncing motion should also be studied. Furthermore, well-planned experimental investigations should be conducted to supplement the computational study.

Author Contributions: Conceptualization, R.L., P.K.D., M.A.P. and P.D.L.; methodology, R.M., R.L. and P.K.D.; software, R.M.; validation, R.M.; formal analysis, R.M., R.L. and P.K.D.; investigation, R.M., R.L. and P.K.D.; resources, R.L. and P.K.D.; writing—original draft preparation, R.M.; writing—review and editing, R.L., P.K.D., M.A.P. and P.D.L.; supervision, R.L. and P.K.D.; project administration, R.L., P.K.D., M.A.P. and P.D.L.; funding acquisition, R.L., P.K.D., M.A.P. and P.D.L. All authors have read and agreed to the published version of the manuscript.

Funding: This research was funded by DST-RFBR through the project FLB.

Data Availability Statement: The data presented in this study can be made available on request.

Acknowledgments: The authors from India would like to acknowledge the support of National Supercomputing Mission (NSM) for providing HPC facility of PARAM Shakti at IIT Kharagpur, which is implemented by C-DAC and supported by the Ministry of Electronics and Information Technology (MeitY) and Department of Science and Technology (DST), Government of India. The co-authors from Russia are grateful to the state contract with IT SB RAS (Projects 121032200034-4 and 121031800217-8).

Conflicts of Interest: The authors declare no conflict of interest. The funders had no role in the design of the study; in the collection, analyses, or interpretation of data; in the writing of the manuscript; or in the decision to publish the results.

Abbreviations

The following abbreviations are used in this manuscript:

d	Bubble Diameter
Eo	Eotvos Number
g	Acceleration due to gravity
Ga	Galilei Number

s	Normalized bubble-wall initial distance
t	Time
vx	x-velocity of bubble centroid
vy	y-velocity of bubble centroid
vz	z-velocity of bubble centroid
VOF	Volume of fluid
x	x-position of bubble centroid
y	y-position of bubble centroid
z	z-position of bubble centroid
α	Phase fraction
ρ	Density
σ	Surface Tension Coefficient
μ	Dynamic Viscosity

References

1. Bhaga, D.; Weber, M.E. Bubbles in viscous liquids: Shapes, wakes and velocities. *J. Fluid Mech.* **1981**, *105*, 61. [CrossRef]
2. Duineveld, P.C. The rise velocity and shape of bubbles in pure water at high Reynolds number. *J. Fluid Mech.* **1995**, *292*, 325–332. [CrossRef]
3. Vries, A.W.D.; Biesheuvel, A.; Wijngaarden, L.V. Notes on the path and wake of a gas bubble rising in pure water. *Int. J. Multiph. Flow* **2003**, *28*, 1823–1835. [CrossRef]
4. Clift, R.; Grace, J.R.; Weber, M.E. *Bubbles, Drops, and Particles*, 2nd ed.; Dover Publications: Mineola, NY, USA, 2005.
5. Zenit, R.; Magnaudet, J. Path instability of rising spheroidal air bubbles: A shape-controlled process. *Phys. Fluids* **2008**, *20*, 061702. [CrossRef]
6. Sharaf, D.M.; Premlata, A.R.; Tripathi, M.K.; Karri, B.; Sahu, K.C. Shapes and paths of an air bubble rising in quiescent liquids. *Phys. Fluids* **2017**, *29*, 122104. [CrossRef]
7. Chen, L.; Garimella, S.V.; Reizes, J.A.; Leonardi, E. The development of a bubble rising in a viscous liquid. *J. Fluid Mech.* **1999**, *387*, 61–96. [CrossRef]
8. Mougin, G.; Magnaudet, J. Path Instability of a Rising Bubble. *Phys. Rev. Lett.* **2001**, *88*, 014502. [CrossRef]
9. Hua, J.; Lou, J. Numerical simulation of bubble rising in viscous liquid. *J. Comp. Phys.* **2007**, *222*, 769–795. [CrossRef]
10. Farhangi, M.M.; Passandideh-Fard, M.; Moin, H. Numerical study of bubble rise and interaction in a viscous liquid. *Int. J. Comp. Fluid Dyn.* **2010**, *24*, 13–28. [CrossRef]
11. Ghosh, S.; Das, A.K.; Vaidya, A.A.; Mishra, S.C.; Das, P.K. Numerical Study of Dynamics of Bubbles Using Lattice Boltzmann Method. *Ind. Eng. Chem. Res.* **2012**, *51*, 6364–6376. [CrossRef]
12. Tripathi, M.K.; Sahu, K.C.; Govindarajan, R. Dynamics of an initially spherical bubble rising in quiescent liquid. *Nat. Commun.* **2015**, *6*, 6268. [CrossRef] [PubMed]
13. Cano-Lozano, J.C.; Martínez-Bazán, C.; Magnaudet, J.; Tchoufag, J. Paths and wakes of de- formable nearly spheroidal rising bubbles close to the transition to path instability. *Phys. Rev. Fluids* **2016**, *1*, 053604. [CrossRef]
14. Gumulya, M.; Joshi, J.B.; Utikar, R.P.; Evans, G.M.; Pareek, V. Bubbles in viscous liquids: Time dependent behaviour and wake characteristics. *Chem. Eng. Sci.* **2016**, *144*, 298–309. [CrossRef]
15. Shew, W.L.; Pinton, J.F. Dynamical Model of Bubble Path Instability. *Phys. Rev. Lett.* **2006**, *97*, 144508. [CrossRef]
16. Ern, P.; Risso, F.; Fabre, D.; Magnaudet, J. Wake-Induced Oscillatory Paths of Bodies Freely Rising or Falling in Fluids. *Ann. Rev. Fluid Mech.* **2012**, *44*, 97–121. [CrossRef]
17. Zenit, R.; Magnaudet, J. Measurements of the streamwise vorticity in the wake of an oscillating bubble. *Int. J. Multiph. Flow* **2009**, *35*, 195–203. [CrossRef]
18. Magnaudet, J.; Eames, I. The Motion of High-Reynolds-Number Bubbles in Inhomogeneous Flows. *Annu. Rev. Fluid Mech.* **2000**, *32*, 659–708. [CrossRef]
19. Gumulya, M.; Utikar, R.; Evans, G.; Joshi, J.; Pareek, V. Interaction of bubbles rising inline in quiescent liquid. *Chem. Eng. Sci.* **2017**, *166*, 1–10. [CrossRef]
20. Senapati, A.; Singh, G.; Lakkaraju, R. Numerical simulations of an inline rising unequal- sized bubble pair in a liquid column. *Chem. Eng. Sci.* **2019**, *208*, 115159. [CrossRef]
21. Tsao, H.K.; Koch, D.L. Observations of high Reynolds number bubbles interacting with a rigid wall. *Phys. Fluids* **1997**, *9*, 44–56. [CrossRef]
22. Takemura, F.; Takagi, S.; Magnaudet, J.; Matsumoto, Y. Drag and lift forces on a bubble rising near a vertical wall in a viscous liquid. *J. Fluid Mech.* **2002**, *461*, 277–300. [CrossRef]
23. Takemura, F.; Magnaudet, J. The transverse force on clean and contaminated bubbles rising near a vertical wall at moderate Reynolds number. *J. Fluid Mech.* **2003**, *495*, 235–253. [CrossRef]
24. Podvin, B.; Khoja, S.; Moraga, F.; Attinger, D. Model and experimental visualizations of the interaction of a bubble with an inclined wall. *Chem. Eng. Sci.* **2008**, *63*, 1914–1928. [CrossRef]

25. Jeong, H.; Park, H. Near-wall rising behaviour of a deformable bubble at high Reynolds number. *J. Fluid Mech.* **2015**, *771*, 564–594. [CrossRef]
26. Lee, J.; Park, H. Wake structures behind an oscillating bubble rising close to a vertical wall. *Int. J. Multiph. Flow* **2017**, *91*, 225–242. [CrossRef]
27. Barbosa, C.; Legendre, D.; Zenit, R. Conditions for the sliding-bouncing transition for the interaction of a bubble with an inclined wall. *Phys. Rev. Fluids* **2016**, *1*, 032201. [CrossRef]
28. Sugioka, K.; Tsukada, T. Direct numerical simulations of drag and lift forces acting on a spherical bubble near a plane wall. *Int. J. Multiph. Flow* **2015**, *71*, 32–37. [CrossRef]
29. Zhang, Y.; Dabiri, S.; Chen, K.; You, Y. An initially spherical bubble rising near a vertical wall. *Int. J. Heat Fluid Flow* **2020**, *85*, 108649. [CrossRef]
30. Zhang, K.; Li, Y.; Chen, Q.; Lin, P. Numerical Study on the Rising Motion of Bubbles near the Wall. *Appl. Sci.* **2021**, *11*, 10918. [CrossRef]
31. Yan, H.; Zhang, H.; Liao, Y.; Zhang, H.; Zhou, P.; Liu, L. A single bubble rising in the vicinity of a vertical wall: A numerical study based on volume of fluid method. *Ocean Eng.* **2022**, *263*, 112379. [CrossRef]
32. Hasan, S.M.M.; Hasan, A.B.M.T. Migration dynamics of an initially spherical deformable bubble in the vicinity of a corner. *Phys. Fluids* **2022**, *34*, 112119. [CrossRef]
33. Khodadadi, S.; Samkhaniani, N.; Taleghani, M.H.; Bandpy, M.G.; Ganji, D.D. Numerical simulation of single bubble motion along inclined walls: A comprehensive map of outcomes. *Ocean Eng.* **2022**, *255*, 111478. [CrossRef]
34. Tagawa, Y.; Takagi, S.; Matsumoto, Y. Surfactant effect on path instability of a rising bubble. *J. Fluid Mech.* **2014**, *738*, 124–142. [CrossRef]
35. Mukundakrishnan, K.; Quan, S.; Eckmann, D.M.; Ayyaswamy, P.S. Numerical study of wall effects on buoyant gas-bubble rise in a liquid-filled finite cylinder. *Phys. Rev. E* **2007**, *76*, 036308. [CrossRef] [PubMed]
36. Krishna, R.; Urseanu, M.; van Baten, J.; Ellenberger, J. Wall effects on the rise of single gas bubbles in liquids. *Int. Comm. Heat Mass Transf.* **1999**, *26*, 781–790. [CrossRef]
37. Klostermann, J.; Schaake, K.; Schwarze, R. Numerical simulation of a single rising bubble by VOF with surface compression. *Int. J. Numer. Methods Fluids* **2013**, *71*, 960–982. [CrossRef]
38. Deshpande, S.S.; Anumolu, L.; Trujillo, M.F. Evaluating the performance of the two-phase flow solver interFoam. *Comp. Sci. Discov.* **2012**, *5*, 014016. [CrossRef]
39. Pradeep, A.; Sharma, A.K. Numerical investigation of single bubble dynamics in liquid sodium pool. *Sādhanā* **2019**, *44*, 56. [CrossRef]
40. Senthilkumar, S.; Delauré, Y.; Murray, D.; Donnelly, B. The effect of the VOF–CSF static contact angle boundary condition on the dynamics of sliding and bouncing ellipsoidal bubbles. *Int. J. Heat Fluid Flow* **2011**, *32*, 964–972. [CrossRef]
41. Albadawi, A.; Donoghue, D.; Robinson, A.; Murray, D.; Delauré, Y. On the assessment of a VOF based compressive interface capturing scheme for the analysis of bubble impact on and bounce from a flat horizontal surface. *Int. J. Multiph. Flow* **2014**, *65*, 82–97. [CrossRef]
42. Kulkarni, A.A.; Joshi, J.B. Bubble Formation and Bubble Rise Velocity in Gas–Liquid Systems: A Review. *Ind. Eng. Chem. Res.* **2005**, *44*, 5873–5931. [CrossRef]

 *water*

Article

An Impact of the Discrete Representation of the Bubble Size Distribution Function on the Flow Structure in a Bubble Column Reactor

Alexander Chernyshev [1,*], Alexander Schmidt [1] and Veronica Chernysheva [2]

1   Computational Physics Laboratory, Ioffe Institute, Politekhnicheskaya Str. 26, 194021 St. Petersburg, Russia
2   Department of Energy, Peter the Great St. Petersburg Polytechnic University, Politekhnicheskaya Str. 29, 195251 St. Petersburg, Russia
*   Correspondence: alexander.tchernyshev@mail.ioffe.ru

Abstract: The purpose of the present study is to analyze the effect of different discrete representations of the continuous bubble size distribution function on the flow structure in a bubble column reactor. Poly- and monodisperse media were considered, such that the mathematical expectation of the bubble size in the polydisperse case was equal to the bubble size in the monodisperse case at the same volumetric bubble contents. For these computations the normalized variances of the velocity profiles of the carrier and the disperse phases, the volume fraction of the disperse phase, and the specific area of the interfacial surface were determined. The normalized variances were calculated from a reference scenario with a detailed resolution of the bubble size distribution function with ten bubble classes. It was shown that with increase of the average bubble sizes mono- and polydisperse approaches provide converging solutions. A modified hybrid discretization of the bubble size distribution function with four classes of bubbles was shown to predict the flow structure with normalized variance less than 5% over the entire computational domain for all monitored parameters.

Keywords: bubble column reactor; bubbly flow; polydisperse media; numerical simulation; Euler–Euler approach; discrete bubble size distribution function

 **check for updates**

Citation: Chernyshev, A.; Schmidt, A.; Chernysheva, V. An Impact of the Discrete Representation of the Bubble Size Distribution Function on the Flow Structure in a Bubble Column Reactor. *Water* **2023**, *15*, 778. https://doi.org/10.3390/w15040778

Academic Editor: Arthur Mynett

Received: 27 January 2023
Revised: 13 February 2023
Accepted: 14 February 2023
Published: 16 February 2023

1. Introduction

Ascending bubble flows with the gravity force as the main driver are a common feature of many natural and technological processes. Typical industrial applications of such flows are bubble column reactors (BCRs). These apparatus are used in many areas of human activity, such as water treatment and purification [1], greenhouse gas elimination [2], hydrogen and biofuel synthesis [3,4], or production of biochemical substances [5]. All of the above shows that studies of the processes occurring in the BCRs are of great importance.

As a rule, the flows in the BCR are polydisperse; this circumstance plays an important role in the formation of both the global structure of the flows [6,7] and the local properties of the working media of reactors [4,8].

The description of the bubble size distribution (BSD) in the framework of the polydisperse Euler–Euler approach can be implemented in various ways. The simplest one is similar to the monodisperse description, but the bubbles have the same size only locally; herewith they can vary in space. The local size of the bubbles is taken to be equal to the Sauter mean diameter [9], which ensures that the total bubble surface is preserved, and is important in problems with intensive interphase mass transfer. At low relative velocities this approach is sufficient. However, at phase velocity nonequilibrium, averaging the bubble diameter results in errors in the calculation of interphase forces, which in turn affects the overall flow pattern.

A detailed description of polydispersity within the Euler–Euler approach can be provided by the Population Balance Model (PBM) [4,10] with a population balance equation.

This model makes it possible to describe the change in the spectrum of bubble sizes at each spatial point but the computational algorithm becomes noticeably more complicated because of the integral–differential nature of the equation.

Another approach which is a subclass of the PBM is based on the introduction of a set of classes of monodisperse bubbles that differ in size (the MUltiple SIze Group or MUSIG model) [11]. The original model has the same velocity for each bubble class and is called homogeneous MUSIG, while its variant with separate velocity for each class is called the inhomogeneous or heterogeneous MUSIG model [12]. A fixed number of classes and ranges of bubble sizes in the classes are chosen to include all possible bubbles. Sets of the conservation equations are written for each class. Such an approach provides a unified description of the working medium. However, this approach is also quite resource-demanding due to the need to solve additional sets of nonlinear equations.

Due to the significant increase in computational resources required, it seems reasonable to assess the influence of the parameters of the discrete representation of the (BSD) function on the flow structure for the regimes where polydispersity is the determining factor.

There are a number of papers dealing with the discretization parameters of the BSD. An unsteady simulation of the BCR was performed in [13] with only two classes for the 13 mm/s surface velocity case, while a monodisperse approach was sufficient for the 3 mm/s surface velocity. The two classes for the 13 mm/s case were chosen to resolve a change in the sign of the lateral lift force at 6 mm bubble diameter. There are other investigations [14,15] which show that for gas surface velocities below 5 mm/s the monodisperse approach is sufficient.

Numerical simulations of the BSD have been performed in [7,16]. With 17 classes of bubbles it was shown [7] that good prediction can be obtained for most of the bubble size range [0.5 mm, 16.5 mm], while the probability of small bubbles was underestimated by the model. Reducing the number of classes [16] from 17 to 14 provides better correlation with experimental data.

A detailed investigation of the effect of the number of bubble classes was carried out in [17]. It was shown that seven classes of bubbles were sufficient for the detailed description of the flow in the BCR in terms of liquid velocity, bubble velocity and bubble volume fraction. However, for the prediction of the interfacial area, 15 classes were selected in the computations. In the subsequent study [18] it was shown that the heterogeneous MUSIG outperformed the original homogeneous version. The same results were obtained in [19] for the simplified heterogeneous MUSIG model with algebraic equations for bubble slip velocities with 13 classes of bubbles for the bubble size range [1 mm, 25 mm].

It should be noted that all of these studies deal with the uniform BSD and it is not clear how the bubble swarm behaves with bubble sizes ranging from fractions of a millimeter to a few millimeters.

The paper presents the results of a numerical study of the dynamics of poly- and monodisperse flows in a column-type bubble reactor. The mathematical expectation of the bubble size for a polydisperse medium was assumed to be equal to the bubble size in a monodisperse medium. Bubbles with characteristic sizes in the range $d_b = 0.25$–1.0 mm are considered; the volume content of the dispersed phase at the inlet boundary was the same in all cases and was equal to 0.0037. Different discretization parameters of the BSD function were chosen to estimate their influence on the flow structure in the BCR.

According to the results of the numerical simulation, the fields of phase velocities, bubble volume fraction and the interfacial area, which influence the intensity of interphase processes, were analyzed. On the basis of the obtained data, an optimal number of classes, as well as the discretization parameters of the BSD function, were chosen in order to obtain a consistent solution in the whole range of the input parameters considered.

2. Mathematical Model

2.1. Basic Assumptions of the Model

A mathematical model of the flow of a polydisperse bubble medium in a bubble column reactor is presented. The model is based on the Euler–Euler approach to the description of multiphase flows (see, for example, [20]). Within the framework of this approach, each phase is considered as a continuum occupying the entire volume, at each point of which the volume fraction of the phase α is specified. In this case, the densities of each of the phases are calculated as $\alpha \cdot \rho_0$, where ρ_0 is the density of the substance of the corresponding phase. In addition to the description of the polydispersity and the interaction of the bubbles (subscript b) and the carrier phase (subscript l), attention is paid to the analysis of the effect of turbulence on the two-phase flow.

In this study, flows are characterized by values of the dispersed phase volume fraction less than 0.01, which allows bubble–bubble interactions to be to neglected.

2.2. Description of Polydispersity of Bubbly Medium

The description of the polydispersity of the working medium is carried out in the framework of the multiclass heterogeneous model of the BSD, which takes into account the different speeds of each bubble class (the heterogeneous MUSIG model, [12]). The BSD is approximated by a piecewise constant function describing N bubble classes. Each of the classes is characterized by the constant bubble diameter d_{ib}, the volume fraction α_{ib} and the bubble number density n_{ib}, where i is the class number. These quantities are related to each other as $d_{ib} = ((6\alpha_{ib})/(\pi n_{ib}))^{1/3}$. Then the volume fraction of the liquid can be represented as follows:

$$\alpha_l = 1 - \sum_{i=1}^{N} \alpha_{ib}. \tag{1}$$

The initial BSD corresponds to the experimental data and can be expressed by the log-normal distribution for the bubble number density [21]:

$$F(n_b) = \frac{1}{d_b \sigma \sqrt{2\pi}} \cdot exp\left(\frac{-[ln(d_b) - \mu]^2}{2\sigma^2}\right), \tag{2}$$

where d_b is the bubble diameter, and μ and σ are the mathematical expectation and standard deviation with respect to $ln(d_b)$. The mathematical expectation M and the standard deviation D with respect to d_b are determined by the formulas:

$$M = exp(\mu + 0.5 \cdot \sigma^2), \; D = M \cdot \sqrt{[exp(\sigma^2) - 1]}. \tag{3}$$

In Equation (3), the value M corresponds to the most probable bubble for a given distribution and will be referred to below as the characteristic bubble size. The monodisperse case is obtained by integrating over the entire interval of bubble sizes, producing a single class of bubbles with a size equal to the most probable bubble size. In this sense, the monodisperse approach can be treated as a polydisperse approach with only one class of bubbles.

The equations of conservation of mass, momentum and number density of the bubbles are formulated for each class.

2.3. Governing Equations

In the framework of the Euler–Euler approach the Navier–Stokes equations, accounting for the N bubble classes, can be written as follows:

$$\frac{\partial \alpha_{ib}\rho_{ib}^0}{\partial t} + div\left(\alpha_{ib}\rho_{ib}^0 \mathbf{V}_{ib}\right) = D_{i\alpha},$$

$$\frac{\partial \alpha_{ib}\rho_{ib}^0 \mathbf{V}_{ib}}{\partial t} + div\left(\alpha_{ib}\rho_{ib}^0 \mathbf{V}_{ib}\mathbf{V}_{ib} + p\mathbf{E}\right) = S_{ib},$$

$$\frac{\partial \alpha_l\rho_l^0}{\partial t} + div\left(\alpha_l\rho_l^0 \mathbf{V}_l\right) = 0,$$

$$\frac{\partial \alpha_l\rho_l^0 \mathbf{V}_l}{\partial t} + div\left(\alpha_l\rho_l^0 \mathbf{V}_l\mathbf{V}_l + p\mathbf{E} - \boldsymbol{\tau}\right) = S_l.$$

$$(4)$$

The following notation is used: $\mathbf{V}$ is the velocity vector, p is the pressure, $\mathbf{E}$ is the unity tensor, S is the source term responsible for the interphase momentum transport, D is the source term responsible for the bubble dispersion due to turbulence and $\boldsymbol{\tau}$ is the strain rate tensor with its components expressed as:

$$\tau_{mn} = 2 \cdot \mu_{eff} \cdot S_{mn}, \quad S_{mn} = \frac{1}{2}\left(\frac{\partial v_m}{\partial x_n} + \frac{\partial v_n}{\partial x_m}\right), \quad \mu_{eff} = \mu_{lam} + \mu_{turb} + \mu_{BIT}, \quad (5)$$

where μ_{eff} is the effective dynamic viscosity, equal to sum of the dynamic viscosity of the medium μ_{lam}, the turbulent viscosity μ_{turb} and an additional member, responsible for turbulence modification due to presence of bubbles μ_{BIT}.

The density of the gas inside the bubbles ρ_{ib}^0 depends on both the external pressure of the carrier medium and on the pressure created by the surface tension, and can be expressed as follows:

$$\rho_{ib}^0 = \frac{p + 2\Sigma/R_{ib}}{\Re_b T_0}, \quad (6)$$

where Σ is the surface tension (assumed to be equal to 0.072 H/m), R_{ib} is the bubble radius for the ith class, T_0 is the ambient temperature and $\Re_b$ is the specific gas constant.

The set of Equation (4) is supplemented by the conservation equation for the bubble number density for each class i:

$$\frac{\partial n_{ib}}{\partial t} + div(n_{ib}\mathbf{V}_{ib}) = D_{in}. \quad (7)$$

2.4. Interphase Momentum Exchange

The details of the interphase momentum exchange processes are discussed below. Corresponding source terms in the governing equations include: the buoyancy force, the drag (Stokes) force, the lift (Saffman) force, the virtual mass force and the wall lubrication force:

$$S_{ib} = \mathbf{F}_{iB} + \mathbf{F}_{iD} + \mathbf{F}_{iL} + \mathbf{F}_{iWL} + \mathbf{F}_{iVM}, \quad S_l = -\sum_{i=1}^{N}(\mathbf{F}_{iD} + \mathbf{F}_{iL} + \mathbf{F}_{iWL} + \mathbf{F}_{iVM}). \quad (8)$$

2.4.1. Buoyancy Force

The buoyancy force is the main driver of flow development in the column and is expressed as the difference between the Archimedean and gravitational forces:

$$\mathbf{F}_{iB} = \alpha_{ib} \cdot (\rho_{ib} - \rho_l) \cdot \mathbf{g}. \quad (9)$$

2.4.2. Drag Force

The expression for the Stokes drag force acting on bubbles of class i can be written in the following form:

$$\mathbf{F}_{iD} = \frac{3}{8} \cdot \frac{\rho_l}{R_{ib}} \alpha_{ib} C_{iD} \mathbf{V}_{irel} |\mathbf{V}_{irel}|, \quad \mathbf{V}_{irel} = \mathbf{V}_l - \mathbf{V}_{ib}. \tag{10}$$

In the general case, when the Reynolds number of the relative motion of the phases $\mathrm{Re}_{ip} = (\rho_l V_{irel} d_{ib})/\mu_{lam}$ cannot be considered small, it is necessary to introduce a correction function f_{ip} for the Stokes drag coefficient, which takes into account the inertia of the flow around the bubbles, their possible asymmetry, flows in the bubbles, etc. Thus, the drag coefficient can be written as $C_{iD} = 24 \cdot f_{ip}/\mathrm{Re}_{ip}$. During the calculations, several options were considered to determine the function f_{ip}. In Ref. [22], the following expression for the correction function is proposed, based on experimental data:

$$f_{ip} = \begin{cases} 2/3, & \mathrm{Re}_{ip} < 1.5 \\ 0.6208\,\mathrm{Re}_{ip}^{0.22}, & 1.5 < \mathrm{Re}_{ip} < 80 \\ 2 - 4.41\,\mathrm{Re}_{ip}^{-0.5} + 7.75 \cdot 10^{-17}\,\mathrm{Re}_{ip}^{5.756}, & 80 < \mathrm{Re}_{ip} < 1500 \\ 0.10875\,\mathrm{Re}_{ip}, & \mathrm{Re}_{ip} > 1500 \end{cases} \tag{11}$$

This expression describes the dynamics of small bubbles, but does not take into account the possible deformation of the interfaces and the loss of sphericity.

The shape of the bubble may be different from spherical if the pressure changes significantly on scales of bubble size. Assuming that the main reason for the pressure variation is gravity, then the effect of the pressure gradient is determined by the Eötvös number $\mathrm{Eo}_i = g(\rho_l - \rho_{ib})d_{ib}^2/\Sigma$, which is the ratio of the hydrostatic pressure of a liquid column whose height is equal to the size of the bubble to the pressure produced by the surface tension force. At $\mathrm{Eo} < 1$ the bubble remains spherical, at $\mathrm{Eo} > 1$ the bubble loses its sphericity, acquiring at the initial stage the shape of an ellipsoid. In Ref. [23], a correlation was proposed for the drag coefficient C_{iD} based on the Reynolds numbers Re_{ip} and Eo_i:

$$C_{iD} = \sqrt{C_D(\mathrm{Re}_{ip})^2 + C_D(\mathrm{Eo}_i)^2}$$

$$C_D(\mathrm{Re}_{ip}) = \frac{16}{\mathrm{Re}_{ip}} \left(1 + \frac{2}{1 + 16/\mathrm{Re}_{ip} + 3.315/\sqrt{\mathrm{Re}_{ip}}} \right) \tag{12}$$

$$C_D(\mathrm{Eo}_i) = 4 \cdot \mathrm{Eo}_i/(\mathrm{Eo}_i + 9.5), \quad \mathrm{Eo}_i < 5$$

In this study, expression (12) for the drag force was used.

2.4.3. Lift Force

The Saffman force occurs when particles move in a stream with significant velocity gradients. The resulting force is normal to the particle relative velocity vector. Although the value of this force is usually much smaller than the Stokes force, it can have a noticeable effect on the structure of the flow and the distribution of bubbles.

The expression for the Saffman force was initially obtained under a number of assumptions, in particular, for the case of low Reynolds numbers of relative motion (Stokes flow), which are not applicable to the problems considered.

For the range of flow parameters of practical interest, the formula proposed in [24] was chosen. Here the expression for the Saffman force can be written as:

$$\mathbf{F}_{iL} = C_{iL}\alpha_{ib}\rho_l \mathbf{V}_{irel} \times rot\mathbf{V}_l. \tag{13}$$

The coefficient C_{iL} is often taken to be constant, ranged in 0.1–0.5, depending on the type of problem.

Based on a large set of experimental data on the ascent of a single bubble in a liquid layer with a transverse linear velocity gradient, the following expression for the coefficient C_{iL} was proposed in [25]:

$$C_{iL} = min\left[0.288\,tanh(0.121\,\mathrm{Re}_{ip}),\ f(\mathrm{Eo}_i)\right],\ \mathrm{Eo}_i < 4,$$
$$f(\mathrm{Eo}_i) = 0.00105\,\mathrm{Eo}_i^3 - 0.0159\,\mathrm{Eo}_i^2 - 0.0204\,\mathrm{Eo}_i + 0.474. \tag{14}$$

When calculating the Saffman transverse force, expression (14) was used in this study.

2.4.4. Virtual Mass Force

It is well known that the effect of added masses in bubble flows plays an important role. This effect is associated with an increase of the bubble inertia due to the inclusion of the ambient liquid in relative motion. The expression for the added mass force can be written as:

$$\mathbf{F}_{iVM} = 0.5\alpha_{ib}\rho_l\left(\frac{D_b\mathbf{V}_{ib}}{Dt} - \frac{D_l\mathbf{V}_l}{Dt}\right). \tag{15}$$

2.4.5. Wall Lubrication Force

Near the channel walls, the flow is characterized by a significant gradient of the velocity and asymmetry in the flow around a separate bubble, resulting in repulsion of the bubble from the wall. This effect can be important in the formation of the bubble flow structure and can lead to flow stratification. The expression for the wall force can be written as follows [12]:

$$\mathbf{F}_{iWL} = -C_{iWL}\alpha_{ib}\rho_l[\mathbf{V}_{irel} - (\mathbf{V}_{ib}\cdot\mathbf{n}_W)\cdot\mathbf{n}_W]^2\cdot\mathbf{n}_W. \tag{16}$$

Here $\mathbf{n}_W$ is the normal to the nearest wall.
Coefficient C_{iWL} is calculated using the following formula:

$$- C_{iWL} = 0.47\cdot max\left\{0,\ \frac{1}{6.3}\cdot\frac{[1 - y_W/(10\,d_{ib})]}{y_W\cdot[y_W/(10\,d_{ib})]^{0.7}}\right\}. \tag{17}$$

Here y_W is the distance to the nearest wall.

2.5. Turbulence

Typical flows in the BCR are turbulent. The presence of a dispersed phase can lead to both the generation and dissipation of turbulence, depending on the flow parameters. An analysis of the influence of these processes on the flow structure in the BCR is of a great interest. In this paper, the $k - \omega$ SST turbulence model [26] is implemented with additional source terms describing the generation of turbulence due to the relative motion of the bubbles and the carrier phase:

$$\frac{\partial\rho_l k}{\partial t} + div(\rho_l\mathbf{V}_l k) = \tilde{P} - \beta^*\rho_l\omega k + div[(\mu_{lam} + \sigma_k\mu_{turb})grad(k)] + S_{bk},$$
$$\frac{\partial\rho_l\omega}{\partial t} + div(\rho_l\mathbf{V}_l\omega) = \alpha_{k\omega}\frac{\rho_l}{\mu_{turb}}\tilde{P} - \beta\rho_l\omega^2 + div[(\mu_{lam} + \sigma_\omega\mu_{turb})grad(\omega)] +$$
$$2(1 - F_1)\rho_l\sigma_{\omega2}\cdot\frac{1}{\omega}\frac{\partial k}{\partial x_j}\frac{\partial\omega}{\partial x_j} + S_{b\omega}, \tag{18}$$
$$\mu_{turb} = \frac{\rho_l a_1 k}{max(a_1\omega, \Omega F_2)}.$$

Here k is the turbulence kinetic energy and ω is the specific dissipation rate. All constants and closing relations are taken from [26].

2.5.1. Turbulence Generation and Dissipation by Bubbles

The modified model of Troshko and Hassan [27], originally developed for the $k - \epsilon$ turbulence model, is used here to account for the effect of the generation of turbulence by the bubbles. Corresponding source terms in the equations for k and ω can be written as:

$$S_{bk} = \frac{3}{8}\frac{C_{iD}}{R_{ib}}\alpha_{ib}\rho_l|\mathbf{V}_{irel}|^3, \quad S_{b\omega} = 0.8\frac{k}{\mu_{turb}}S_{bk},\tag{19}$$

An additional term responsible for the effect of turbulence generation in the expression for the effective viscosity (5) is calculated based on Sato model [28]:

$$\mu_{BIT} = 1.2 \cdot \alpha_{ib}\rho_l R_{ib}|\mathbf{V}_{irel}|.\tag{20}$$

2.5.2. Bubble Path Dispersion

The dispersion of bubbles due to turbulent velocity fluctuations in the carrier phase is accounted for by introducing an additional diffusion term in the equations for the volume fraction of bubbles and the bubble number density [29]:

$$D_{i\alpha} = \frac{1}{Sc}div\left(\frac{\mu_{eff}}{\rho_l}grad(\alpha_{ib})\right), \quad D_{in} = \frac{1}{Sc}div\left(\frac{\mu_{eff}}{\rho_l}grad(n_{ib})\right).\tag{21}$$

Here Sc is the Schmidt number with $Sc = 0.83$; the value is based on the comparison with experimental data of other authors [29,30].

Thus, the set of the conservation equations and closing relations for flows in the BCR are formulated.

2.6. Numerical Method

An algorithm and a computer code are developed for the numerical implementation of the model described above.

The algorithm is based on a common approach using the finite volume method and unstructured grids. Special attention is paid to the discretization of the convection and diffusion terms, both of which are discretized with second-order accuracy. The convection term is discretized using a second-order upwind discretization that satisfies the total variation diminishing (TVD) criterion. Three types of TVD limiters were tested: minmod, vanLeer and superbee. The minmod limiter was used in the computations due to the high stability of the resulting computational scheme.

The source terms resulting from the interphase interactions usually have nonlinear components which can affect the stability of the solution. To overcome this effect, a linearization of implicitly calculated source terms was introduced, especially for the drag force.

The coupling of the pressure and velocity fields was performed by the modified SIMPLE (Phase Coupled-SIMPLE) algorithm, taking into account the effects of multiphase. The PC-SIMPLE includes not only the major liquid phase fluxes in the calculation of pressure corrections but also the minor fluxes of the bubble phase. Rhie–Chow interpolation is used to avoid non-physical oscillations when updating the pressure and velocity fields.

The resulting discretized equations are solved by iterative methods such as Gauss–Zeidel with successive over-relaxations. The pressure correction equation is solved by the preconditioned conjugate gradient method with incomplete zero-filling Cholesky decomposition as preconditioner.

A detailed description of the basis of the numerical method can be found in [31].

The developed mathematical model, numerical methods and computer code have been extensively tested on the flows under study [20,32,33].

3. Problem Statement

3.1. Simulation Domain and Conditions

Since the main goal of this study is to analyze the effects of the description of the polydispersity of the working medium of the BCR, a series of calculations were carried out using various parameters of the piecewise constant representation of the BSD. Different numbers of classes N and different characteristic sizes of bubbles M were considered at their constant flow rate. Flows in axisymmetric BCRs due to Archimedes' force are considered using a simplified axisymmetric formulation. In Refs. [34,35] it was shown that such an approach allows the main features of the flows under study to be captured.

The analysis is performed for flows in a vertical cylindrical BCR with a coaxial aerator at the bottom and the top open.

Figure 1 shows a schematic diagram of an axisymmetric BCR; reactor diameter is $D = 0.07$ m and reactor height is $H = 0.65$ m. The carrier phase is water. The disperse phase, in the form of polydisperse gas bubbles, enters the column through a coaxial axisymmetric bottom mounted aerator ($d = 0.05$ m).

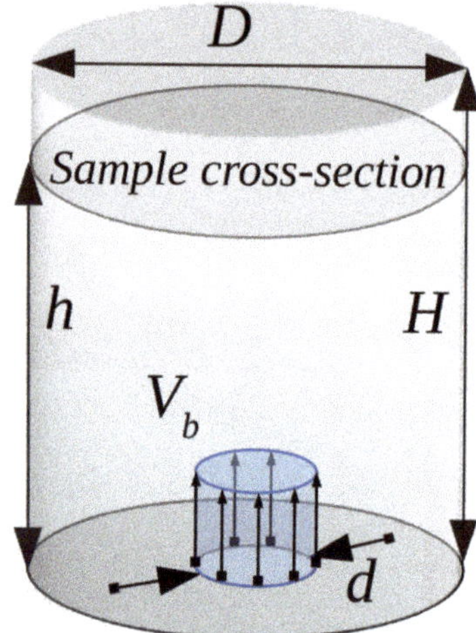

Figure 1. Schematics of a bubble column reactor.

A cross-section at a height of $h = 0.48$ m, depicted in the scheme, is used as a reference cross-section for the presentation of flow parameters.

Normal conditions for ambient gas at the free surface are assumed.

Simulations were carried out on a mesh of 50 cells along the radial coordinate and 90 cells along the axial coordinate, with the mesh denser toward the solid boundaries to provide detailed boundary layer resolution to meet both multiphase and turbulence requirements. Independent mesh sensitivity studies were performed and the presented mesh was selected as adequate. The time step Δt was chosen equal to 10^{-2} s with a normalized convergence criterion of 10^{-6} for each independent variable.

3.2. A Discrete Approximation of the BSD Function

The initial BSD can be described by the lognormal function (2). To determine the most efficient variant of the piecewise constant approximation of this function, three sets of approximations are used, denoted by **R1**, **R2** and **R3**. The entire interval of bubble size variation is replaced by the range $[d_{min}, d_{max}]$, where $d_{min} = exp(\mu - 3\sigma)$, $d_{max} = exp(\mu + 3\sigma)$, which corresponds to 99.7% of all bubbles. The **R1** distribution approximation is uniform with respect to the variable d_b in the range $[d_{min}, d_{max}]$; the entire range is divided into N classes, each class having the same width with respect to d_b. The **R2** distribution approximation is uniform with respect to the variable $ln(d_b)$; the range $[ln(d_{min}), ln(d_{max})]$ is divided into N classes, the width of each class with respect to $ln(d_b)$ is the same. The **R3** approximation is a hybrid **R2–R1**: the first $N/2$ classes are uniformly distributed with respect to the variable $ln(d_b)$ in the range $[ln(d_{min}), \mu]$; the subsequent $N/2$ classes are uniformly distributed with respect to the variable d_b in the range $[exp(\mu), d_{max}]$.

The distribution approximation of type **R1** is used for validation purposes. A set of different numbers of classes is used with $N = [1 \dots 10]$, covering a range from a monodisperse approach to a detailed polydisperse one. The validation is performed for the characteristic bubble size $M = 0.5$ mm.

The lognormal distributions under consideration correspond to four characteristic bubble sizes M: 0.25 mm, 0.375 mm, 0.5 mm and 1.0 mm. Five combinations of the number of classes and distribution approximation options are introduced: ($N = 1$, **R1**), which corresponds to monodisperse bubbles, ($N = 4$, **R3**), which corresponds to a compact economic distribution, ($N = 7$, **R1**) and ($N = 7$, **R2**), which allow a qualitative description of the polydisperse medium, and ($N = 10$, **R3**), which provides a detailed description of the polydisperse bubble flow.

The discretized BSDs for the characteristic bubble size $d_b = 0.25$ mm are shown in Figure 2, with the exception of the monodisperse one, since it is elementary. It is worth noting some features of these distributions.

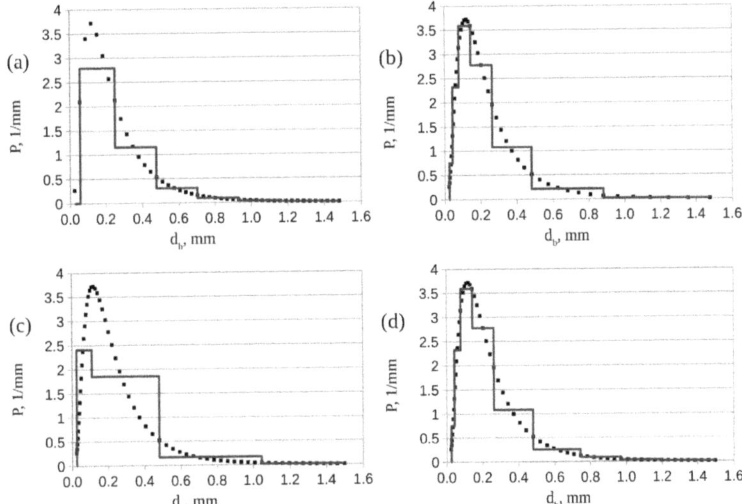

Figure 2. BSD function and piecewise approximations. The characteristic bubble size is $d_b = 0.25$ mm. (**a**) $N = 7$, **R1**; (**b**) $N = 7$, **R2**; (**c**) $N = 4$, **R3**; (**d**) $N = 10$, **R3**.

It can be seen that the distribution ($N = 7$, **R1**) does not describe the features of the distribution of small bubbles, providing for this domain only one class for bubbles from 0 to 0.25 mm. The distribution ($N = 7$, **R2**) resolves small bubbles at the expense of details for large bubbles. The distribution ($N = 4$, **R3**), despite some roughness, singles out a class for small and medium bubbles. The distribution ($N = 10$, **R3**) is the most detailed, describing both small and large bubbles well over the entire range.

4. Results and Discussion

4.1. Algorithm Verification

To assess the convergence of the algorithm by the number of classes N, the following parameters are used: the interfacial area S_{int}, the bubble volume fraction α_b, the bubble velocity V_b and the liquid velocity V_l. The bubble volume fraction, interfacial area and bubble velocity are calculated as follows:

$$\alpha_b = \sum_{i=1}^{N} \alpha_{ib},$$

$$S_{int} = \sum_{i=1}^{N} n_{ib} \cdot \pi d_{ib}^2, \tag{22}$$

$$V_b = \sum_{i=1}^{N} (\alpha_{ib} V_{ib}) / \alpha_b.$$

The overall averaged values are used to monitor the convergence to a stable solution as N increases, as well as the deviation of solutions with number of classes N and N-1.

Figure 3 shows the averaged values of the variables mentioned above and the normalized deviation of the variable values compared to the solution obtained for N-1 bubble classes as a function of the number of classes. Due to the averaging procedure, these values represent the entire flow behavior. Figure 3 shows the averaged values of the above variables as a function of the number of classes and the normalized deviation of the variable values compared to the solution obtained for N-1 bubble classes. Due to the averaging procedure, the values represent the entire flow behavior.

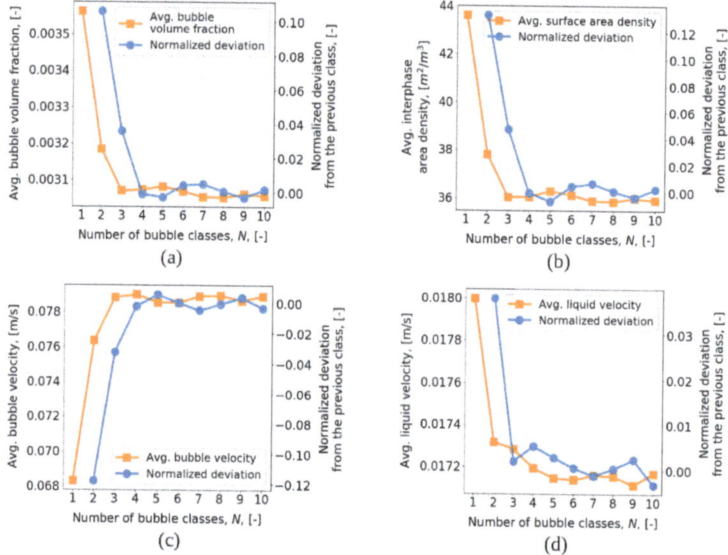

Figure 3. Distributions of flow parameters as a function of the number of bubble classes. The characteristic bubble size d_b = 0.25 mm; (**a**) the bubble volume fraction; (**b**) the interfacial area density; (**c**) the bubble velocity; (**d**) the liquid velocity.

It can be seen that the monodisperse approach overpredicts the total gas holdup within a column as well as the interfacial density. With increasing N the solution tends to converge to the polydisperse solution, with normalized deviation tending to the vicinity of zero at $N \geq 5$ for the bubble volume fraction, the interfacial area density and the bubble velocity. The differences between the variants are less than 1% at $N = 4$, indicating that sufficient detail is captured to estimate the integral flow parameters. However, the fluid velocity, despite having the lowest normalized deviation, also has the lowest rate of deviation decay, approaching 0 at $N > 6$. Thus, a converged solution in terms of N can be obtained at $N = 7$.

4.2. An Impact of Characteristic Bubble Size and Parameters of Approximation of the BSD Function on the Flow Structure

Some results of the BCR flow simulation are presented below. The radial profiles of the bubble volume fraction α_b (Figure 4), the interfacial surface area S_{int} (Figure 5), the bubble velocity V_b (Figure 6) and the carrier phase velocity V_l (Figure 7) are shown.

The data correspond to the reference cross-section (see Figure 1) for three characteristic bubble sizes (0.25 mm, 0.5 mm and 1 mm) and four variants of approximations: ($N = 1$, **R1**), ($N = 4$, **R3**), ($N = 7$, **R1**) and ($N = 10$, **R3**).

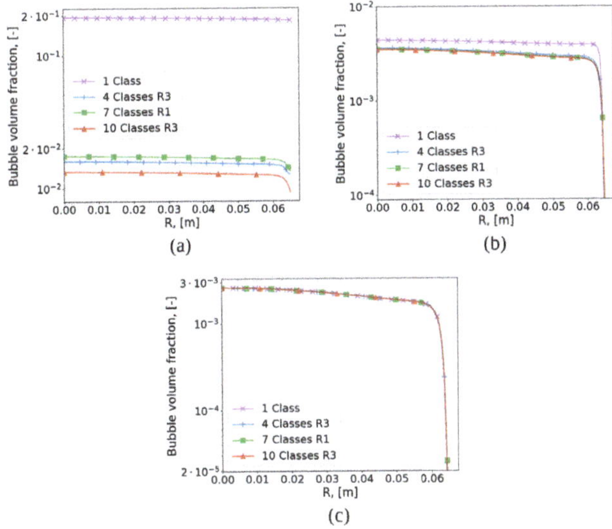

Figure 4. Radial profiles of the bubble volume fraction in the reference cross-section (see Figure 1) per different characteristic bubble sizes: (**a**) $M = 0.25$ mm; (**b**) $M = 0.5$ mm; (**c**) $M = 1.0$ mm.

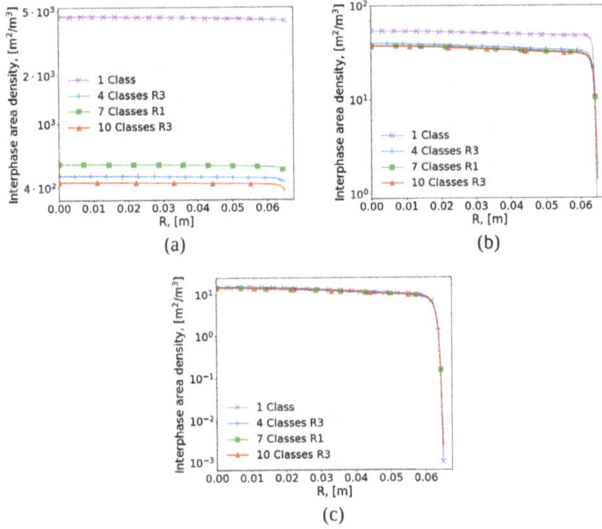

Figure 5. Radial profiles of the interfacial area density in the reference cross-section (see Figure 1) for different characteristic bubble sizes: (**a**) $M = 0.25$ mm; (**b**) $M = 0.5$ mm; (**c**) $M = 1.0$ mm.

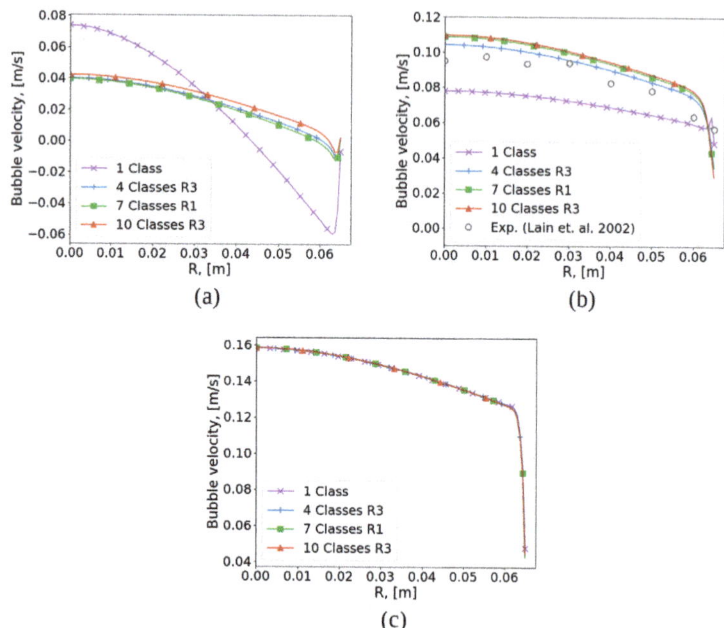

Figure 6. Radial profiles of bubble velocity in the reference cross-section (see Figure 1) for different characteristic bubble sizes: (**a**) $M = 0.25$ mm; (**b**) $M = 0.5$ mm; (**c**) $M = 1.0$ mm.

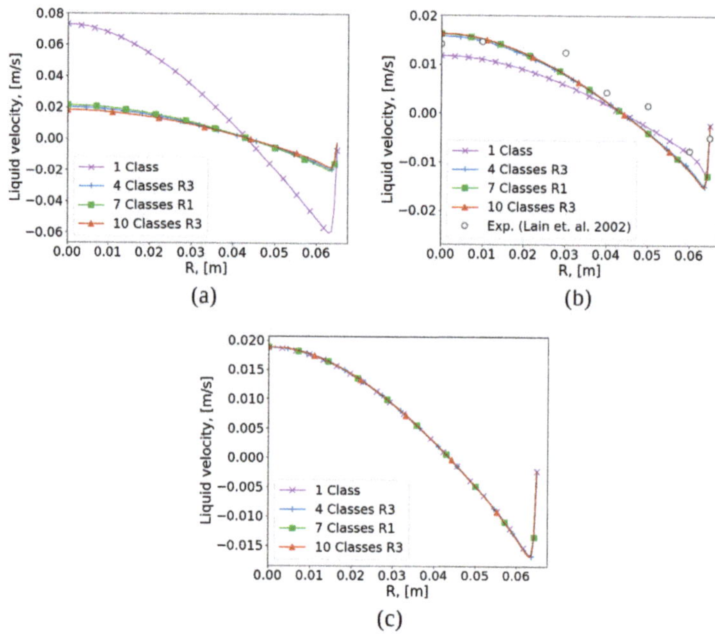

Figure 7. Radial profiles of liquid velocity in the reference cross-section (see Figure 1) for different characteristic bubble sizes: (**a**) $M = 0.25$ mm; (**b**) $M = 0.5$ mm; (**c**) $M = 1.0$ mm.

It can be seen that for small bubbles (d_b = 0.25 mm) there is a significant difference between the monodisperse and polydisperse approaches in all studied parameters. As the bubble size increases (d_b = 1.00 mm), the profiles of the volume fraction and the velocity of the carrier phase coincide for the two approaches. It should also be noted that, although the variant (N = 4, **R3**) has fewer bubble classes than (N = 7, **R1**), it gives a solution close to (N = 10, **R3**) for the volume fraction and the interfacial area, in contrast to (N = 7, **R1**). A comparison with the experimental data [22] for d_b = 0.5 mm shows that the option (N = 4, **R3**) is preferable.

The main emphasis is placed on the comparison of the obtained solutions with the "reference" solution for the case (N = 10, **R3**) as the most detailed representation of the BSD function. The comparison was carried out using the normalized variance of the obtained solution and the "reference" solution in the entire domain for a set of characteristic values of the polydisperse flow, such as the velocities of the carrier and dispersed phases, the volume fraction of the bubbles, and the interfacial area. The normalized variance was estimated according to the formula:

$$\sigma_{norm}(\phi) = \sqrt{\left[\sum_{i=1}^{N_{cells}} (\phi_i - \phi_i^*)^2 / N_{cells}\right]} / max(|\phi^*|), \tag{23}$$

where ϕ is the comparable variable for the solution under study and ϕ^* is the same variable in the same cell for the "reference" solution.

Below are plots of the weighted variances, σ_{norm}, as a function of case number and bubble diameter, d_b. For the case numbers, 1 stands for (N = 1, **R1**), 2 stands for (N = 4, **R3**), 3 stands for (N = 7, **R1**) and 4 stands for (N = 7, **R2**) (Figures 8–11).

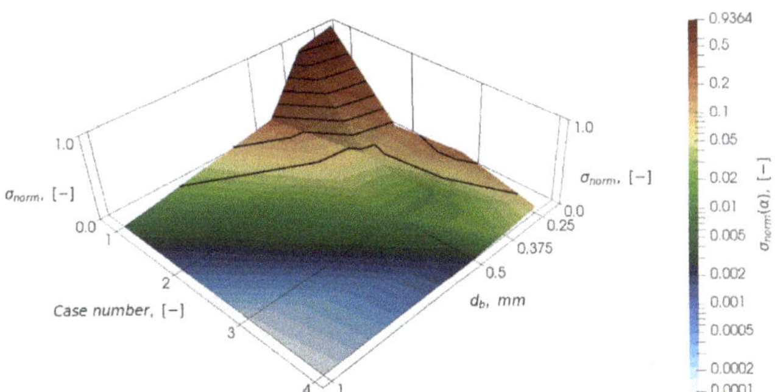

Figure 8. Distribution of a weighted variance σ_{norm} for the bubble volume fraction, α_b.

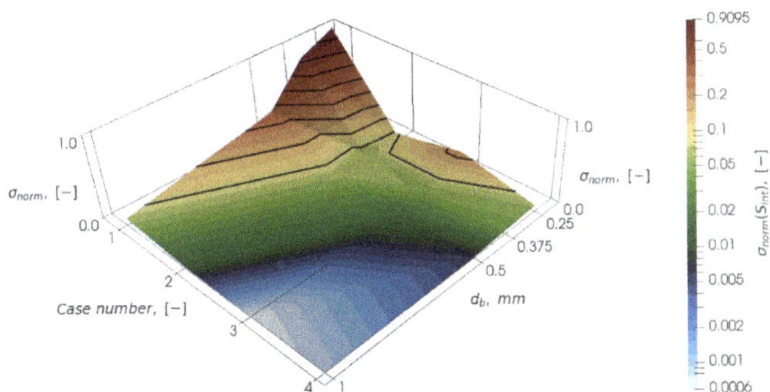

Figure 9. Distribution of a weighted variance σ_{norm} for the interfacial area density, S_{int}.

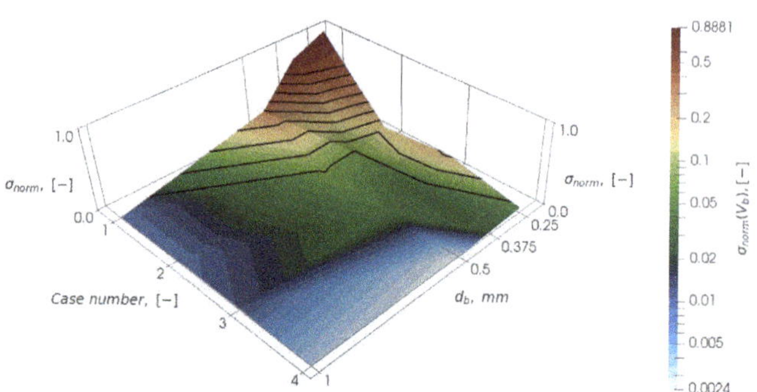

Figure 10. Distribution of a weighted variance σ_{norm} for the bubble velocity, V_b.

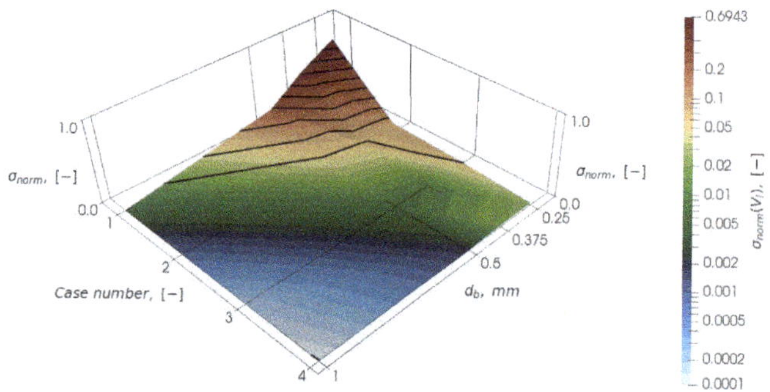

Figure 11. Distribution of a weighted variance σ_{norm} for the liquid velocity, V_l.

Based on the analysis of all four distributions, a general conclusion can be drawn: as the characteristic size of the bubbles increases, the difference between all approaches is leveled and, even for the monodisperse approach, the difference with the reference one turns out to be small: it is 5% for the interfacial area and about 1% for other quantities for a bubble diameter of 1 mm. However, for a bubble diameter of 0.5 mm, the difference between the monodisperse and polydisperse approaches becomes significant (up to 20%). The difference increases with decreasing bubble diameter.

For 0.5 mm bubbles the normalized variance decreases monotonically and the solution approaches the reference one with increase of N (variation of the parameters of the BSD function approximation): from monodisperse ($N = 1$, **R1**), then through distributions ($N = 4$, **R3**), ($N = 7$, **R1**) and ending with ($N = 7$, **R2**).

However, for the smallest bubble considered, with a diameter of 0.25 mm, this behavior changes. Using the distribution ($N = 4$, **R3**), a solution can be obtained that is close to the variant ($N = 7$, **R2**) and has a smaller difference from the "reference" one than ($N = 7$, **R1**). This can be explained by the fact that, in the case of the distribution ($N = 7$, **R1**), small bubbles corresponding to the left side of the BSD function are not resolved at all in the simulation. However, their influence is important since they have the ability to accumulate in stagnation or recirculation zones, significantly influencing the flow parameters and changing the flow structure. Although the distribution ($N = 4$, **R3**) is coarser than ($N = 7$, **R1**), it still describes the behavior of small bubbles.

5. Conclusions

The numerical simulation of the dynamics of polydisperse and monodisperse flows in the BCR was performed to clarify the effect of the parameters of the piecewise constant approximation of the BSD function.

It was found that with increasing of bubble size the normalized variances of the bubble volume fraction, the liquid velocity, and the specific density of the interfacial surface calculated for the polydisperse and monodisperse descriptions of the BCR working medium decrease. In particular, for bubbles with a characteristic size of $M = 1.0$ mm, the normalized variance is about 1% for the velocity and 5% for the density of the interfacial surface.

The discrete approximation of the BSD function with the uniform distribution of bins with respect to $ln(d_b)$ (**R2**) provides a better solution, in terms of normalized variances, than that with the uniform bins with respect to d_b (**R1**), due to the correct resolution of the domain of small bubbles (with $d_b < M$).

A hybrid approximation of the BSD function **R3** has been proposed which is uniform with respect to $ln(d_b)$ for bubbles with sizes smaller than M and uniform with respect to d_b for sizes larger than M. Such an approximation is potentially preferable with substantially large bubble sizes up to ($d_b > 6$ mm).

The hybrid approximation of the BSD (**R3**) with four classes of bubbles provides better results than the original **R1** approximation with seven classes and is comparable to the **R2** approximation with seven classes of bubbles. The obtained results were compared with the results for the **R3** approximation with 10 classes of bubbles as well as with the experimental data, and demonstrated qualitative and quantitative agreement. This makes the proposed approach promising for flow pattern prediction and robust due to low computational resource requirements.

Author Contributions: Conceptualization, A.C. and A.S.; methodology, A.C. and A.S.; software, A.C.; validation, A.C. and V.C.; formal analysis, A.S.; investigation, A.S.; writing—original draft preparation, A.C., A.S. and V.C.; writing—review and editing, A.C. and A.S.; visualization, A.C. and V.C.; supervision, A.S.; project administration, A.S.; funding acquisition, A.S. All authors have read and agreed to the published version of the manuscript.

Funding: This research received no external funding.

Institutional Review Board Statement: Not applicable.

Informed Consent Statement: Not applicable.

Data Availability Statement: Data available upon request.

Conflicts of Interest: The authors declare no conflict of interest.

Abbreviations

The following abbreviations are used in this manuscript:

BCR	Bubble column reactor
BSD	Bubble size distribution
MUSIG	Multiple size group
TVD	Total variation diminishing
SIMPLE	Semi-implicit method for pressure-linked equations

References

1. Yang, X.; Liu, Z.; Manhaeghe, D.; Yang, Y.; Hogie, J.; Demeestere, K.; van Hulle, S.W.H. Intensified ozonation in packed bubble columns for water treatment: Focus on mass transfer and humic acids removal. *Chemosphere* **2021**, *283*, 131217. [CrossRef] [PubMed]
2. Inkeri, E.; Tynjälä, T. Modeling of CO$_2$ Capture with Water Bubble Column Reactor. *Energies* **2020**, *13*, 5793. [CrossRef]
3. Perez, B.J.L.; Jimenez, J.A.M.; Bhardwaj, R.; Goetheer, E.; van Sint Annaland, M.; Gallucci, F. Methane pyrolysis in a molten gallium bubble column reactor for sustainable hydrogen production: Proof of concept & techno-economic assessment. *Int. J. Hydrogen Energy* **2021**, *46*, 4917–4935. [CrossRef]
4. Vik, C.B.; Solsvik, J.; Hillestad, M.; Jakobsen, H.A. Interfacial mass transfer limitations of the Fischer-Tropsch synthesis operated in a slurry bubble column reactor at industrial conditions. *Chem. Eng. Sci.* **2018**, *192*, 1138–1156. [CrossRef]
5. Kantarci, N.; Borak, F.; Ulgen, K.O. Bubble column reactors. *Process Biochem.* **2005**, *40*, 2263–2283. [CrossRef]
6. Lau, Y.M.; Thiruvalluvan Sujatha, K.; Gaeini, M.; Deen, N.G.; Kuipers, J.A.M. Experimental study of the bubble size distribution in a pseudo-2D bubble column. *Chem. Eng. Sci.* **2013**, *98*, 203–211. [CrossRef]

7. Besagni, G.; Inzoli, F. Prediction of Bubble Size Distributions in Large-Scale Bubble Columns Using a Population Balance Model. *Computations* **2019**, *7*, 17. [CrossRef]
8. Pakhomov, M.A.; Terekhov, V.I. Simulation of the turbulent structure of a flow and heat transfer in an ascending polydisperse bubble flow. *Tech. Phys.* **2015**, *60*, 1268–1276. [CrossRef]
9. Yao, W.; Morel, C. Volumetric interfacial area prediction in upward bubbly two-phase flow. *Int. J. Heat Mass Transf.* **2004**, *47*, 307–328. [CrossRef]
10. Wang, T.; Wang, J.; Jin, Y. Population Balance Model for Gas-Liquid Flows: Influence of Bubble Coalescence and Breakup Models. *Ind. Eng. Chem. Res.* **2005**, *44*, 7540–7549. [CrossRef]
11. Lo, S. *Application of Population Balance to CFD Modelling of Bubbly Flow via the MUSIG Model*; AEA Technology: Rotherham, UK, 1996.
12. Krepper, E.; Lucas, D.; Frank, T.; Prasser, H.-M.; Zwart, P.J. The inhomogeneous MUSIG model for the simulation of polydispersed flows. *Nucl. Eng. Des.* **2008**, *238*, 1690–1702. [CrossRef]
13. Ziegenhein, T.; Rzehak, R.; Lucas, D. Transient simulation for large scale flow in bubble columns. *Chem. Eng. Sci.* **2015**, *122*, 1–13. [CrossRef]
14. Olmos, E.; Gentric, C.; Vial, C.; Wild, G.; Midoux, N. Numerical simulation of multiphase flow in bubble column reactors. Influence of bubble coalescence and break-up. *Chem. Eng. Sci.* **2001**, *56*, 6359–6365. [CrossRef]
15. Fard, M.G.; Stiriba, Y.; Gourich, B.; Vial, C.; Grau, F.X. Euler–Euler large eddy simulations of the gas–liquid flow in a cylindrical bubble column. *Nucl. Eng. Des.* **2020**, *369*, 110823. [CrossRef]
16. Zhang, X.B.; Yan, W.C.; Luo, Z.H. Numerical simulation of local bubble size distribution in bubble columns operated at heterogeneous regime. *Chem. Eng. Sci.* **2021**, *231*, 116266. [CrossRef]
17. Wu, Y.; Liu, Z.; Li, B.; Gan, Y. Numerical simulation of multi-size bubbly flow in a continuous casting mold using population balance model. *Powder Technol.* **2022**, *396*, 224–240. [CrossRef]
18. Wu, Y.; Liu, Z.; Li, B.; Xiao, L.; Gan, Y. Numerical simulation of multi-size bubbly flow in a continuous casting mold using an inhomogeneous multiple size group model. *Powder Technol.* **2022**, *402*, 117368. [CrossRef]
19. Bhole, M.R.; Joshi, J.B.; Ramkrishna, D. CFD simulation of bubble columns incorporating population balance modeling. *Chem. Eng. Sci.* **2008**, *63*, 2267–2282. [CrossRef]
20. Chernyshev, A.S.; Schmidt, A.A. Investigation of the evolution of the bubble size distribution in the ascending and descending flows. *J. Phys. Conf. Ser.* **2017**, *899*, 032009. [CrossRef]
21. Lain, S.; Broder, D.; Sommerfeld, M. Experimental and numerical studies of the hydrodynamics in a bubble column. *Chem. Eng. Sci.* **1999**, *54*, 4913–4920. [CrossRef]
22. Lain, S.; Broder, D.; Sommerfeld, M.; Goz, M.F. Modelling hydrodynamics and turbulence in a bubble column using the Euler–Lagrange procedure. *Int. J. Multiph. Flow* **2002**, *28*, 1381–1407. [CrossRef]
23. Roghair, I.; Lau, Y.M.; Deen, N.G.; Slagter, H.M.; Baltussen, M.W.; van Sint Annaland, M.; Kuipers, J.A.M. On the drag force of bubbles in bubble swarms at intermediate and high Reynolds numbers. *Chem. Eng. Sci.* **2011**, *66*, 3204–3211. [CrossRef]
24. Auton, T.R. The lift force on a spherical body in a rotational flow. *J. Fluid Mech.* **1987**, *183*, 199–218. [CrossRef]
25. Tomiyama, A.; Tamai, H.; Zun, I.; Hosokawa, S. Transverse migration of single bubbles in simple shear flows. *Chem. Eng. Sci.* **2002**, *57*, 1849–1858. [CrossRef]
26. Menter, F.R.; Kuntz, M.; Langtry, R. Ten Year of Industrial Experience with the SST Turbulence Model. In Proceedings of the 4th International Symposium on Turbulence, Heat and Mass Transfer, Antalya, Turkey, 12–17 October 2003; pp. 625–632.
27. Troshko, A.A.; Hassan, Y.A. A two-equation turbulence model of turbulent bubbly flows. *Int. J. Multiph. Flow* **2001**, *27*, 1965–2000. [CrossRef]
28. Sato, Y.; Sekoguchi, R. Liquid velocity distribution in two-phase bubble flow. *Int. J. Multiph. Flow* **1975**, *2*, 79–95. [CrossRef]
29. Sokolichin, A.; Eigenberger, G.; Lapin, A. Simulation of Buoyancy Driven Bubbly Flow: Established Simplifications and Open Questions. *AIChE J.* **2004**, *50*, 24–45. [CrossRef]
30. Moraga, F.J.; Larreteguy, A.E.; Drew, D.A.; Lahey, R.T., Jr. Assessment of turbulent dispersion models for bubbly flows in the low Stokes number limit. *Int. J. Multiph. Flow* **2003**, *29*, 655–673. [CrossRef]
31. Versteeg, H.K.; Malalasekera, W. *An Introduction to Computational Fluid Dynamics. The Finite Volume Method*, 2nd ed.; Pearson Education Limited: Harlow, UK, 2007; pp. 134–211.
32. Chernyshev, A.S.; Schmidt, A.A. Numerical modeling of gravity-driven bubble flows with account of polydispersion. *J. Phys. Conf. Ser.* **2016**, *754*, 032005. [CrossRef]
33. Chernyshev, A.S.; Schmidt, A.A. Numerical simulation of gravity-driven mono- and polydisperse bubbly flows in a three-dimensional column in the framework of the Euler–Euler approach. *J. Phys. Conf. Ser.* **2020**, *1697*, 012236. [CrossRef]
34. Chen, P.; Sanyal, J.; Dudukovic, M.P. Numerical simulation of bubble columns flows: Effect of different breakup and coalescence closures. *Chem. Eng. Sci.* **2005**, *60*, 1085–1101. [CrossRef]
35. Allio, A.; Buffo, A.; Marchisio, D.; Savoldi, L. Two-fluid modelling for poly-disperse bubbly flows in vertical pipes: Analysis of the impact of geometrical parameters and heat transfer. *Nucl. Eng. Technol.* **2022**, *in press*. [CrossRef]

Article

Spreading of Impacting Water Droplet on Surface with Fixed Microstructure and Different Wetting from Superhydrophilicity to Superhydrophobicity

Sergey Starinskiy [1,2,*], Elena Starinskaya [1,2], Nikolay Miskiv [1,2], Alexey Rodionov [1,2], Fedor Ronshin [1,2], Alexey Safonov [1], Ming-Kai Lei [3] and Vladimir Terekhov [1,2]

1   S.S. Kutateladze Institute of Thermophysics SB RAS, Lavrentyev Ave. 1, 630090 Novosibirsk, Russia
2   Novosibirsk State University, Pirogova Str. 2, 630090 Novosibirsk, Russia
3   School of Materials Science and Engineering, Dalian University of Technology, Dalian 116024, China
*   Correspondence: starikhbz@mail.ru

Abstract: The spreading of the water droplets falling on surfaces with a contact angle from 0 to 160° was investigated in this work. Superhydrophilicity of the surface is achieved by laser treatment, and hydrophobization is then achieved by applying a fluoropolymer coating of different thicknesses. The chosen approach makes it possible to obtain surfaces with different wettability, but with the same morphology. The parameter t^* corresponding to the time when the capillary wave reaches the droplet apex is established. It is shown that for earlier time moments, the droplet height change does not depend on the type of used substrate. A comparison with the data of other authors is made and it is shown that the motion of the contact line on the surface weakly depends on the type of the used structure if its characteristic size is less than 10 μm.

Keywords: droplet impact; spreading; superhydrophobicity; superhydrophilicity; wettability; water droplet; laser ablation; HW CVD

Citation: Starinskiy, S.; Starinskaya, E.; Miskiv, N.; Rodionov, A.; Ronshin, F.; Safonov, A.; Lei, M.-K.; Terekhov, V. Spreading of Impacting Water Droplet on Surface with Fixed Microstructure and Different Wetting from Superhydrophilicity to Superhydrophobicity. *Water* **2023**, *15*, 719. https://doi.org/10.3390/w15040719

Academic Editor: Yaoming Ma

Received: 12 January 2023
Revised: 6 February 2023
Accepted: 9 February 2023
Published: 11 February 2023

1. Introduction

The development of surface treatment methods at the micro- and nanoscale opens the way to the evolution of biomechanical technologies. Particular attention is paid to the issue of reproducing the hierarchical topologies observed in nature, possessing the principle of self-cleaning (lotus effect), retention of liquid droplets (rose petal effect), and reduction in hydrodynamic resistance (shark skin) [1–3]. The experience accumulated by researchers formed the basis of several approaches to the creation of materials called superhydrophobic, superhydrophilic, and biphilic [4–8].

The actual changes in the wettability properties are caused by two factors—the chemical composition of the surface and its topology. Depending on their combination, according to the classical studies of Wenzel [9] and Cassie-Baxter [10], surface wettability can be described in two modes. The first one (Wenzel mode) implies that the liquid is in full contact with the surface. In this case, the topology development entails increasing hydrophobicity for hydrophobic materials and hydrophilicity for hydrophilic ones by the expression $cos\,\theta_r = r \cdot cos\,\theta$, where r is the roughness value, θ is the contact angle (CA) of the smooth surface, and θ_r is the contact angle of the rough surface. Thus, the initially hydrophilic silicon surface acquires superhydrophilic properties after texture creation [11]. In the second mode (Cassie–Baxter), an air layer, called a plastron, is trapped in the cavities formed by the microtexture. It is the presence of air pockets that ensures the enhancement of hydrophobic properties up to reaching the superhydrophobic state. The static contact angle can be estimated by the following expression: $cos\theta_r = \sum_n f_n cos\theta_n$, where f_n is the total area of each interface under the droplet per unit projected area. These simple models are very illustrative and have good applicability; however, they have some limitations [12–14], for example, in describing the rose petal effect when a superhydrophobic state with high adhesive strength is realized on a hierarchical micro/nanostructure [15].

Materials with both superhydrophilic [16,17], superhydrophobic [18,19], and biphilic [20] properties are very promising in terms of passive (energy-free) control of the liquid droplets' interaction with a solid wall. This affects important applications such as spray cooling [21], fuel combustion [22], additive technology [23], coating [24], deicing [25], biofouling, and surface contamination [26]. One of the actual challenges is the study of multiphase phenomena [27,28], in particular, the behavior of liquid droplets [29] including interphase phenomena. However, there is currently no unequivocal understanding of the material wettability effect on the dynamics of the falling droplet spreading on the surface. Thus, the data available in the literature on the dynamics of the falling droplets spreading on superhydrophobic surfaces with similar contact angles under close conditions can differ appreciably [30–32]. According to the results of Pachchigar et al. [33], maximum droplet spreading on a structured fluoropolymer with contact angles of 105–154° is independent of surface morphology for Weber numbers $We = \frac{\rho D_0 v^2}{\sigma} < 40$, where ρ is the density, D_0 is the droplet diameter, v is the droplet velocity, and σ is the surface tension. On the other hand, Pan et al. [34] observed a significant difference in the dispersion of the falling droplets, although they used materials with a contact angle range (77–145°) close to the work of Pachchigar et al. [33]. Moreover, a difference in the spreading dynamics was found in the investigation of Lv et al. [35], where the influence of nanostructure on the droplet bounce dynamics from surfaces with close roughness at the microlevel was clearly shown. This agrees with the results of other authors [36,37], which show that for the same static contact angles, a droplet on the surface can be in both the lotus (without pinning) and rose petal (with pinning) states, which affect the liquid spreading on the surface. These results are in agreement with the MDPD calculations carried out by Du et al. [38], where it is shown that sufficiently high contact angles (~145°) can be achieved in both the Wenzel and Cassie–Baxter modes.

Despite the apparent simplicity, the fall of droplets on a solid wall is a complex two-phase process. Upon primary contact with the wall at the liquid–gas interface, capillary-surface waves are generated, which leads to the formation of pyramidal structures [39] with a characteristic wavelength $\sigma/\rho V$, where ρ is density, σ is surface tension, and V is velocity. Further movement of the liquid leads to the formation of a dish-like or torus-shaped topology. According to Renardy et al. [39], surface dry-out is determined by the ratio $We = 1590/Re^{1.49} + 3.62$. The formation of a dry cavity leads to the capture of an air bubble during the retraction stage. The higher the falling velocity, the higher the side lamella velocity. To describe the maximum spreading β_{max}, various analytical models are used, which, as a rule, do not take into account the surface characteristics of the wall [40]. However, many authors have noticed the incorrectness of this approach for low We numbers in conditions before splashing. Thus, it was shown by Ukiwe and Kwok [41] that the experimental results are much better described by taking into account the contact angle. For modes involving the development of instability and subsequent splashing, as a rule, the boundary between deposition and splashing is determined by the parameter $K = We^{0.5} Re^{0.25}$. However, according to Roisman et al. [42], even in this case, it is necessary to take into account the morphology of the analytical surface, and it was proposed to redefine the Reynolds number, taking into account the surface roughness. After reaching the maximum spreading, a reverse flow occurs, which cannot be independent of the receding contact angle. Much attention was paid to this issue in the work of Wang and Fang [32], and quite complex analytical approaches were proposed for constructing retraction curves. However, the authors took into account only the contact angles, but not the topology of the surface.

The influence of surface topological characteristics on the falling droplet process continues to be intensively studied as there are now reliable ways to control surface roughness [2,43] or to create periodic structures with different spatial distributions [30,44–47]. A noticeably lower amount of work is devoted to the study of the dynamics of liquid spreading over surfaces with different wettability but the same morphology, except the results obtained using different liquids [43,48–51]. Zhang et al. [52] considered the effect of surface wettability on the water droplet spreading in a wide range of Weber numbers We = 0–3000 (although the data in the work start from We = 80). Hydrophilic, hydrophobic, and superhydrophobic surfaces were investigated with CA = 30–150°. Samples were obtained by plasma treatment and functionalization methods with hydrophobic agents. According to the proposed mechanism, the wettability property mainly affects the rise in the lamella edge and the subsequent air leakage, which has a significant influence on droplet spreading and splashing characteristics. Sun et al. [53] examined falling droplets on surfaces with CA = 5 and 134°. Different wettability was achieved by UV exposure to titanium oxide, suggesting that the surface morphology remains unchanged. It was shown that for superhydrophilic surfaces, the spreading dynamics also depend on the material texture, which determines the spreading velocity; in particular, the detachment of secondary droplets is possible only in the case when the spreading velocity is above the falling droplet velocity.

Further analysis of the effect of transitions between Cassie–Baxter and Wenzel modes is required as several authors have shown a very pronounced effect on the velocity of liquid movement along the wall [45,54], even though more and more attention has recently been paid in the literature to the issue of a water droplet falling on surfaces with the rose petal effect [2,55,56]. Most papers consider the effect of simultaneous changes in surface structure and wettability, while these parameters have significantly different effects on the falling liquid droplet spreading. However, there is still no complete understanding of the influence of the surface topology and wettability on the dynamics of falling droplet spreading, and the data available in the literature are quite scattered. There are no data on the flow dynamics for materials with structures similar to rose petals but different contact angles. Thus, the purpose of this work is to fill the research gap and investigate the surface wettability effect on the liquid droplet spreading on surfaces with the same morphology. For the first time, we studied water droplets falling on surfaces with a contact angle from <5° (superhydrophilic) to 155° (superhydrophobic) with a fixed surface topology close in structure to a rose petal. Experiments were performed in a wide range of Weber numbers We = 0.3–33 for different droplet sizes in the range of D_0 = 2–3 mm. For a detailed study of the spreading dynamics, simulations were performed using the lattice Boltzmann method, which allows us to analyze the velocity field in time.

2. Experimental Setup

In this work, the change in the surface wettability was achieved in a two-stage process. During the first stage, the surface was irradiated in the air by pulses of the basic harmonic Nd:YAG laser (home-made) with a wavelength of 1064 nm and a pulse duration of 11 ns. The average energy density in the beam was 3.6 J/cm^2; thus, conditions favorable for the formation of a special hierarchical structure of the laser spot were created on the surface [11]. The material was irradiated in the mode of beam scanning along the surface with an area of 12 × 18 mm. The laser spot was 0.4 mm^2, the number of laser pulses per spot was about 60, and the total number of surface preparations was 30,000. The overlapping of laser spots was 60% and controlled by laser pulse frequency and scanning velocity. The initial surface of monocrystalline silicon with natural oxidation had a contact angle of ~55°. Laser processing of the silicon surface led to the formation of a self-organized periodic structure consisting of alternating hillocks and hollows with a characteristic spatial size of about 10 μm. In addition to the micron-sized periodic structure, there was a second level of nanometer roughness or porosity formed by ablation products returning to the surface. The contact angle on the laser-processed surface was less than 5°, i.e., the substrate became superhydrophilic. The texture can retain its properties for a very long time under various external influences, particularly during pool boiling [57].

In the second stage, the laser-treated samples were hydrophobized by applying a fluoropolymer coating of different thicknesses by the Hot Wire Chemical Vapor Deposition

(HW CVD) method [58]. In the HWCVD method, a hot catalytic metal wire mesh is used to activate the precursor gas. The experimental setup for depositing coatings was described in detail by Safonov et al. [58]. Hexafluoropropylene oxide C_3F_6O was used as the precursor gas of the fluoropolymer film. Silicon substrates were placed in a cooled substrate holder in a vacuum chamber. At a distance of 30 mm above the substrates, there was a catalytic activator in the form of a mesh made of a 0.5 mm diameter helically coiled nichrome wire with a spacing of 20 mm. The mesh temperature was fixed at 580 °C and monitored by volt-ampere measurements (Mastech MS8050, Huayi Mastach Co., Shenzhen, China). The precursor gas pressure in the deposition chamber was 0.5 Torr and the gas flow rate was 20 sccm. The substrate temperature was about 30 °C during deposition. The thickness of the fluoropolymer coating was controlled by varying the deposition time within the range from 30 s to 750 s. The fluoropolymer coating does not affect the material topology at both micro- and nanolevels [59]. The thickness of the fluoropolymer coating did not exceed 50 nm and was controlled by the deposition duration. This technique already allows us to achieve stable superhydrophobic properties with a contact angle greater than 160° [59]. The microstructures of the obtained samples and the rose petal were very similar to each other (Figure 1).

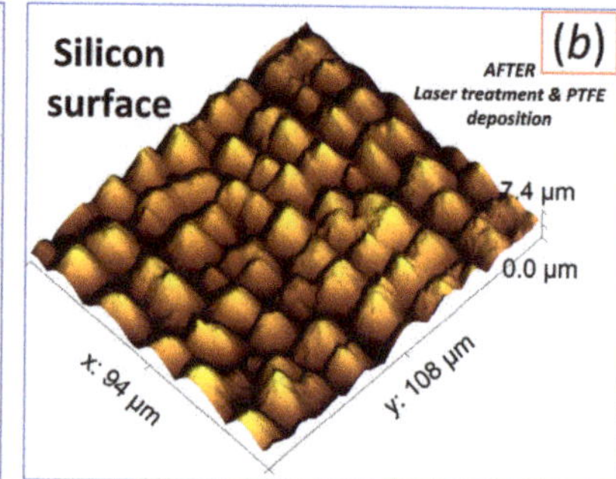

Figure 1. Comparison of the surface topology of the rose petal (**a**) (adapted from [15]) and samples synthesized in the present work (**b**).

Both advancing contact angle (ACA) and receding contact angle (RCA) were measured using the sessile drop method on a KRUSS DSA 100 (KRUSS GmbH, Hamburg, Germany). The liquid droplet is placed on the substrate using an automatic dosing system. In the first stage, the advancing contact angle is measured when liquid is pumped into the droplet. At the initial moment, the contact angle increases as the contact line is pinned. Then, the depinning of the contact line takes place, the contact angle becomes constant, and the measurements are taken at this moment. In the second stage, the receding contact angle is measured in a similar way as the liquid droplet is reduced in volume. The results are presented in Table 1. The obtained contact angles make it possible to fully characterize the surface wetting hysteresis.

Table 1. Wettability of samples as a function of fluoropolymer layer thickness (PTFE).

Sample	ACA, °	RCA, °	PTFE Thickness, nm
Surface 1	<5	<5	0
Surface 2	22	5	2
Surface 3	50	17	5
Surface 4	90	20	12
Surface 5	145	22	25
Surface 6	160	159	50

A scheme of the experimental setup to study the dynamics of the spreading of Milli-Q water droplets falling on the synthesized surfaces is shown in Figure 2a. Visualization was performed with a Phantom VEO710 high-speed camera (Vision Research, Inc. Wayne, NJ, USA). Typical images of the process are shown in Figure 2b. The main analyzed parameters were droplet height H and contact line diameter D (Figure 2b). Occasionally, during the spreading of the droplet, the central part was hidden behind the droplet side parts, in which case H was measured as the maximum height of the lamella. The experiments were performed for distilled water droplets 2–3 mm in size falling from a height of 1–60 mm (Weber numbers We = 0–33 and Reynolds numbers Re = 250–3000), which corresponds to deposition, spreading, partial rebound, and rebound dependent on wettability surface regimes [31,50]. In other words, we worked under no splash conditions. The Bond number Bo ~ 1, i.e., the dynamics of droplet spreading was predominantly determined by surface tension.

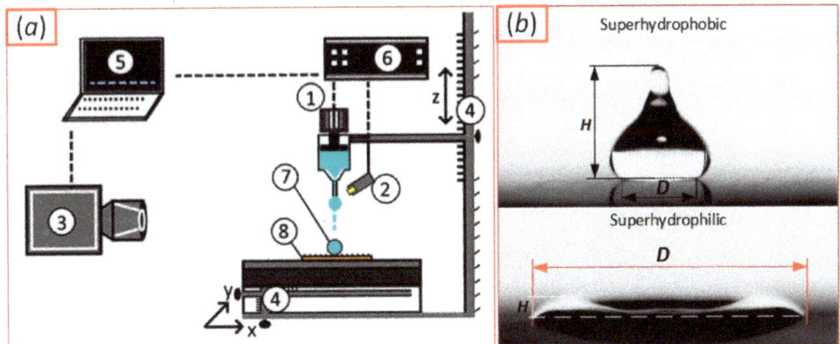

Figure 2. (**a**) Experimental setup for measuring the droplet impact and spreading characteristics on the substrate surface: 1—injection pump, 2—droplet break-away sensor, 3—high-speed video camera Phantom VEO710, 4—3-axis positioning table, 5—computer, 6—microcontroller unit, 7—droplet, 8—substrate. (**b**) Typical snapshots of droplet spreading on superhydrophilic and superhydrophobic materials.

3. LBM Simulation

In this work, the simulation was performed using the multiple-relaxation-time lattice Boltzmann method (MRT-LBM) [60]. LBM is currently an effective method for studying fluid flows; its advantages are most clearly seen in modeling multiphase flows and surface phenomena. There are several approaches to describe the interphase interactions in the LBM; in this work, we chose the pseudopotential model [61,62], in which the interparticle interactions are represented as a force based on the potential, which depends on the density of the medium. When modeling in the present work, the medium was represented as a one-component medium with separation into liquid and vapor phases. The conditions under which the equilibrium properties of vapor and liquid are close to the properties of air and water in the experiment were determined by the given temperature.

It is important to note that in this work, we used a two-dimensional model, simplifying the problem significantly. However, there is a successful experience of the 2D approach to solving the problem of droplet surface interaction dynamics [63–66]. In addition, the goal of modeling is to obtain qualitative data on the velocity distribution, which was difficult to obtain in the experiment. Thus, we expect to provide important information for understanding the physical mechanism of the interaction of a drop with a surface despite the accepted 2D simplification.

For a simple description of the physical phenomena considered in this work, we used a two-dimensional formulation, which, despite the obvious limitations, allows us to qualitatively describe the process of droplet spreading [63]. The basic LBM equation, which uses a multiple-relaxation-time collision operator, is as follows:

$$f_\alpha(x + e_\alpha, t + \Delta t) = f_\alpha(x, t) - \overline{L}_{\alpha\beta}\left(f_\beta - f_\beta^{eq}\right)\Big|_{(x,t)} + \Delta t\left(S_\alpha - \frac{1}{2}\overline{L}_{\alpha\beta}S_\beta\right)\Big|_{(x,t)} \tag{1}$$

where $f_\alpha(x, t)$ is the density distribution function; index eq is its equilibrium value; t is time, x is a coordinate, e_α is a discrete set of velocity vectors in the α direction; S_α is the term describing the action of various forces, $\overline{L} = M^{-1}LM$ is the collision matrix, in which M is the orthogonal transformation matrix and L is the diagonal matrix. Using the above transformation matrix, the distribution function is transformed to the moment space $m = Mf$ and $m^{eq} = Mf^{eq}$.

For the D2Q9 lattice used in this paper, the matrices M and L and vectors e_α and f^{eq} are given by Bouzidi et al. [67]. The solution of Equation (1) is carried out in two stages. The first stage is the transition to the space of moments and carrying out the process of collisions.

$$m^* = m - L(m - m^{eq}) + \Delta t\left(I - \frac{L}{2}\right)S \tag{2}$$

The second stage is to return to the distribution function $f^* = M^{-1}m^*$ and conduct the distribution process:

$$f_\alpha(x + e_\alpha, t + \Delta t) = f_\alpha^*(x, t). \tag{3}$$

After that, the macroscopic quantities are defined as $\rho = \sum_\alpha f_\alpha$ and $\rho V = \sum_\alpha e_\alpha f_\alpha$. Phase separation was modeled according to the pseudopotential approach. According to this model, a force acts in a gas or liquid [62]:

$$F = -G\psi(x)\sum_\alpha w_\alpha \psi(x + e_\alpha)e_\alpha, \tag{4}$$

where G is the interaction force, $\psi(x)$ is the potential, and w_α is the weighting factor in the α direction. The interaction potential was determined according to [62]

$$\psi(\rho) = \sqrt{\frac{2(p_{EOS} - \rho c_s^2)}{Gc^2}}, \tag{5}$$

where the pressure is determined from the state equation, in our case, Carnahan–Starling:

$$p_{EOS} = \varrho RT\frac{1 + \frac{b\rho}{4} + \left(\frac{b\rho}{4}\right)^2 - \left(\frac{b\rho}{4}\right)^3}{\left(1 - \frac{b\rho}{4}\right)^3} - a\rho^2, \tag{6}$$

where parameters a and b are functions of critical temperature and pressure, respectively. Following the approach of Li et al. [64], we chose $R = 1$, $b = 4$, and $a = 0.25$. The introduction of force (4) into Equation (2) was carried out according to the exact difference method [68], and gravity was included in the same way. Such an approach to the integration of interfacial interaction significantly suppresses non-physical currents arising in the case of the original method [61,62] due to the non-isotropy of discrete operators. It allows us to achieve liquid-to-gas density ratios of ~1000, i.e., to simulate a water/air system at normal conditions.

Aspects of the construction of boundary conditions in the LBM for multiphase applications using the pseudopotential approach were studied in detail by Khajepor [69]. The top and bottom boundaries of the domain were considered to be solid surfaces, while the periodic boundary condition was applied to the side boundaries.

The interaction of a fluid with a solid surface was modeled by a similar approach:

$$F = -G\psi(x)\sum_\alpha w_\alpha \psi(\varrho_w)H(x + e_\alpha)e_\alpha, \tag{7}$$

where H is a function taking the value $H = 1$ at the points of the solid surface and $H = 0$ at any other points. ϱ_w is a parameter characterizing the wetting degree and determining the value of the contact angle. Note that, unlike most works using such an approach, in the present work, the parameter characterizing the wetting degree ϱ_w was not a constant value both in space and time. As shown below, the surfaces considered in the work have a significant contact angle hysteresis, so the value of the contact angle changes significantly when the liquid flows over the surface and during the reverse process. Thus, the initial state ϱ_w was taken from the considerations of reproducing the advancing contact angle according to the experiment and then, depending on some "critical" pressure on the wall, "switched" to the receding contact angle.

In all calculations, the distribution corresponding to a circular droplet of a given diameter located near the surface and having a given velocity was set as the initial condition. The boundary conditions on the upper and bottom surfaces were assumed to be solid surfaces with a given ϱ_w, and the side faces were assumed to be periodic. Preliminary calculations were performed to determine the dependence of the contact angle value on ϱ_w, the dependences of vapor/liquid density and surface tension on temperature, and to verify mesh convergence. For most cases, a grid of 768×768 cells was used, while, in lattice units, the drop diameter was 164 cells, the relaxation time was 0.515, and the initial drop velocity was 0.021. Especially for the superhydrophilic surface, the horizontal resolution of the grid was increased up to 2048 unity.

4. Results and Discussion

Figure 3a shows images of a 2.3 mm diameter water droplet falling on surfaces with different wettability at a velocity of 0.3 m/s at the moment of the collision, which corresponds to the dimensionless criteria of $We = 3$, $Re = 775$, and $Bo = 0.73$. Based on the sweep of the droplet falling dynamics, several stages can be distinguished. The first one is the inertial stage with a duration of less than 3 ms. Thus, a capillary wave is formed upon droplet contact with the surface, which propagates across the droplet surface and deforms the droplet into a pyramidal structure [39]. At this stage, the shape of the droplet is determined only by the Weber number (falling velocity) and weakly depends on the type of surface used. The only exception is a superhydrophilic surface for which, at small Weber numbers ($We < 3$), the contact line velocity may exceed the lateral spreading velocity of the droplet. However, this has an insignificant effect on the dynamics of the droplet height H.

The inertial stage is followed by the viscous spreading stage until the maximum lateral droplet size is reached. For all the samples studied except for the superhydrophilic one, the droplet behavior is qualitatively almost identical. For superhydrophilicity, we see the gradual spreading of liquid. For other cases, we observe the formation of a toroidal structure, i.e., the central part of the droplet is lower relative to its sides. The influence of surface wettability can be seen in the dynamic contact angle (Figure 3, $t = 4.6$ ms). For the superhydrophilic case at the beginning of the viscous flow stage, the achievement of a local maximum on the dependence $H(t)$ is observed. This process is based on the same mechanism as the droplet emission induced by ultrafast spreading on a superhydrophilic surface [53]. Like Sun et al. [53], we observe droplet emission from the surface of a superhydrophilic material at small Weber numbers (We < 0.1). The viscous stage ends when the droplet on the substrate reaches its maximum lateral size, which is determined by the size of the region where the droplet is pinned to the surface. Next, the retraction stage is realized. For hydrophilic surfaces, the dynamics of liquid motion at this stage are qualitatively similar, although the difference in the dynamic contact angle is already more pronounced (Figure 3, $t = 7.2$ ms). For the hydrophobic surface, the return flow coincides for some time with the flow for superhydrophobic material. However, while, for the superhydrophobic surface, a decrease in the contact line up to the detachment is observed, for the hydrophobic one, there is a pinning of the droplet to the surface. After the contact line pinning on hydrophilic surfaces, the contact line diameter changes weakly, and complete relaxation occurs significantly later, after at least 500 ms. At this stage, gradually decaying oscillations can be seen for the droplet height value. The frequency of oscillations is determined by the contact angle, although the general behavior of the droplets is almost identical. Attenuation is slowest on a hydrophobic surface with high adhesive properties. After attenuation, the contact angle becomes close to 145°, as was recorded in the measurements on the KRUSS DSA 100. The observed nonmonotonicity in the frequency of oscillations depending on the contact angle is one of the directions for further research.

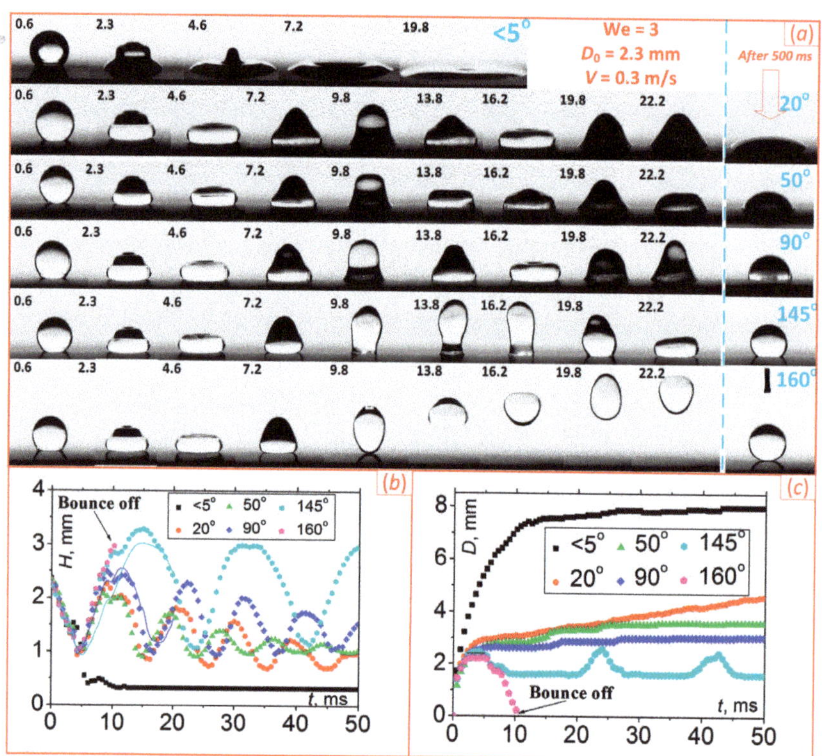

Figure 3. (**a**) Snapshots of droplet D_0 = 2.3 mm falling on surface with different contact angles for We = 3, numbers are time in ms; the dotted line separates the snapshots after 500 ms (**b**) Variation in droplet height and (**c**) contact line diameter with time. Contact angle θ < 5° (■), 20° (●), 50° (▲), 90° (◆), 145° (●), 155° (●). Solid lines in (**b**) are LBM modeling for surface 4 and 5.

We do not observe any fundamental difference in droplet behavior depending on the size, as can be seen from Figure 4, which compares the change in droplet height dynamics with time for three sizes. Time t^*, at which inertial flow is realized, is found—for this stage $t < t^*$, the behavior of the change in droplet height with time for superhydrophilic and superhydrophobic materials is the same (as well as for intermediate contact angles). Moreover, values of t^* do not depend on droplet falling velocity and are determined only by droplet size at least for We < 30 (Figure 4b). This point allows the comparison of data with other authors' data and calculated result verification for different droplet sizes. It is expected that for the dependence of t^* on D_0, the power is 3/2 as the inertial-capillary number $t_c = \sqrt{\frac{\rho R_0^3}{\sigma}}$ (Figure 5a) [31]. This expression is obtained in the Hertz problem on the deformation of the ball against the surface [70]. This moment corresponds to reaching the maximum droplet spreading or the minimum droplet height. We cannot register exactly the moment of achieving the minimum droplet height center in the experiment, because it is closed by the lamella (see Figure 3, $t = 4.6$ ms), which is why in Figure 4a, we observe a plateau in the area of transition from viscous spreading to return flow at the stage of $t = 4$–5 ms (for $D_0 = 2.3$ mm). The dependence of the time moment of droplet rebound from the superhydrophobic surface t_b on D_0 also has the same power of $3/2$ [70] (Figure 5).

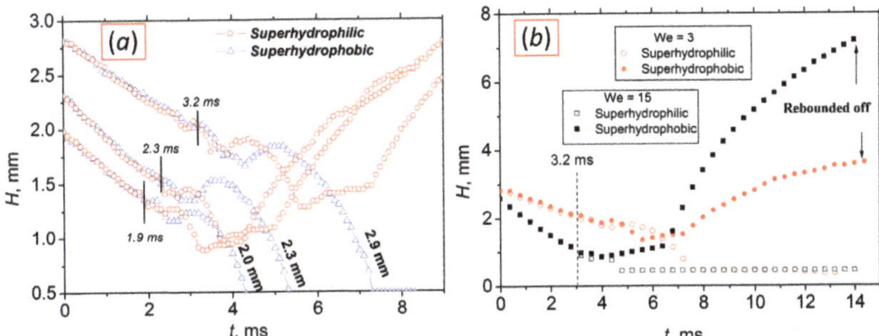

Figure 4. (**a**) Evolutions of height of droplets with different diameters impacted on superhydrophobic and superhydrophilic surfaces We = 3. (**b**) Evolutions of height of droplets for We = 3 and 15 $D_0 = 2.9$ mm.

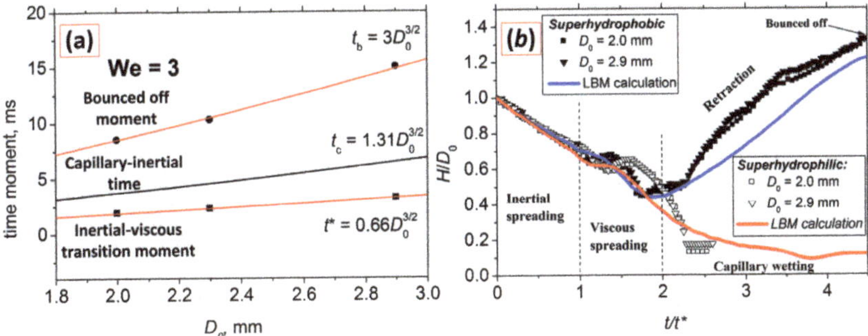

Figure 5. (**a**) Dependence of time t^*, capillary-inertial time t_c, and droplet rebound moment t_b on the diameter. (**b**) Comparison of experimental results for $D_0 = 2$ and 2.9 mm in relative coordinates, We = 3.

Further, we use the value of t^* to generalize the obtained data and compare it with the other authors. A similar approach was used for time t_c [71]. For superhydrophilic surfaces, a local minimum (for We < 10) or a transition to the plateau (We > 10) is fixed at this time, while time t_c and t_b are weakly applicable. In particular, as it has been noted above, for small Weber numbers (We < 3) at sufficiently high spreading velocities (exceeding the droplet velocity) in the region $t < t_c$, a local increase in droplet height is observed [72] and may lead to detachment of secondary droplets, which we have recorded as with Sun et al. [53] at We ~ 0.1.

Figure 6 shows a direct comparison of the water droplet dynamics calculation for a diameter of 2.9 mm on the superhydrophilic and superhydrophobic surfaces with the experimental results. A qualitative and quantitative agreement is observed (Figure 5b). In the calculation, as well as in the experiment, there is a local maximum associated with the limited lateral velocity of the droplet spreading, both on the superhydrophobic and superhydrophilic surfaces. The experimental and calculated times of the liquid droplet rebound from the superhydrophobic surface are also in good agreement. We associate some discrepancies that we observe at the viscous spreading and retraction stages with the limited simulation capabilities for the case of 2D simulation. Figure 5b successfully shows the idea of introducing the time t^*. This is the time, observed for the droplet height, during which the wave reaches the surface. Based on the time value thus determined, it is possible to generalize the data without additional information about the properties of the liquid used or the initial size of the droplets.

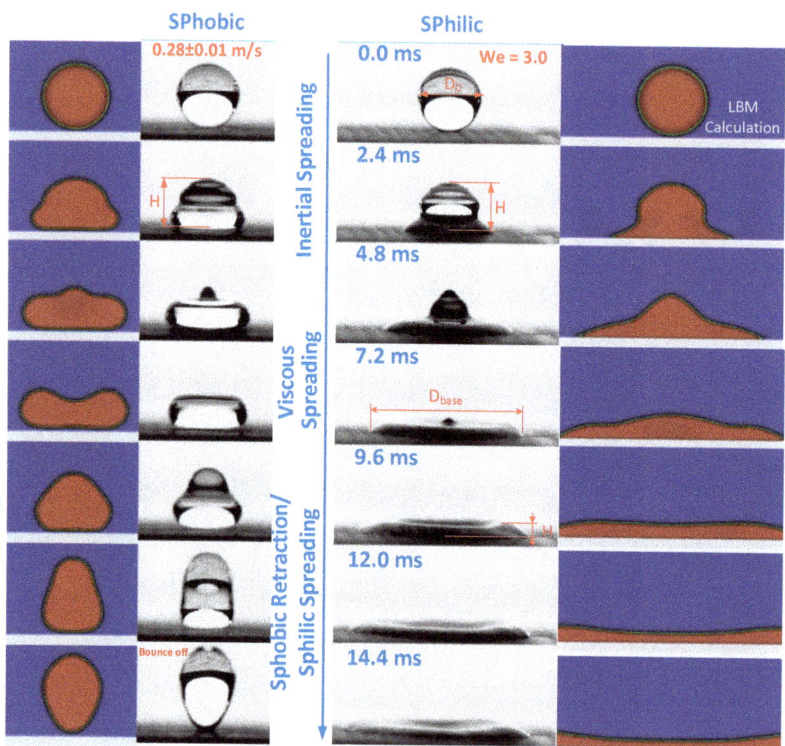

Figure 6. Comparison of the calculated and experimental dynamics of the water droplet interaction (D_0 = 2.9 mm) with superhydrophobic and superhydrophilic surfaces. The colors shows qualitatively different densities to recognize the interface.

From Figures 5b and 6, one can also trace all the characteristic stages of the process of interaction of a droplet with a surface both in the superhydrophilic and superhydrophobic cases, as well as compare them with each other. The first stage is inertial, limited by the propagation time of the surface wave to the droplet apex. It can be seen that the upper part of the droplet on fundamentally different surfaces remains similar, while the droplet height is the same for superhydrophobic and superhydrophilic surfaces (Figures 6 and 7). In the second stage, the viscous one, a rapid (especially in the hydrophilic case) spreading of the droplet along the surface is observed. Finally, the final stage for superhydrophobic and superhydrophilic surfaces is fundamentally different. In the first case, the spreading is replaced by runoff and droplet rebound from the surface; in the second case, the droplet completely spreads over the surface.

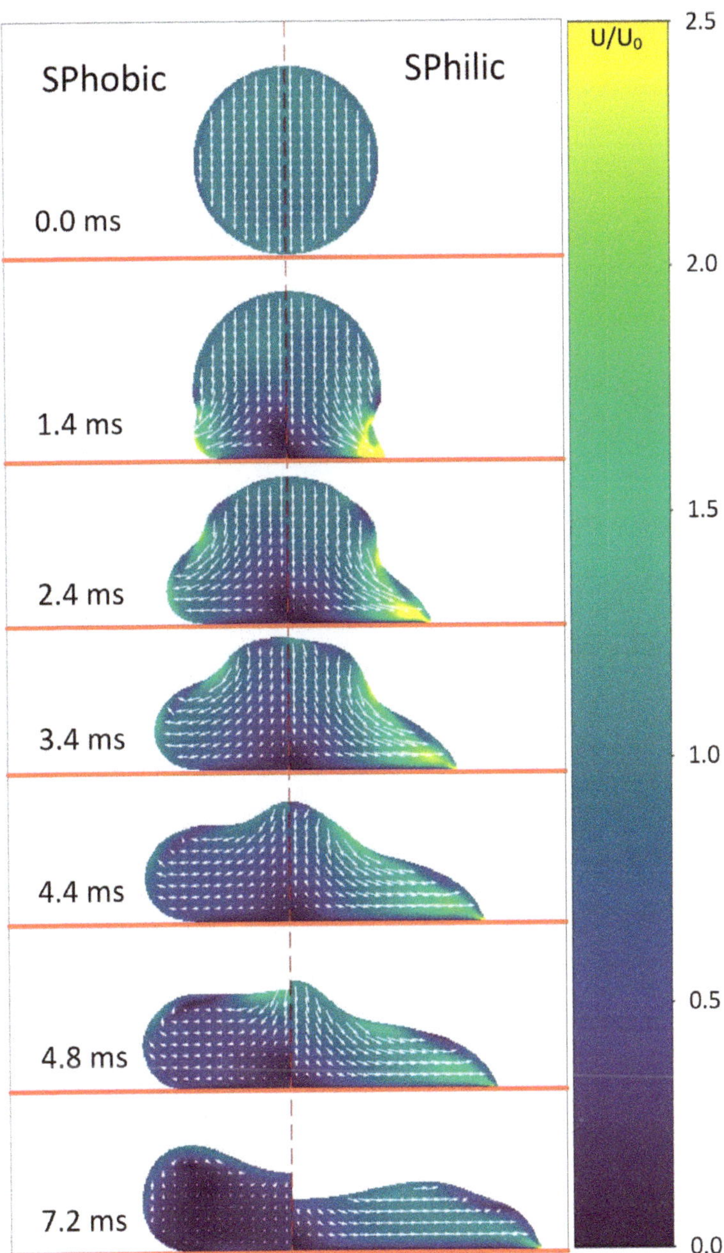

Figure 7. Velocity field in the droplet with D_0 = 2.9 mm falling on the superhydrophobic and superhydrophilic surfaces at We = 3.

The good agreement between calculation and experiment for the inertial spreading stage allows us to analyze in detail the dynamics of liquid flow (see also Figure 3b for surfaces 4 and 5). Thus, Figure 7 shows the velocity field in a droplet in contact with surfaces with different wettability. It can be seen that wave motion on the interface appears at the moment when the droplet touches the surface and reaches the droplet apex at t^*, which is the physical meaning of this parameter. It is seen that for $t = 3.4$ ms $> t^* = 3.2$ ms, there is a divergence in droplet height associated with wave deformation of the interface. Further spreading over the superhydrophobic and superhydrophilic surface becomes perfectly different throughout the volume of the droplet. By the time t_c, the velocity of the droplet on the superhydrophobic surface converges to zero at all points except the interface. However, for superhydrophilic surfaces, t_c has no physical meaning.

Using the t^* parameter, we compare the dynamics of falling droplets spreading on different surfaces with the data available in the literature (Figure 8). The list of references and experimental conditions are given in Table 2.

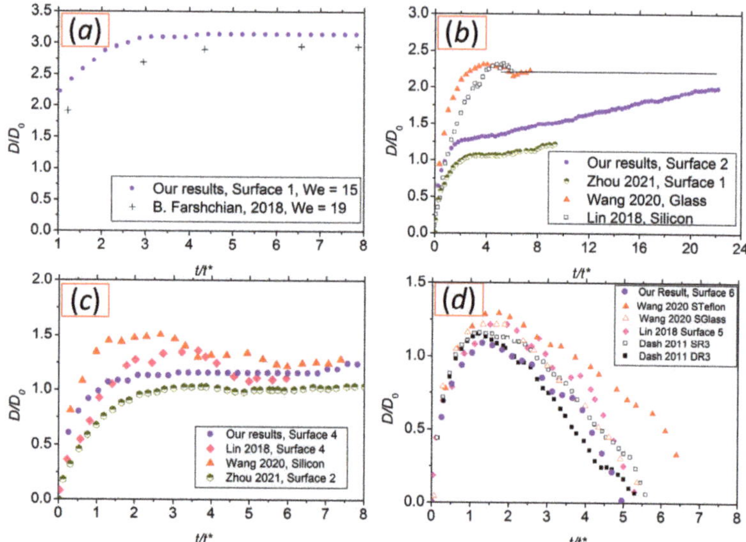

Figure 8. Comparison of experimental data on the dynamics of water droplets spreading on the (**a**) Superhydrophilic surfaces at $We = 15$–19, (**b**) surfaces with $CA = 20°$–$40°$ at $We = 1.5$–4, (**c**) surfaces with $CA = 84°$–$110°$, $We = 1.5$–4, (**d**) superhydrophobic surfaces $CA = 140°$–$160°$, $We = 2$–4.

We were unable to find data on the water droplets spreading dynamics on superhydrophilic surfaces at $We = 3$ in the literature; however, comparison with the data of Farshchian et al. [73] shows good agreement at $We \sim 17$ (Figure 8a). Figure 8b shows a comparison of contact line dynamics for droplets falling at $We = 1.5$–3 on our textured surface (Surface 2) with a contact angle of $20°$, a surface with a carbon nanotube forest with a contact angle of $29°$ [74] and smooth silicon and glass surfaces from other works [31,32]. The droplet spreading on the textured surfaces is identical; the difference in the maximum droplet spreading can be explained by a small difference in the contact angles and the We number. At the same time, there is a significant difference in the contact line motion on smooth and textured surfaces at the stage of viscous spreading ($t/t^* > 1$): first, there are no oscillations of the contact line, due to liquid flowing into the texture; second, reaching the equilibrium state requires considerable time, which may be related to the limited rate of water penetration into the material structure. For high contact angles (i.e., for thick fluoropolymer coating), lateral droplet movement over textured surfaces is limited to the pinning region reached by the end of the inertial spreading stage (Figure 8c), while oscillations occur on the smooth surface. However, the contact line diameters are close to each other for all surfaces considered at $t/t^* > 6$. The data for falling droplets on superhydrophobic surfaces are qualitatively similar (Figure 8d). The maximum droplet spreading and detachment distances from the surface at approximately the same time are reached under the condition that it is wetted in the Cassie–Baxter state. It was shown by Wang and

Fang [32] that the transition between Cassie–Baxter and Wenzel wetting states changes the dynamics of liquid motion at the retraction stage (STeflon Figure 8d), which can lead to later droplet detachment and partial adhesion of liquid to the surface.

Table 2. References for comparing the dynamics of water droplet spreading on surfaces with different wettability.

Reference	We	D_0, mm	Surface	ACA, °	RCA, °	SCA, °
S. Dash, et al. 2011, [30]	2.8	2.2	Single-roughness surface (SR3)	155	122	144
S. Dash, et al. 2011, [30]	2.8	2.2	Double-roughness surface (SR3)	165	155	166
S. Lin, et al. 2018, [31]	2	2.3	Fractal-like network of hydrophobized silica shells on clean glass slides (surface 5)	163	159	161
F. Wang, et al. 2020, [32]	4	2.5	Sanding Teflon (STeflon)	146	137	-
F. Wang, et al. 2020, [32]	4	2.5	Superhydrophobic solution NeverWet on a piece of clean glass (SGlass)	158	153	-
S. Lin, et al. 2018, [31]	2	2.3	Silanized silicon wafers (surface 4)	111	100	106
F. Wang, et al. 2020, [32]	4 17	2.5	Silicon	92	74	-
F. Wang, et al. 2020, [32]	4	2.5	Glass	46	21	-
S. Lin, et al. 2018, [31]	4	2.5	Silicon	31	-	27
B. Farshchian, et al. 2018, [73] *	19	2.3	Plasma-treated nanoparticles on PMMA	-	-	9
W. Ding, et al. 2022, [47] **	10.5	2	Salinized smooth microcones on silicon surface (SP8H20)	-	-	93
W. Ding, et al. 2022, [47] **	10.5	2	Salinized smooth microcones on silicon surface (SP8H20)	-	-	134
W. Ding, et al. 2022, [47] **	10.5	2	Salinized rough microcones on silicon surface (RP8H27)	-	-	159
M. Zhou, et al. 2021, [74]	1.5 17.7	2	Carbon nanotube forest treated by plasma (Substrate 1)	-	-	29
M. Zhou, et al. 2021, [74]	1.5 17.7	2	Carbon nanotube forest treated by plasma (Substrate 2)	-	-	84
M. Zhou, et al. 2021, [74]	1.5 17.7	2	Carbon nanotube forest treated by plasma (Substrate 3)	-	-	147

Notes: * We were unable to find data to compare superhydrophilic surfaces for a droplet falling from $We = 3$, so we give an example for $We = 19$. ** In Ding et al. [47], a change in wettability was achieved either by changing the texture or the liquid. Contact angles 134° and 159° were obtained for a water–ethanol mixture of 35%, and the contact angle of 93° was obtained for a mixture of 67%.

Of particular interest is the droplet behavior the Surface 5, which, on the one hand, has high adhesion characteristics and, on the other hand, has a large value of the contact angle of ~145°. The droplet behaves on a superhydrophobic surface for a considerable time up to $t/t^* \sim 3.5$ until the contact line pinning occurs. As a result, the droplet fails to detach from the surface. It is difficult to determine exactly at what point the droplet pinning to the surface occurs; however, unlike CA ~ 90°, the adhesion area is slightly smaller than the droplet diameter, i.e., it can be realized both at the inertial stage and later, as it was described by Lee et al. [45]. The authors of the paper report two types of transitions depending on the We number between the Cassie–Wenzel wetting states at different times. In the first case, liquid pinning is realized at the inertial stage. In the second case (at lower Weber numbers), wetting is observed at a stage much later in time during recoil and before the rebound. In addition, the rebounding droplet appears pinned to the surface through a small and central wetted area. We still suggest that fixation (i.e., the Cassie–Wenzel transition) occurs at the first stage of contact of the droplet with the surface. This is supported by the absence of dependence on the fixation moment on the We. The fixing moment for Surface 5 for the droplet diameter of 2.3 mm is reached at $t \sim 8$ ms (see

Supplementary information). In addition, the spreading data for Surfaces 4, 5, and 6 for $We = 11$ are presented in Figure 9. It should be noted that, in contrast to Figure 8, data on the ordinate axis result in the maximum droplet spreading D_m, and on the abscissa axis, the moment of the droplet rebound from superhydrophobic surface is $t_b = 10.2$ ms. This is carried out for a direct comparison with the results of Ding et al. [47], who investigated the liquid droplet spreading that falls on a silicon surface with a nanostructured microcone. The authors note that the fluid flowing into the structure leads to a fundamental difference in the behavior of the droplet on the retraction stage in comparison with a smooth surface, which is observed in our experiments. In the work of Ding et al. [47], the change in the contact angles was achieved either by changing the surface texture (a nanoporous structure on a microcone was created) or by mixing water with ethanol. The authors argue that the higher the static contact angle, the earlier the rebound from the superhydrophobic surface will occur, which is consistent with the results shown in Figure 8d. In addition, despite differences in liquid and structure types (Figure 9), the dynamics of water droplets spreading on our surfaces with different surface energies (but the same morphology) are identical to the results of Ding et al. [47] for similar contact angles. Our results in the range of $We = 1.5$–33 also agree very well with the work of M. Zhou [74], where the droplet ($D_0 = 2$ mm) falling on a carbon nanotube forest was considered, and the contact angle of the surfaces was varied by the plasma treatment intensity (Figure 8b,c and Figure 9b). Note that the data from this work generalize well using the dependence in Figure 5a, which gives a value of $t^* = 1.9$ ms. Figure 9b shows that after $t/t^* = 1$, the contact line dynamics on hydrophilic and superhydrophobic surfaces begin to differ. For comparison, Figure 8b also shows the water droplet spreading data on a smooth silicon surface with natural oxidation and it shows that the contact line dynamics is performed differently from the case of a droplet falling on a textured surface. A similar result is obtained for $We \sim 33$. Thus, it may be concluded that such a factor as the type of hierarchical structure has an insignificant effect on the falling droplets spreading on surfaces with high adhesion characteristics in the investigated range of $We = 0.3$–33, at least if the characteristic size of non-uniformities is less than 10 μm. However, on the other hand, according to Tang et al. [75], the change in texture roughness retained the character dynamics as spreading on a smooth surface, i.e., contact line fluctuations were observed (Wang 2020, Silicon in Figure 9).

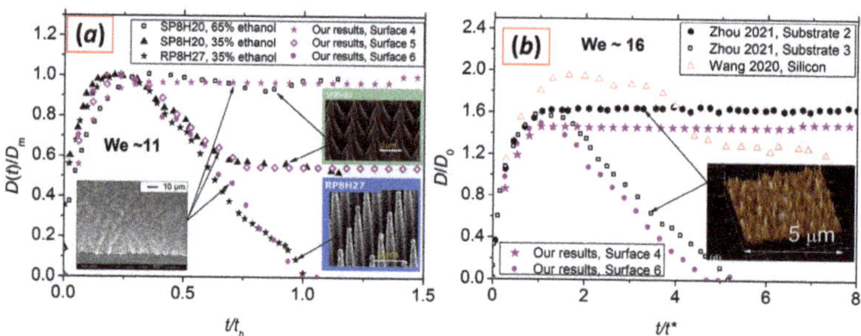

Figure 9. (**a**) Comparison with data of Ding et al. [47] for $We = 11$; the abscissa axis is normalized to the rebound moment t_b, the ordinate axis is normalized to the maximum spreading distance; (**b**) Comparison of our results for $We = 15$ with those of Zhou et al. [74] for $We = 17.7$ and Wang et al. [32] for $We = 17$. The insets show the microstructure of our surface and the surfaces from [32] and [74].

5. Conclusions

1. For the first time, we systematically studied the dynamics of falling water droplets on surfaces with identical hierarchical structures but different wettability in a wide range of contact angles 5–161° for $We = 0.3$–33.
2. We proposed a generalizing parameter—the time value $t^* = 0.66\,D_0^{3/2}$—corresponding to the transition between inertial and viscous flow regimes. We compared the dynam-

ics of water droplets falling at different velocities and onto different surfaces. It was shown that the parameter t^* does not depend on the *We* number in all investigated conditions.

3. Analyzing the velocity fields obtained by the LBM, it was found that the inertial spreading regime $<t^*$ corresponds to the moment of capillary-surface waves reaching the droplet apex for all surfaces in the considered conditions. The inertial-capillary number t_c corresponds to the zeroing of velocity for the superhydrophobic surface. However, for superhydrophilic surfaces, t_c has no physical meaning.

4. It was shown that surfaces with absolutely different hierarchical structures can provide the identity of the contact line dynamics for falling droplets, regardless of the liquid used, where the contact angles equality is the necessary condition.

5. It was found that the droplet spreading over surfaces with high adhesion force (or exhibiting the rose petal effect or, in other words, having very large contact angle hysteresis) is fundamentally different from droplets spreading over a smooth surface despite the equality of contact angles. For the first time, it was shown that the surface structure does not affect the dynamics of the falling droplet spreading if we deal with the rose petal effect, i.e., the key factor is the liquid to the surface adhesion force.

Supplementary Materials: The following supporting information can be downloaded at https://www.mdpi.com/article/10.3390/w15040719/s1, Figure S1: Snapshots of droplet $D_0 = 2.3$ mm falling on surface with different contact angle for We = 0.3, Figure S2: Snapshots of droplet $D_0 = 2.3$ mm falling on surface with different contact angle for We = 11, Figure S3: Snapshots of droplet $D_0 = 2.3$ mm falling on surface with different contact angle for We = 15, Figure S4: Snapshots of droplet $D_0 = 2.3$ mm falling on surface with different contact angle for We = 22, Figure S5: Snapshots of droplet $D_0 = 2.3$ mm falling on surface with different contact angle for We = 33.

Author Contributions: Conceptualization, M.-K.L.; methodology, V.T. and F.R.; validation, S.S.; investigation, N.M., A.R., and E.S.; resources, S.S. and A.S.; data curation, S.S. and A.S.; writing—original draft preparation, S.S.; writing—review and editing, S.S., F.R., and E.S.; visualization, N.M.; supervision, S.S.; project administration, S.S. All authors have read and agreed to the published version of the manuscript.

Funding: The reported study was funded by RFBR and NSFC, project number 21-52-53025 GFEN_a; the equipment for the study was provided as part of the financial support of a grant from the Government of the Russian Federation to support scientific research conducted under the guidance of leading scientists (mega-grant No. 075-15-2021-575).

Data Availability Statement: Not applicable.

Conflicts of Interest: The authors declare no conflict of interest.

References

1. Yu, C.; Liu, M.; Zhang, C.; Yan, H.; Zhang, M.; Wu, Q.; Liu, M.; Jiang, L. Bio-Inspired Drag Reduction: From Nature Organisms to Artificial Functional Surfaces. *Giant* **2020**, *2*, 100017. [CrossRef]
2. Weng, W.; Tenjimbayashi, M.; Hu, W.H.; Naito, M. Evolution of and Disparity among Biomimetic Superhydrophobic Surfaces with Gecko, Petal, and Lotus Effect. *Small* **2022**, *18*, 2200349. [CrossRef]
3. Liu, Y.; Li, G. A New Method for Producing "Lotus Effect" on a Biomimetic Shark Skin. *J. Colloid Interface Sci.* **2012**, *388*, 235–242. [CrossRef] [PubMed]
4. Liravi, M.; Pakzad, H.; Moosavi, A.; Nouri-Borujerdi, A. A Comprehensive Review on Recent Advances in Superhydrophobic Surfaces and Their Applications for Drag Reduction. *Prog. Org. Coatings* **2020**, *140*, 105537. [CrossRef]
5. Betz, A.R.; Jenkins, J.; Kim, C.J.; Attinger, D. Boiling Heat Transfer on Superhydrophilic, Superhydrophobic, and Superbiphilic Surfaces. *Int. J. Heat Mass Transf.* **2013**, *57*, 733–741. [CrossRef]
6. Drelich, J.; Marmur, A. Physics and Applications of Superhydrophobic and Superhydrophilic Surfaces and Coatings. *Surf. Innov.* **2014**, *2*, 211–227. [CrossRef]
7. Drelich, J.; Chibowski, E.; Meng, D.D.; Terpilowski, K. Hydrophilic and Superhydrophilic Surfaces and Materials. *Soft Matter* **2011**, *7*, 9804–9828. [CrossRef]
8. Starinskaya, E.; Miskiv, N.; Terekhov, V.; Safonov, A.; Li, Y.; Lei, M.; Starinskiy, S. Evaporation Dynamics of Sessile and Suspended Almost-Spherical Droplets from a Biphilic Surface. *Water* **2023**, *15*, 273. [CrossRef]

9. Wenzel, R.N. Resistance of Solid Surfaces To Wetting By Water. *Ind. Eng. Chem.* **1936**, *28*, 988–994. [CrossRef]
10. Cassie, A.B.D.; Baxter, S. Wettability of Porous Surfaces. *Trans. Faraday Soc.* **1944**, *40*, 546. [CrossRef]
11. Starinskiy, S.V.; Rodionov, A.A.; Shukhov, Y.G.; Safonov, A.I.; Maximovskiy, E.A.; Sulyaeva, V.S.; Bulgakov, A.V. Formation of Periodic Superhydrophilic Microstructures by Infrared Nanosecond Laser Processing of Single-Crystal Silicon. *Appl. Surf. Sci.* **2020**, *512*, 145753. [CrossRef]
12. Yildirim Erbil, H.; Elif Cansoy, C. Range of Applicability of the Wenzel and Cassie-Baxter Equations for Superhydrophobic Surfaces. *Langmuir* **2009**, *25*, 14135–14145. [CrossRef]
13. McHale, G. Cassie and Wenzel: Were They Really so Wrong? *Langmuir* **2007**, *23*, 8200–8205. [CrossRef]
14. Milne, A.J.B.; Amirfazli, A. The Cassie Equation: How It Is Meant to Be Used. *Adv. Colloid Interface Sci.* **2012**, *170*, 48–55. [CrossRef]
15. Feng, L.; Zhang, Y.; Xi, J.; Zhu, Y.; Wang, N.; Xia, F.; Jiang, L. Petal Effect: A Superhydrophobic State with High Adhesive Force. *Langmuir* **2008**, *24*, 4114–4119. [CrossRef]
16. Zheng, Y.; Chen, G.; Zhao, X.; Sun, W.; Ma, X. Falling Liquid Film Periodical Fluctuation over a Superhydrophilic Horizontal Tube at Low Spray Density. *Int. J. Heat Mass Transf.* **2020**, *147*, 118938. [CrossRef]
17. Zhu, Y.; Liu, J.; Hu, Y.; Yang, D.Q.; Sacher, E. Dynamic Behaviours and Drying Processes of Water Droplets Impacting on Superhydrophilic Surfaces. *Surf. Eng.* **2021**, *37*, 1301–1307. [CrossRef]
18. Hu, Z.; Chu, F.; Wu, X. Double-Peak Characteristic of Droplet Impact Force on Superhydrophobic Surfaces. *Extrem. Mech. Lett.* **2022**, *52*, 101665. [CrossRef]
19. Hu, Z.; Chu, F.; Lin, Y.; Wu, X. Contact Time of Droplet Impact on Inclined Ridged Superhydrophobic Surfaces. *Langmuir* **2022**, *38*, 1540–1549. [CrossRef] [PubMed]
20. Satpathi, N.S.; Malik, L.; Ramasamy, A.S.; Sen, A.K. Drop Impact on a Superhydrophilic Spot Surrounded by a Superhydrophobic Surface. *Langmuir* **2021**, *37*, 14195–14204. [CrossRef] [PubMed]
21. Tilton, D.E.; Kearns, D.A.; Tilton, C.L. Liquid Nitrogen Spray Cooling of a Simulated Electronic Chip. *Adv. Cryog. Eng.* **1994**, *39*, 1779–1786. [CrossRef]
22. Moreira, A.L.N.; Moita, A.S.; Panão, M.R. Advances and Challenges in Explaining Fuel Spray Impingement: How Much of Single Droplet Impact Research Is Useful? *Prog. Energy Combust. Sci.* **2010**, *36*, 554–580. [CrossRef]
23. Singh, M.; Haverinen, H.M.; Dhagat, P.; Jabbour, G.E. Inkjet Printing-Process and Its Applications. *Adv. Mater.* **2010**, *22*, 673–685. [CrossRef]
24. Kwon, H.-M.; Paxson, A.T.; Varanasi, K.K.; Patankar, N.A. Rapid Deceleration-Driven Wetting Transition during Pendant Drop Deposition on Superhydrophobic Surfaces. *Phys. Rev. Lett.* **2011**, *106*, 036102. [CrossRef]
25. Schutzius, T.M.; Jung, S.; Maitra, T.; Eberle, P.; Antonini, C.; Stamatopoulos, C.; Poulikakos, D. Physics of Icing and Rational Design of Surfaces with Extraordinary Icephobicity. *Langmuir* **2015**, *31*, 4807–4821. [CrossRef] [PubMed]
26. Choi, C.; Kim, M. Wettability Effects on Heat Transfer. In *Two Phase Flow, Phase Change and Numerical Modeling*; InTech: Shanghai, China, 2011; pp. 313–340. Available online: https://cdn.intechopen.com/pdfs/20823/InTech-Wettability_effects_on_heat_transfer.pdf (accessed on 10 February 2021).
27. Xu, Q.; Liu, Q.; Zhu, Y.; Luo, X.; Tao, L.; Guo, L. Entropy and Exergy Analysis of the Steam Jet Condensation in Crossflow of Subcooled Water. *Ann. Nucl. Energy* **2023**, *180*, 109485. [CrossRef]
28. Xu, Q.; Yuan, X.; Liu, C.; Wang, X.; Guo, L. Signal Selection for Identification of Multiphase Flow Patterns in Offshore Pipeline-Riser System. *Ocean Eng.* **2023**, *268*, 113395. [CrossRef]
29. Castrejón-Pita, J.R.; Muñoz-Sánchez, B.N.; Hutchings, I.M.; Castrejón-Pita, A.A. Droplet Impact onto Moving Liquids. *J. Fluid Mech.* **2016**, *809*, 716–725. [CrossRef]
30. Dash, S.; Kumari, N.; Garimella, S.V. Characterization of Ultrahydrophobic Hierarchical Surfaces Fabricated Using a Single-Step Fabrication Methodology. *J. Micromechanics Microengineering* **2011**, *21*, 105012. [CrossRef]
31. Lin, S.; Zhao, B.; Zou, S.; Guo, J.; Wei, Z.; Chen, L. Impact of Viscous Droplets on Different Wettable Surfaces: Impact Phenomena, the Maximum Spreading Factor, Spreading Time and Post-Impact Oscillation. *J. Colloid Interface Sci.* **2018**, *516*, 86–97. [CrossRef]
32. Wang, F.; Fang, T. Retraction Dynamics of Water Droplets after Impacting upon Solid Surfaces from Hydrophilic to Superhydrophobic. *Phys. Rev. Fluids* **2020**, *5*, 033604. [CrossRef]
33. Pachchigar, V.; Ranjan, M.; Sooraj, K.P.; Augustine, S.; Kumawat, D.; Tahiliani, K.; Mukherjee, S. Self-Cleaning and Bouncing Behaviour of Ion Irradiation Produced Nanostructured Superhydrophobic PTFE Surfaces. *Surf. Coatings Technol.* **2021**, *420*, 127331. [CrossRef]
34. Pan, Y.; Shi, K.; Duan, X.; Naterer, G.F. Experimental Investigation of Water Droplet Impact and Freezing on Micropatterned Stainless Steel Surfaces with Varying Wettabilities. *Int. J. Heat Mass Transf.* **2019**, *129*, 953–964. [CrossRef]
35. Lv, C.; Hao, P.; Zhang, X.; He, F. Drop Impact upon Superhydrophobic Surfaces with Regular and Hierarchical Roughness. *Appl. Phys. Lett.* **2016**, *108*, 141602. [CrossRef]
36. Guo, Y.; Zhang, X.; Wang, X.; Xu, Q.; Geng, T. Wetting and Tribological Properties of Superhydrophobic Aluminum Surfaces with Different Water Adhesion. *J. Mater. Sci.* **2020**, *55*, 11658–11668. [CrossRef]
37. Sun, S.; Li, H.; Guo, Y.; Mi, H.-Y.; He, P.; Zheng, G.; Liu, C.; Shen, C. Superefficient and Robust Polymer Coating for Bionic Manufacturing of Superwetting Surfaces with "Rose Petal Effect" and "Lotus Leaf Effect". *Prog. Org. Coatings* **2021**, *151*, 106090. [CrossRef]

38. Du, Q.; Zhou, P.; Pan, Y.; Qu, X.; Liu, L.; Yu, H.; Hou, J. Influence of Hydrophobicity and Roughness on the Wetting and Flow Resistance of Water Droplets on Solid Surface: A Many-Body Dissipative Particle Dynamics Study. *Chem. Eng. Sci.* **2022**, *249*, 117327. [CrossRef]

39. Renardy, Y.; Popinet, S.; Duchemin, L.; Renardy, M.; Zaleski, S.; Josserand, C.; Drumright-Clarke, M.A.; Richard, D.; Clanet, C.; Quéré, D. Pyramidal and Toroidal Water Drops after Impact on a Solid Surface. *J. Fluid Mech.* **2003**, *484*, 69–83. [CrossRef]

40. Eggers, J.; Fontelos, M.A.; Josserand, C.; Zaleski, S. Drop Dynamics after Impact on a Solid Wall: Theory and Simulations. *Phys. Fluids* **2010**, *22*, 062101. [CrossRef]

41. Ukiwe, C.; Kwok, D.Y. On the Maximum Spreading Diameter of Impacting Droplets on Well-Prepared Solid Surfaces. *Langmuir* **2005**, *21*, 666–673. [CrossRef]

42. Roisman, I.V.; Lembach, A.; Tropea, C. Drop Splashing Induced by Target Roughness and Porosity: The Size Plays No Role. *Adv. Colloid Interface Sci.* **2015**, *222*, 615–621. [CrossRef]

43. Zhang, H.; Zhang, X.; Yi, X.; Du, Y.; He, F.; Niu, F.; Hao, P. How Surface Roughness Promotes or Suppresses Drop Splash. *Phys. Fluids* **2022**, *34*, 022111. [CrossRef]

44. Li, X.; Ma, X.; Lan, Z. Dynamic Behavior of the Water Droplet Impact on a Textured Hydrophobic/Superhydrophobic Surface: The Effect of the Remaining Liquid Film Arising on the Pillars' Tops on the Contact Time. *Langmuir* **2010**, *26*, 4831–4838. [CrossRef] [PubMed]

45. Lee, C.; Nam, Y.; Lastakowski, H.; Hur, J.I.; Shin, S.; Biance, A.L.; Pirat, C.; Kim, C.J.; Ybert, C. Two Types of Cassie-to-Wenzel Wetting Transitions on Superhydrophobic Surfaces during Drop Impact. *Soft Matter* **2015**, *11*, 4592–4599. [CrossRef] [PubMed]

46. Cetiner, A.; Evren, B.; Budakli, M.; Arik, M.; Ozbek, A. Spreading Behavior of Droplets Impacting over Substrates with Varying Surface Topographies. *Colloids Surfaces A Physicochem. Eng. Asp.* **2020**, *606*, 125385. [CrossRef]

47. Ding, W.; Dorao, C.A.; Fernandino, M. Improving Superamphiphobicity by Mimicking Tree-Branch Topography. *J. Colloid Interface Sci.* **2022**, *611*, 118–128. [CrossRef]

48. Foltyn, P.; Ribeiro, D.; Silva, A.; Lamanna, G.; Weigand, B. Influence of Wetting Behavior on the Morphology of Droplet Impacts onto Dry Smooth Surfaces. *Phys. Fluids* **2021**, *33*, 063305. [CrossRef]

49. Schiaffino, S.; Sonin, A.A. Molten Droplet Deposition and Solidification at Low Weber Numbers. *Phys. Fluids* **1997**, *9*, 3172–3187. [CrossRef]

50. Almohammadi, H.; Amirfazli, A. Droplet Impact: Viscosity and Wettability Effects on Splashing. *J. Colloid Interface Sci.* **2019**, *553*, 22–30. [CrossRef]

51. Zhang, Z.; Liu, D.; Zhang, Y.; Xue, T.; Huang, Y.; Zhang, G. Fabrication and Droplet Impact Performance of Superhydrophobic Ti6Al4V Surface by Laser Induced Plasma Micro-Machining. *Appl. Surf. Sci.* **2022**, *605*, 154661. [CrossRef]

52. Zhang, H.; Zhang, X.; Yi, X.; He, F.; Niu, F.; Hao, P. Effect of Wettability on Droplet Impact: Spreading and Splashing. *Exp. Therm. Fluid Sci.* **2021**, *124*, 110369. [CrossRef]

53. Sun, R.; Bai, H.; Ju, J.; Jiang, L. Droplet Emission Induced by Ultrafast Spreading on a Superhydrophilic Surface. *Soft Matter* **2013**, *9*, 9285. [CrossRef]

54. Antonini, C.; Villa, F.; Bernagozzi, I.; Amirfazli, A.; Marengo, M. Drop Rebound after Impact: The Role of the Receding Contact Angle. *Langmuir* **2013**, *29*, 16045–16050. [CrossRef] [PubMed]

55. Xu, P.; Coyle, T.W.; Pershin, L.; Mostaghimi, J. From Lotus Effect to Petal Effect: Tuning the Water Adhesion of Non-Wetting Rare Earth Oxide Coatings. *J. Eur. Ceram. Soc.* **2020**, *40*, 1692–1702. [CrossRef]

56. Roy, D.; Pandey, K.; Banik, M.; Mukherjee, R.; Basu, S. Dynamics of Droplet Impingement on Bioinspired Surface: Insights into Spreading, Anomalous Stickiness and Break-Up. *Proc. R. Soc. A Math. Phys. Eng. Sci.* **2019**, *475*, 20190260. [CrossRef]

57. Serdyukov, V.; Starinskiy, S.; Malakhov, I.; Safonov, A.; Surtaev, A. Laser Texturing of Silicon Surface to Enhance Nucleate Pool Boiling Heat Transfer. *Appl. Therm. Eng.* **2021**, *194*, 117102. [CrossRef]

58. Safonov, A.I.; Sulyaeva, V.S.; Gatapova, E.Y.; Starinskiy, S.V.; Timoshenko, N.I.; Kabov, O.A. Deposition Features and Wettability Behavior of Fluoropolymer Coatings from Hexafluoropropylene Oxide Activated by NiCr Wire. *Thin Solid Films* **2018**, *653*, 165–172. [CrossRef]

59. Starinskiy, S.V.; Bulgakov, A.V.; Gatapova, E.Y.; Shukhov, Y.G.; Sulyaeva, V.S.; Timoshenko, N.I.; Safonov, A.I. Transition from Superhydrophilic to Superhydrophobic of Silicon Wafer by a Combination of Laser Treatment and Fluoropolymer Deposition. *J. Phys. D Appl. Phys.* **2018**, *51*, 255307. [CrossRef]

60. d'Humières, D. Multiple–Relaxation–Time Lattice Boltzmann Models in Three Dimensions. *Philos. Trans. R. Soc. London. Ser. A Math. Phys. Eng. Sci.* **2002**, *360*, 437–451. [CrossRef]

61. Shan, X.; Chen, H. Lattice Boltzmann Model for Simulating Flows with Multiple Phases and Components. *Phys. Rev. E* **1993**, *47*, 1815–1819. [CrossRef]

62. Shan, X.; Chen, H. Simulation of Nonideal Gases and Liquid-Gas Phase Transitions by the Lattice Boltzmann Equation. *Phys. Rev. E* **1994**, *49*, 2941–2948. [CrossRef] [PubMed]

63. Shen, S.; Bi, F.; Guo, Y. Simulation of Droplets Impact on Curved Surfaces with Lattice Boltzmann Method. *Int. J. Heat Mass Transf.* **2012**, *55*, 6938–6943. [CrossRef]

64. Li, Q.; Luo, K.H.; Li, X.J. Lattice Boltzmann Modeling of Multiphase Flows at Large Density Ratio with an Improved Pseudopotential Model. *Phys. Rev. E* **2013**, *87*, 053301. [CrossRef] [PubMed]

65. Ma, R.; Zhou, X.; Dong, B.; Li, W.; Gong, J. Simulation of Impacting Process of a Saturated Droplet upon Inclined Surfaces by Lattice Boltzmann Method. *Int. J. Heat Fluid Flow* **2018**, *71*, 1–12. [CrossRef]
66. Liu, X.; Cheng, P.; Quan, X. Lattice Boltzmann Simulations for Self-Propelled Jumping of Droplets after Coalescence on a Superhydrophobic Surface. *Int. J. Heat Mass Transf.* **2014**, *73*, 195–200. [CrossRef]
67. Bouzidi, M.; D'Humières, D.; Lallemand, P.; Luo, L.-S. Lattice Boltzmann Equation on a Two-Dimensional Rectangular Grid. *J. Comput. Phys.* **2001**, *172*, 704–717. [CrossRef]
68. Kupershtokh, A.L.; Medvedev, D.A.; Karpov, D.I. On Equations of State in a Lattice Boltzmann Method. *Comput. Math. Appl.* **2009**, *58*, 965–974. [CrossRef]
69. Khajepor, S.; Cui, J.; Dewar, M.; Chen, B. A Study of Wall Boundary Conditions in Pseudopotential Lattice Boltzmann Models. *Comput. Fluids* **2019**, *193*, 103896. [CrossRef]
70. Richard, D.; Clanet, C.; Quéré, D. Contact Time of a Bouncing Drop. *Nature* **2002**, *417*, 811. [CrossRef]
71. Liu, X.; Zhang, X.; Min, J. Spreading of Droplets Impacting Different Wettable Surfaces at a Weber Number Close to Zero. *Chem. Eng. Sci.* **2019**, *207*, 495–503. [CrossRef]
72. Roux, D.C.D.; Cooper-White, J.J. Dynamics of Water Spreading on a Glass Surface. *J. Colloid Interface Sci.* **2004**, *277*, 424–436. [CrossRef]
73. Farshchian, B.; Pierce, J.; Beheshti, M.S.; Park, S.; Kim, N. Droplet Impinging Behavior on Surfaces with Wettability Contrasts. *Microelectron. Eng.* **2018**, *195*, 50–56. [CrossRef] [PubMed]
74. Zhou, M.; Qu, J.; Ji, Y. Dynamic Characteristics of Droplet Impingement on Carbon Nanotube Array Surfaces with Varying Wettabilities. *Appl. Surf. Sci.* **2021**, *554*, 149359. [CrossRef]
75. Tang, C.; Qin, M.; Weng, X.; Zhang, X.; Zhang, P.; Li, J.; Huang, Z. Dynamics of Droplet Impact on Solid Surface with Different Roughness. *Int. J. Multiph. Flow* **2017**, *96*, 56–69. [CrossRef]

 *water*

Article

The Effect of the Angle of Pipe Inclination on the Average Size and Velocity of Gas Bubbles Injected from a Capillary into a Liquid

Anastasia E. Gorelikova [1,2,*], Vyacheslav V. Randin [1,2,†] , Alexander V. Chinak [1] and Oleg N. Kashinsky [1]

1   Heat Mass Transfer Problem Laboratory Power Engineering, Kutateladze Institute of Thermophysics SB RAS, 630090 Novosibirsk, Russia
2   Department of Physics, Novosibirsk State University, 630090 Novosibirsk, Russia
*   Correspondence: gorelikova.a@gmail.com
†   Deceased.

Abstract: This work is devoted to an experimental study of the effect of coalescence on the average diameter and velocity of gas bubbles in an inclined pipe. The measurements were carried out for agas flow rate of 3.3 and 5 mL/min at pipe inclination angles of 30–60°. The study of gas bubble diameters was performed using a shadow photography method. The values of the average diameter and velocity of the bubbles were obtained depending on the angle of inclination of the pipe. A map of regime parameters was constructed at which gas bubbles form a stable structure—a chain of bubbles with an equal diameter.

Keywords: inclined pipe; bubbles; coalescence; bubble velocity; chain of bubbles

1. Introduction

Gas bubbles floating in a stationary or moving liquid can have a significant effect on heat exchange due to the mixing of the wall layers of the liquid. The paper [1] shows that, in a downward bubbly flow, a change in the size of the dispersed phase can lead to both intensification and deterioration of the heat transfer as compared to a single-phase flow at constant rates of liquid and gas flow at the channel inlet. Adding small gas bubbles to the flow leads to "laminarization" in the wall region and deterioration in the heat transfer by about 25% as compared to a single-phase flow. Large bubbles lead to a higher turbulence level in the near-wall region, an increase in the average friction, and an intensification of the heat transfer up to 50%. In the paper [2], the addition of air bubbles led to a significant increase in the heat transfer rate (up to 300%) in a downstream bubbly flow in a sudden pipe expansion.

The dynamics of bubbles can be nonlinear and complex [3]. The movement of bubbles is influenced by the geometry of the channel, the internal diameter of the capillary on which the bubble is formed [4], and the characteristics of the liquid [5].

The paper [6] presented an experimental investigation of highly deformed bubbles. The dynamics of bubble formation were investigated via experiments with ultrapure water and silicone fluids, with gas flow rates of 5–300 mL/min. The behavior of bubbles in the immediate vicinity of the capillary was considered. It was shown that the coalescing dynamics are affected by capillary inertia; the viscosity has a negligible effect on coalescence. A phase diagram for coaxial coalescence and no coalescence was presented in terms of the Weber number and the Morton number to describe the effects of inertia and fluid properties on the dynamics of bubble formation.

Much attention has been paid to the study of vertical pipes and channels [7–9]. The efficiency of heat and mass transfer and gas-phase residence time largely depends on the dynamics of gas bubbles. According to the paper [10], gas–liquid interaction in the bubble

Citation: Gorelikova, A.E.; Randin, V.V.; Chinak, A.V.; Kashinsky, O.N. The Effect of the Angle of Pipe Inclination on the Average Size and Velocity of Gas Bubbles Injected from a Capillary into a Liquid. *Water* **2023**, *15*, 560. https://doi.org/10.3390/w15030560

Academic Editor: Wencheng Guo

Received: 21 November 2022
Revised: 26 January 2023
Accepted: 28 January 2023
Published: 31 January 2023

flow can have a significant impact on the size and velocity of bubbles, the area of interaction between gas and liquid, and the number of small bubbles.

Studies of multiphase systems with horizontal channels have been widely presented [11,12]. In the paper [13], an experimental investigation of bubbly flow in an annulus pipe was presented. It was shown that the presence of various values of total dissolved solid materials can affect the thermal conductivity of water and the bubble formation characteristics.

Much less attention has been paid to inclined pipes and channels, despite the fact that the angle of inclination can make a significant contribution to the nature of gas–liquid flows and affect the bubbles that move in stationary liquid.

The paper [14] presented an experimental study of the velocity of a free-floating gas slug in glass tubes with diameters of 11.8–30 mm using a time-of-flight method combined with high-speed video shooting and numerical processing of sequential images. The presented results indicate that the bubble velocity depends on the angle of inclination of the pipe and is non-monotonic in nature.

The paper [15] presented the results of an experimental study of the rise of single bubbles in an inclined flat channel. The fluid velocity was 0–0.2 m/s, the volume of one bubble varied in the range from 1 to 80 mL, and the channel depth was 8, 4 or 1.5 mm. It was shown that with the vertical arrangement of the channel, the bubble velocity primarily depends on the depth of the channel. When the angle of inclination of the channel changes from 0° to 90°, the velocity of the bubble monotonously increases and reaches a maximum with the vertical arrangement of the channel. This differs from the results obtained for projectile flows in round inclined pipes with large cross sections, where the velocity reaches a maximum at angles of inclination close to 45°.

The paper [16] studied the dynamics of growth and collapse of vapor bubbles for a laser-induced bubble on or near the wall. The difference in the shapes of bubbles near and on the wall was demonstrated. Deformation of a spherical bubble floating near the wall was noted. It was shown that the lifetime of the bubble near the wall was longer than the lifetime of the bubble on the wall.

In [17], liquid bubbles with a relative spherical diameter in the range of 4.75–9.14 mm floating near an inclined surface (inclination angle of 30°) were investigated. The authors discuss an increase by up to 8 times in the local heat transfer coefficient in comparison to the case of free convection, in which the average heat transfer value increases by 2 times compared to in the case without the addition of the gas phase.

In the paper [18], a study of three-dimensional trajectories of bubbles in a stationary liquid with working volumes of $300 \times 300 \times 1500$ mm^3 and $300 \times 150 \times 500$ mm^3 was carried out. Bubbles with a diameter of 3–5 mm were considered (depending on the capillary and fluid flow, the diameter of the bubbles changed slightly). The dependences of the effect of gas flow rate, height of the liquid column, and nozzle diameter on bubble ascent trajectory and deformation of the bubble surface are shown.

The paper [19] presented a bubble coalescence model consisting of two steps. First, an existing model of coalescence of bubbles of the same diameter in a turbulent flow was expanded to the case of bubbles of different diameters. In the second step, the obtained expression for the coalescence rate is used to obtain the dependences of the kinetic equations on the bubble diameter, which can be estimated using CFD packages (CFD, Computational Fluid Dynamics). As a result, a compact expression is obtained to describe the evolution of bubble sizes.

The paper [20] is devoted to the development of three-dimensional models of multiphase models for calculations and analysis of processes in nuclear reactors (Pressurized Water Nuclear Reactor (PWR)). A wide overview of methods is given, including various boiling models, Direct Numerical Simulation (DNS), calculations of adiabatic flows, etc. Despite the fact that models have been developed since 1980, and significant progress has been made in forecasting flows, the authors point out the need for further improvement in methods and for comparison of the results obtained with experimental data.

It is impossible not to note the convenience and relatively low cost of computer modeling of two-phase flows in comparison to experimental works. However, when using various software packages, the question of the applicability of the algorithms, validation of the data obtained, and appropriateness of the models developed for calculating the physical quantities of gas–liquid flows becomes acute. In addition, due to the complexity of the flow structure, it is often difficult to obtain results without using already existing empirical data.

Despite the increasing interest in the influence of the angle of inclination on the movement of bubbles and on gas–liquid flows, there is a significant shortage of experimental data. In the empirical work, attention is mostly paid to large gas consumption and large-diameter bubbles, and angles of inclination of the channels close to the vertical or horizontal position are chosen.

This work is devoted to the experimental study of the effect of coalescence on the average size and velocity of gas bubbles in an inclined pipe. The values of the average diameter and average velocity of the bubbles were obtained depending on the angle of inclination of the pipe. A map of regime parameters was constructed at which gas bubbles form a stable structure—a chain of bubbles with an equal diameter.

2. Experimental Setup and Technique

The scheme of the experimental setup is shown in Figure 1. Gas (air) was supplied from the compressor through the gas flow meter (1) and introduced into the liquid through one capillary of 0.2 mm inner diameter (2). The gas flow rate was monitored using a mass flow controller, Aalborg GFC17 (flow range 0–10 mL/min, accuracy $\pm$ 0.1 mL/min). The test section was a 1.2 m long tube made of Plexiglas with an inner diameter of 32 mm. Bubbles were filmed by a Nikon Zfc camera (4) through an optical section, and the flow was illuminated by an LED matrix (5). The distance from the location of the gas-phase introduction to the measurement point was from 100 to 600 mm. The channel inclination angle θ was counted from the vertical line, with the value $\theta = 0°$ corresponding to the vertical position of the channel and $\theta = 90°$ to the horizontal position. The measurements were carried out for the gas flow rates of 3.3 and 5.0 mL/min at pipe inclination angles of 30–60°. This range was chosen because, at similar channel inclination angles, there was a significant increase in heat exchange from the channel wall for gas–liquid bubble flows [21].

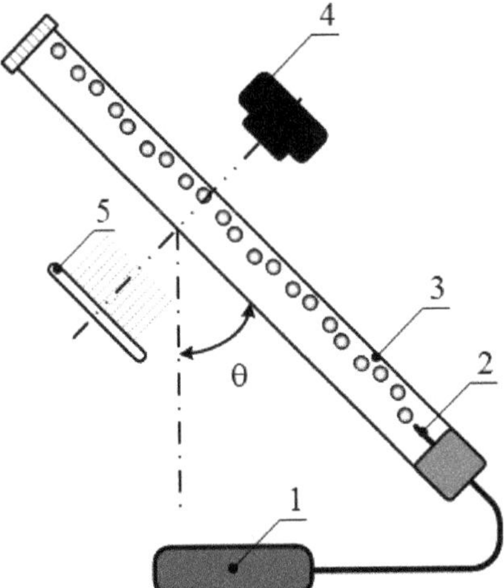

Figure 1. Experimental setup.: 1—flow rate meter; 2—capillary; 3—test section; 4—camera; 5—LED matrix.

Distilled water was used as the working fluid in the experiments, and air was used as the working gas. The exact physical properties of the working fluid [22] are presented in Table 1. All experiments were carried out at the standard pressure (1 atm) and room temperature (20 °C ± 1).

Table 1. Physical properties of the liquid used in this study (at 20 °C).

Fluid	Density, ϱ (kg/m^3)	Viscosity, μ (mPa·s)	Surface Tension, σ (N/m)
Distilled water	998	1	0.072

Figure 2 shows an example of the software processing of the received images. The processing was carried out in two stages. In the first stage, the videos were split into separate frames and translated into the "Grayscale" format. According to the level of illumination and the gradient illumination, the boundaries of the objects were found, and the frames were binarized. In the second stage of processing, the properties of the objects—size and location—were found. Correlation analysis of successive frames was carried out to determine the speed and trajectories of the bubbles.

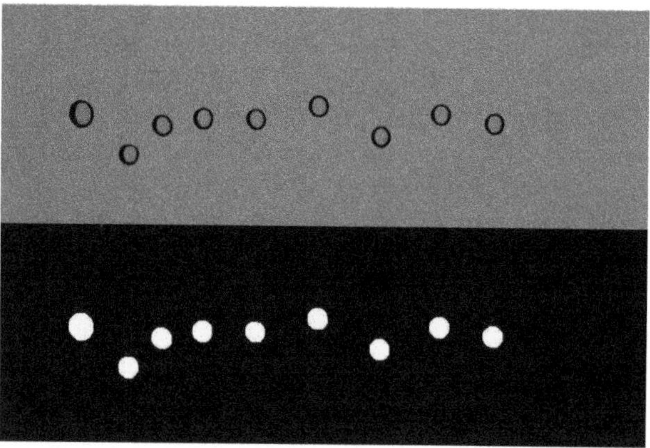

Figure 2. Example of image preprocessing for calculating the diameter and velocity of gas bubbles.

The resulting images were processed numerically, similarly to the method described in [23]. The diameter of gas bubbles was calculated from the area of the bubble in the image as an equivalent diameter according to the formula:

$$D = \sqrt{4S/\pi} \tag{1}$$

where S is the area of the bubble in the image.

The size of the main part of the bubbles in the experiments did not exceed 2–3 mm. Bubbles of this size are close in shape to a sphere, which allows the use of this approximation.

The absolute accuracy of determining the diameter of bubbles ΔD is calculated using the formula:

$$\Delta D = \sqrt{\left(\frac{\partial f}{\partial S}\Delta S\right)^2} \tag{2}$$

Using the formula for determining the equivalent diameter of the bubble (1) as a function in Equation (2), we obtain an expression for the relative accuracy of determining the diameter δD:

$$\Delta D = \sqrt{\left(\frac{\partial f}{\partial S}\Delta S\right)^2} = \frac{\partial\left(\sqrt{\frac{4S}{\pi}}\right)}{\partial S}\Delta S = \frac{1}{2}\sqrt{\frac{4}{\pi S}}\Delta S \frac{1}{2}\sqrt{\frac{4S}{\pi S^2}}\Delta S = \frac{1}{2}\sqrt{\frac{4S}{\pi}}\frac{\Delta S}{S} = \frac{1}{2}D\delta S \quad (3)$$

$$\frac{\Delta D}{D} = \delta D = \frac{1}{2}\delta S \quad (4)$$

The accuracy of determining the position of the bubble boundary was ± 1 pixel. According to the calibration frames, 1 mm is equal to 22 pixels, and the relative accuracy of determining the diameter for bubbles of 0.3–7 mm is 0.01–0.1. The value of the gas flow rate obtained during video processing converged with the values of the mass flow controller with an accuracy of 0.05.

The velocity of the bubbles was calculated from the frame-by-frame shift.

To validate the measuring system, a series of experiments were carried out with a vertical orientation of the pipe ($\theta = 0°$) and a gas flow rate $Q_g = 3.3$ mL/min.

The rate of bubble ascent is influenced by the buoyancy force

$$F_B = \frac{\pi D_b^3}{6}(\rho_L - \rho_G)g \quad (5)$$

and the drag force

$$F_D = \frac{\pi D_b^2}{4}C_d\frac{\rho_L}{2}V_b^2 \quad (6)$$

For a large volume of a quiescent liquid, the bubble rise velocity is

$$V_b = \sqrt{\frac{4D_b(\rho_L - \rho_G)g}{3\rho_L C_d}} \quad (7)$$

where C_d is the drag coefficient. Bubbles with a small diameter (up to 3–4 mm) do not experience pulsation of shape, and their movement can be described as the movement of a rigid sphere. When liquid flows around a rigid sphere, the drag coefficient is equal to [24]:

$$\begin{cases} C_d = 24/Re_b, \; at \; Re_b < 2 \\ C_d = 18.5/(Re_b)^{0.6}, \; at \; 2 < Re_b < 500 \\ C_d = 0.44, \; at \; 500 \le Re_b \end{cases} \quad (8)$$

where the Reynolds number Re_b is calculated relative to the velocity V_b and diameter D_b of the gas bubble. Measuring values were compared with the calculated bubble rise velocity for the same bubble diameters (Figure 3).

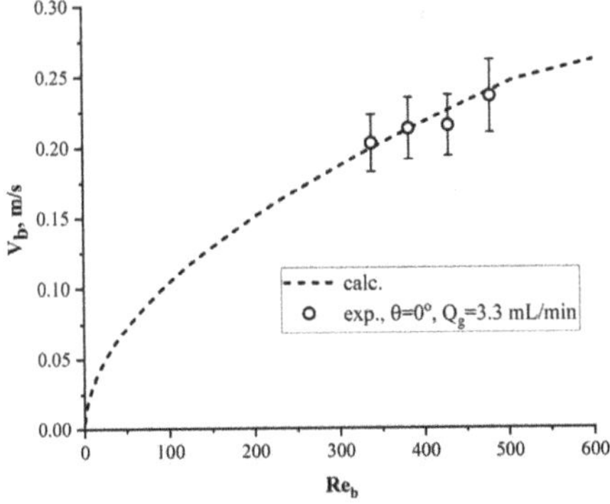

Figure 3. Comparisons of bubble rise velocity in vertical pipe ($\theta = 0°$) in experiments, and calculated bubble rise velocity.

3. Results and Discussion

3.1. Bubble Chain Movement Mode

Figure 4 shows characteristic images of bubbles with equal gas flow rates (Q_g = 3.3 mL/min) and equal distances from the gas input point to the shooting point (L = 200 mm) for different angles of inclination of the pipe (θ = 30°–60°).

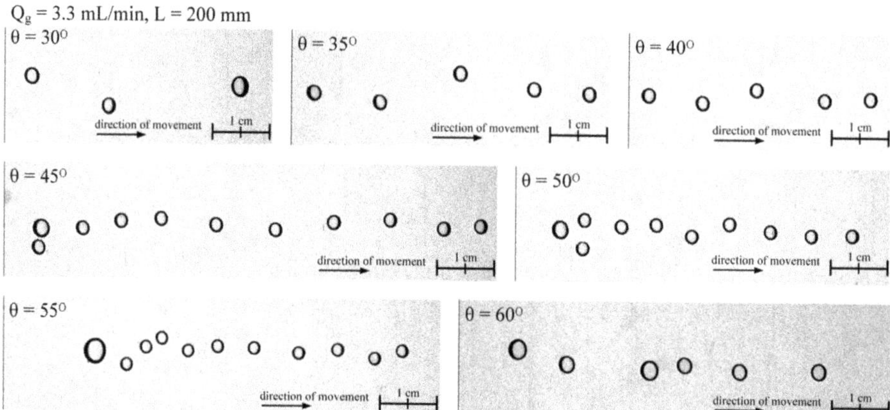

Figure 4. Characteristic images of bubbles. Gas flow rates Q_g = 3.3 mL/min; distances from the gas input point to the shooting point L = 200 mm; angle of inclination of the pipe θ = 30°–60°.

At channel inclination angles θ = 30°–35°, the diameter and the nature of the movement of bubbles largely depended on their departure diameter and coalescence near the capillary. The bubbles oscillated perpendicular to the direction of motion and moved one at a time or in small, unstructured groups.

At pipe inclination angles from 40° to 55°, the bubbles formed a chain. The perpendicular velocity component was significantly reduced due to friction against the upper wall of the pipe. Between 5 and 15 bubbles in the chain had the same diameter, and the last one had a significantly increased diameter in comparison with the others. The long chains of bubbles were formed at angles of inclination θ = 45°–50°. With a further increase in the angle of inclination of the pipe, the distance between the bubbles decreased. Reducing the distance between the bubbles led to the coalescence of the bubbles. This resulted in the destruction of the chain of bubbles at an angle θ = 60°.

Figure 5 shows an example of bubble movement in modes without cluster formation (pipe inclination angle θ = 35°) and with cluster formation (pipe inclination angle θ = 50°) at a gas flow rate Q_g = 3.3 mL/min. For cluster modes of bubble movement, the average values of bubble size and velocity were characteristically close to the values of size and velocity for a single bubble in a chain (deviation of no more than 5%). For modes without cluster formation, there were several typical bubble diameters that moved with different velocities.

Figure 6 shows a map of the modes of movement of the bubbles, as well as the area of formation of a chain of bubbles.

For gas flow rates from 3.0 to 5.0 mL/min and distances from the gas input point to the shooting point from 100 to 600 mm at pipe inclination angles of 30°–35°, the formation of a chain of bubbles did not occur. The oscillatory movements of the bubbles did not allow them to form a stable cluster structure.

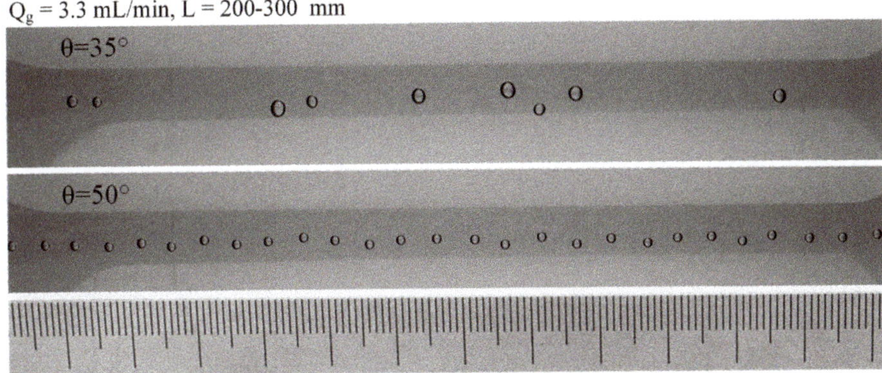

Figure 5. Bubble movement in modes without cluster formation ($\theta = 35°$) and with cluster formation ($\theta = 50°$); gas flow rate $Q_g = 3.3$ mL/min.

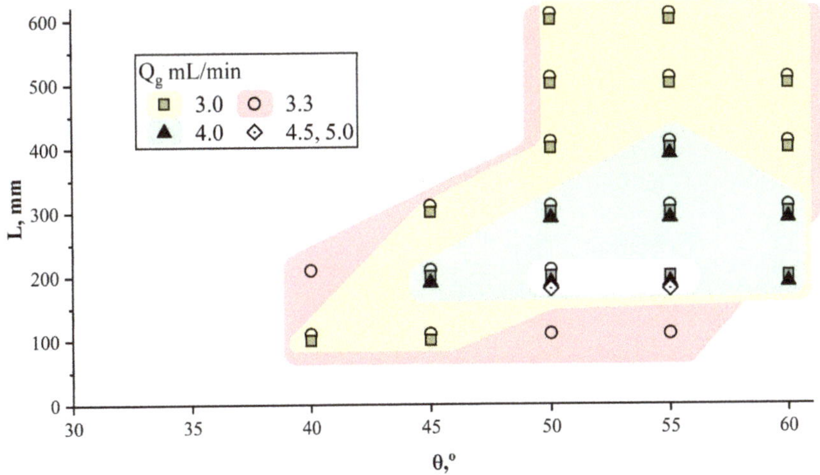

Figure 6. Map of the modes of movement of bubbles. The points correspond to the parameters for which the bubble chain mode appeared.

Chains of bubbles began to form at channel angles of 40° for gas flow rates 3.0 and 3.3 mL/min. Increasing the angle of inclination of the pipe increased the friction of the bubbles against the upper wall of the pipe. This led to a decrease in the transverse vibrations of the bubbles and allowed the chain of bubbles to form.

For gas flow rates of 3.0 and 3.3 mL/min, with an increase in the angle of inclination, the distance at which the bubble chain mode could be observed also increased up to 600 mm.

For gas flow rates of 4.0 mL/min, the maximum distance at which the bubble chain mode was observed first increased with an increase in the angle of inclination, and then decreased for channel inclination angles of more than 50°. For gas flow rates of 4.5 and 5.0 mL/min, clusters of bubbles were formed only at the distance of $L = 200$ mm and at angles of 45° and 50°.

For gas flow rates of more than 5 mL/min, the bubble chain mode was not observed. The coalescence of bubbles near the capillary and those moving along the channel led to an increase in the diameter of individual bubbles, and the greater distance between individual bubbles did not allow them to form a chain.

3.2. Average Size of Gas Bubbles in the Inclined Tube

The dependence of the average size of gas bubbles as a function of the pipe inclination angle is shown in Figure 7.

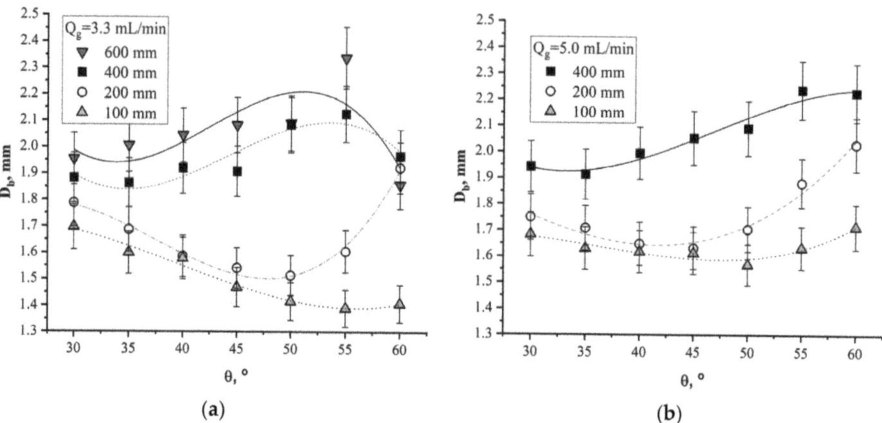

Figure 7. Dependence of the average size of gas bubbles as a function of the pipe inclination angle. Gas flow rate (**a**) Q_g = 3.3 mL/min; (**b**) Q_g = 5.0 mL/min.

The average size of the gas bubbles was D_b = 1.4–2.3 mm for the gas flow rate Q_g = 3.3 mL/min and D_b = 1.5–2.2 mm for the gas flow rate Q_g = 5.5 mL/min.

At a distance of 100 mm for the gas flow rate Q_g = 3.3 mL/min (Figure 7a), the average diameter decreased with an increase in the angle of inclination of the pipe, since the departure diameter decreased due to the angle of inclination of the capillary. At a distance of 200 mm and pipe inclination angles up to 50°, the average diameter decreased due to the decrease in the departure diameter. However, at angles of inclination of more than 50°, bubbles were actively grouped when moving along the upper wall, and the average diameter increased due to coalescence. At distances of 400 and 600 mm, the effect of bubble coalescence along the upper wall of the inclined pipe compensated for the decrease in the departure diameter in the range of channel inclination angles from 30° to 50°. The average diameter of gas bubbles at the flow rate of 5 mL/min (Figure 7b) demonstrated similar behavior.

3.3. Average Velocity of Gas Bubbles in the Inclined Tube

The dependence of the average velocity of gas bubbles as a function of the pipe inclination angle is shown in Figure 8.

As the angle of inclination increased, the projection of the Archimedes force along the axis of movement of the bubbles decreased. Due to this, the bubbles were pressed against the upper wall of the inclined pipe and moved more slowly due to the friction generated by their movement against the wall.

At a distance of 100 mm for the gas flow rate Q_g = 3.3 mL/min (Figure 8a), the velocity of the bubbles decreased throughout the selected range of angles. At a distance L = 200 mm for angles of 55° and 60°, an increase in the velocity of the bubbles was observed. This was due to the fact that the bubbles were actively coalescing when moving along the upper wall of the pipe, and the average velocity of the bubbles increased due to the Archimedes force. The bubble chain mode for angles greater than 55 was not obtained. For the gas flow rate Q_g = 5.0 mL/min (Figure 8b), the dependences of the average velocity of the bubbles for different angles of inclination of the pipe demonstrated similar behavior. An increase in the average velocity for the same distance L was associated with the increase in the average diameter of the bubbles as the gas flow rate increased. The nonlinearity in the average velocity graphs for the distance $L = 400$ mm was associated with the coalescence of bubbles when moving along the upper wall of an inclined pipe. This was also typical for gas flow rate $Q_g = 3.3$ mL/min.

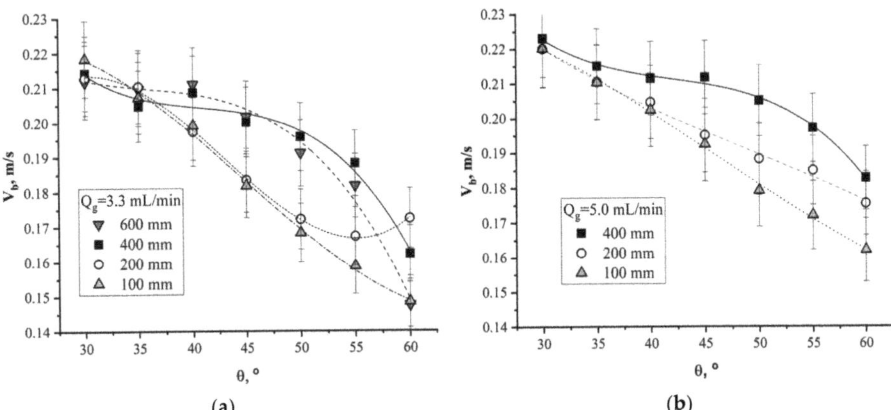

Figure 8. Dependence of the average velocity of gas bubbles as a function of the pipe inclination angle. Gas flow rate (**a**) $Q_g = 3.3$ mL/min; (**b**) $Q_g = 5.0$ mL/min.

The movement of bubbles was caused by buoyancy forces, surface tension, and the force of friction against the upper wall of the inclined channel. One of the most important dimensionless characteristics of the ascent of gas bubbles is the Reynolds number $Re_b = \rho V_b D_b / \mu$, where ϱ is the density of the fluid, μ is the dynamic viscosity of the fluid, V_b is the mean velocity of bubbles, and D_b is the mean diameter of the bubbles. Figure 9 shows the values of the Reynolds number for gas bubbles, depending on the distance between the gas input point and the point of measurement.

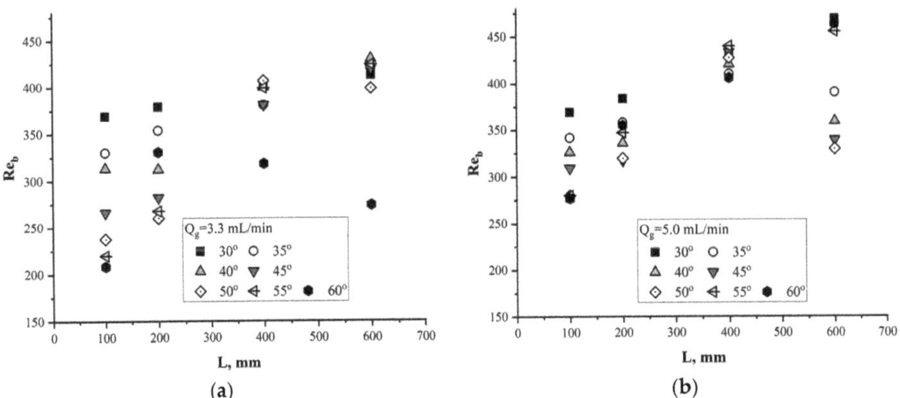

Figure 9. The dependence of the Reynolds number Re_b on the distance between the gas input point and the camera shooting point. Angle of inclination of the pipe $\theta = 30°$–$60°$, gas flow rate (**a**) $Q_g = 3.3$ mL/min; (**b**) $Q_g = 5.0$ mL/min.

As the distance from the gas input point to the measurement point increased from 100 to 600 mm, the graphs for all pipe inclination angles (except 60°) tended toward the value of the Reynolds number $Re_b = 390$–450 for a gas flow rate $Q_g = 3.3$ mL/min (Figure 9a). The behavior of the Reynolds number at the channel angle of 60° differed significantly from its behavior at other angles. From this angle of inclination, the velocity of bubbles due to coalescence and the growth of the average diameter were significantly reduced, leading to a decrease in the Reynolds number.

For the gas flow rate Q_g = 5.0 mL/min (Figure 9b), the graphs tended toward the value of the Reynolds number Re_b = 390–450 at a distance from the gas input point to the camera shooting point L = 400 mm. With an increase in the distance to L = 600 mm, due to coalescence, the differences in the diameters and velocity of the bubbles increased, leading to a discrepancy in the graphs of the Reynolds number.

4. Conclusions

An experimental study of bubble size from a single capillary in an inclined tube was presented.

A map of bubble movement modes was built. It is shown for which values of the gas flow rate, the angle of inclination of the pipe, and the distance from the gas input point to the measurement point the individual bubbles form a chain of bubbles. For gas flow rates of 3.0 and 3.3 mL/min, chains of bubbles were formed in a wide range of pipe inclination angles and distances from the gas input point to the shooting point. When the gas flow rate increased to 4.0 mL/min or more, the range of formation of chains of bubbles decreased due to the coalescence of bubbles as they moved along the upper wall of the inclined pipe. At gas flow rates of more than 5.0 mL/min, chains of bubbles were not formed.

The values of the average diameters were obtained depending on the angle of inclination of the pipe. The average diameter of the gas bubbles was D_b = 1.4–2.3 mm for the gas flow rate Q_g = 3.3 mL/min and D_b = 1.5–2.2 mm for the gas flow rate Q_g = 5.5 mL/min. With an increase in the angle of inclination of the channel, the average diameter of the gas bubbles decreased. The nonlinearity of the graphs of the average diameter was caused by the influence of the angle of inclination on the departure diameter, the coalescence of bubbles near the capillary, and the coalescence of bubbles when moving along the upper wall of the inclined pipe.

Dependences of the average velocity of bubbles on the angle of inclination of the pipe were obtained. With an increase in the angle of inclination of the pipe, the average velocity of the bubbles decreased. This was due to the friction on the upper wall and an increase in the influence of the Archimedes force. With an increase in the angle of inclination of the channel from 30° to 60°, the average velocity of bubbles decreased from ~0.22 m/s to ~0.15 m/s for the gas flow rate Q_g = 3.3 mL/min and to ~0.18 m/s for the gas flow rate Q_g = 5.0 mL/min. The nonlinearity in the average velocity graphs for distances L = 400 mm and greater was associated with the coalescence of bubbles when moving along the upper wall of an inclined pipe.

The results can be used to verify CFD packages and for a deeper analysis of the results of complex gas–liquid flows, including under the conditions of nuclear reactors.

Author Contributions: Conceptualization, A.E.G., V.V.R., A.V.C. and O.N.K.; methodology, A.E.G., V.V.R. and A.V.C.; software, A.V.C.; investigation, A.E.G.; resources, A.E.G., A.V.C. and O.N.K.; data curation, A.E.G. and A.V.C.; writing—original draft preparation, A.E.G., A.V.C. and O.N.K.; writing—review and editing, A.E.G., A.V.C. and O.N.K.; project administration, O.N.K. All authors have read and agreed to the published version of the manuscript.

Funding: This research was funded by RSF, grant number 22-21-20029 and Ministry of Science and Innovation Policy of the Novosibirsk region, https://rscf.ru/project/22-21-20029/ (accessed on 27 January 2023).

Institutional Review Board Statement: Not applicable.

Informed Consent Statement: Not applicable.

Conflicts of Interest: The authors declare no conflict of interest.

References

1. Lobanov, P.D. Wall Shear Stress and Heat Transfer of Downward Bubbly Flow at Low Flow Rates of Liquid and Gas. *J. Eng. Thermophys.* **2018**, *27*, 232–244. [CrossRef]
2. Lobanov, P.; Pakhomov, M.; Terekhov, V. Experimental and Numerical Study of the Flow and Heat Transfer in a Bubbly Turbulent Flow in a Pipe with Sudden Expansion. *Energies* **2019**, *12*, 2735. [CrossRef]
3. Feng, Z.C.; Leal, L.G. Nonlinear Bubble Dynamics. *Annu. Rev. Fluid Mech.* **1997**, *29*, 201–243. [CrossRef]
4. Liow, J.L. Quasi-equilibrium bubble formation during topsubmerged gas injection. *Chem. Eng. Sci.* **2000**, *55*, 4515. [CrossRef]
5. Jamialahmadi, M.; Zehtaban, M.R.; Müller-Steinhagen, H.; Sarrafi, A.; Smith, J.M. Study of Bubble Formation Under Constant Flow Conditions. *Chem. Eng. Res. Des.* **2001**, *79*, 523–532. [CrossRef]
6. Li, M.; Hu, L. Experimental investigation of the behaviors of highly deformed bubbles produced by coaxial coalescence. *Exp. Therm. Fluid Sci.* **2020**, *117*, 110114. [CrossRef]
7. Houston, S.D.; Cornwell, K. Heat transfer to sliding bubbles on a tube under evaporating and non-evaporating conditions. *Int. J. Heat Mass Transf.* **1996**, *39*, 211–214. [CrossRef]
8. Dabiri, S.; Tryggvason, G. Heat transfer in turbulent bubbly flow in vertical channels. *Chem. Eng. Sci.* **2015**, *122*, 106–113. [CrossRef]
9. Chakraborty, S.; Das, P.K. Characterisation and classification of gas-liquid two-phase flow using conductivity probe and multiple optical sensors. *Int. J. Multiph. Flow* **2020**, *124*, 103193. [CrossRef]
10. Zhou, Y.; Kang, P.; Huang, Z.; Yan, P.; Sun, J.; Wang, J.; Yang, Y. Experimental measurement and theoretical analysis on bubble dynamic behaviors in a gas-liquid bubble column. *Chem. Eng. Sci.* **2020**, *211*, 115295. [CrossRef]
11. Razzaque, M.M.; Afacan, A.; Liu, S.; Nandakumar, K.; Masliyah, J.H.; Sanders, R.S. Bubble size in coalescence dominant regime of turbulent air–water flow through horizontal pipes. *Int. J. Multiph. Flow* **2003**, *29*, 1451–1471. [CrossRef]
12. Kadri, U.; Henkes, R.A.W.M.; Mudde, R.F.; Oliemans, R.V.A. Effect of gas pulsation on long slugs in horizontal gas–liquid pipe flow. *Int. J. Multiph. Flow* **2011**, *37*, 1120–1128. [CrossRef]
13. Sarafraz, M.M.; Shadloo, M.S.; Tian, Z.; Tlili, I.; Alkanhal, T.A.; Safaei, M.R.; Goodarzi, M.; Arjomandi, M. Convective Bubbly Flow of Water in an Annular Pipe: Role of Total Dissolved Solids on Heat Transfer Characteristics and Bubble Formation. *Water* **2019**, *11*, 1566. [CrossRef]
14. Pokusaev, B.G.; Kazenin, D.A.; Karlov, S.P.; Ermolaev, V.S. Motion of a gas slug in inclined tubes. *Theor. Found. Chem. Eng.* **2011**, *45*, 640–645. [CrossRef]
15. Tihon, J.; Pěnkavová, V.; Vejražka, J. Wall shear stress induced by a large bubble rising in an inclined rectangular channel. *Int. J. Multiph. Flow* **2014**, *67*, 76–87. [CrossRef]
16. Wang, H.; Zhang, C.; Xiong, H. Growth and Collapse Dynamics of a Vapor Bubble near or at a Wall. *Water* **2021**, *13*, 12. [CrossRef]
17. Donnelly, B.; Meehan, B. O'Reilly; Nolan, K.; Murray, D.B. The dynamics of sliding air bubbles and the effects on surface heat transfer. *Int. J. Heat Mass Transf.* **2015**, *91*, 532–542. [CrossRef]
18. Augustyniak, J.; Perkowski, D.M. Compound analysis of gas bubble trajectories with help of multifractal algorithm. *Exp. Therm. Fluid Sci.* **2021**, *124*, 110351. [CrossRef]
19. Kamp, A.M.; Chesters, A.K.; Colin, C.; Fabre, J. Bubble coalescence in turbulent flows: A mechanistic model for turbulence-induced coalescence applied to microgravity bubbly pipe flow. *Int. J. Multiph. Flow* **2001**, *27*, 1363–1396. [CrossRef]
20. Lahey, R.T.; Baglietto, E.; Bolotnov, I.A. Progress in multiphase computational fluid dynamics. *Nucl. Eng. Des.* **2021**, *374*, 111018. [CrossRef]
21. Chinak, A.V.; Gorelikova, A.E.; Kashinsky, O.N.; Pakhomov, M.A.; Randin, V.V.; Terekhov, V.I. Hydrodynamics and heat transfer in an inclined bubbly flow. *Int. J. Heat Mass Transf.* **2018**, *118*, 785–801. [CrossRef]
22. Haynes, W.M. *CRC Handbook of Chemistry and Physics*; CRC: Boca Raton, FL, USA, 2014.
23. Fu, Y.; Liu, Y. Development of a robust image processing technique for bubbly flow measurement in a narrow rectangular channel. *Int. J. Multiph. Flow* **2016**, *84*, 217–228. [CrossRef]
24. Bhunia, A.; Pais, S.C.; Kamotani, Y.; Kim, I.-H. Bubble formation in a coflow configuration in normal and reduced gravity. *AIChE J.* **1998**, *44*, 1499–1509. [CrossRef]

 *water*

Article

Evaporation Dynamics of Sessile and Suspended Almost-Spherical Droplets from a Biphilic Surface

Elena Starinskaya [1,2,*], Nikolay Miskiv [1,2], Vladimir Terekhov [1], Alexey Safonov [1], Yupeng Li [3], Ming-Kai Lei [3] and Sergey Starinskiy [1,2]

[1]   S.S. Kutateladze Institute of Thermophysics SB RAS, Lavrentyev Ave. 1, 630090 Novosibirsk, Russia
[2]   Novosibirsk State University, Pirogova Str. 2, 630090 Novosibirsk, Russia
[3]   School of Materials Science and Engineering, Dalian University of Technology, Dalian 116024, China
[*]   Correspondence: prefous-lm@yandex.ru

Abstract: Research in the field of the evaporation of liquid droplets placed on surfaces with special wetting properties such as biphilic surfaces is of great importance. This paper presents the results of an experimental study of the heat and mass transfer of a water droplet during its evaporation depending on the direction of the gravitational force. A special technique was developed to create unique substrates, which were used to physically simulate the interaction of liquid droplets with the surface at any angle of inclination to the horizontal. It was found that the suspended and sessile droplets exhibited fundamentally different evaporation dynamics. It was shown that the suspended droplets had a higher temperature and, at the same time, evaporated almost 30% faster.

Keywords: droplet; biphilic surface; evaporation; sessile; pendant; heat and mass transfer

Citation: Starinskaya, E.; Miskiv, N.; Terekhov, V.; Safonov, A.; Li, Y.; Lei, M.-K.; Starinskiy, S. Evaporation Dynamics of Sessile and Suspended Almost-Spherical Droplets from a Biphilic Surface. *Water* **2023**, *15*, 273. https://doi.org/10.3390/w15020273

Academic Editor: Yaoming Ma

Received: 7 December 2022
Revised: 31 December 2022
Accepted: 6 January 2023
Published: 9 January 2023

1. Introduction

The evaporation of droplets from solid surfaces is an important fundamental process used in various applications such as inkjet printing [1], the controlled deposition of self-assembled surface coatings [2], DNA extraction [3], disease diagnosis and drug development [4,5], spray painting and surface coating with solid particles [6,7], the creation of self-cleaning and water-repellent surfaces [8], removing condensed droplets from rear-view mirrors or windshields [9], increasing yields through the effective application of foliar fertilizers [10], and surface cooling due to a phase transition [11–21].

Many studies have been devoted to the investigation of droplet evaporation on heated surfaces [22–25]. Tran et al. [26] studied the behavior of droplets upon impact with superheated surfaces, noting that, at surface temperatures above the boiling point of the liquid, droplet evaporation differs greatly from natural evaporation on a substrate under normal atmospheric conditions [26]. Gao et al. [9] experimentally studied the evaporation of sessile droplets of various volumes on heated hydrophilic and hydrophobic surfaces under constant heat fluxes. At low heat fluxes, the droplet size strongly affected the evaporation time and this effect decreased with increasing heat fluxes. Many researchers carried out their experiments under normal atmospheric conditions [8,27–34].

The dynamics of droplet evaporation from a solid surface depend on many factors including wettability, evaporation flux at the interface and triple line, substrate temperature, external fields, and thermocapillarity [35,36]. In terms of heat transfer, this process is a complex interaction among convection in the gas and liquid phases, evaporation at the contact line, vapor diffusion, cooling at the liquid–gas interface, and possible Marangoni effects [27]. The droplet evaporation process is very complex and although significant progress has been made, this process is still not fully understood, in particular, with regard to coupled heat transfer near the triple contact line [37]. Droplet evaporation can occur in a constant contact angle mode, as shown in [38]. In some cases, the evaporation of droplets occurs in a mixed mode with a simultaneous change in the contact angle and radius of the contact line [39]. Yu et al. [28] and Fang et al. [29] observed that the evaporation of droplets on a hydrophobic surface first occurred in a constant contact radius mode with a change in the contact angle (CCR pinning mode) and the subsequent stage was dominated by a constant contact angle mode with a change in the contact area of the droplet with the surface (CCA depinning mode). Subsequently, Yu et al. [40] studied sessile water droplets evaporating on PDMS and Teflon surfaces and found that in all experiments on hydrophobic surfaces, evaporation started with the CCR mode, switched (after a short time) to the CCA mode, and ended with a mixed mode. Most researchers came to the same conclusion, that is, evaporation on hydrophobic surfaces occurs in two stages: in the initial stage, the pinning mode occurs and then the depinning mode prevails [28,29]. However, different results have also been reported in the literature. It has been shown [40,41] that during evaporation, the depinning mode prevails on hydrophobic surfaces and the pinning mode occurs on hydrophilic surfaces. Birdi and Vu [42] and Uno et al. [30] found that droplet evaporation occurs in the CCR mode on a hydrophilic surface and the CCA mode on a hydrophobic surface. In addition, Shin et al. [8] investigated water evaporation on various wetted surfaces and found that both the contact angle and contact area change at the end of droplet evaporation.

It is of great interest to study the evaporation of liquid droplets placed on surfaces whose structures can lead to significant changes in the droplet evaporation process. Such surfaces include biphilic surfaces, which have regions with different wetting properties [43,44]. Over the past few years, a number of studies have shown that micro- and nanostructured surfaces can increase the overall heat transfer coefficient, e.g., in pool boiling [45], flow boiling [46], and film boiling [47]. At the same time, although the study of the role of structured surfaces may be important, there are still many unresolved issues related to fluid flow and heat transfer, even when pure fluids come into contact with smooth surfaces [11]. However, the investigation of the evaporation process of a suspended droplet on a superhydrophobic surface is experimentally very difficult due to the limitation of droplet spatial stabilization. The fixation of a droplet in a specific place can be performed on a biphilic surface. Creating a superhydrophobic surface with a hydrophilic region that will hold the droplet in place and prevent it from rolling will help to solve the problem of evaporation on surfaces with a high contact angle. We have not found similar studies in the literature, but they could be very useful for controlling deposition processes, understanding droplet evaporation processes on surfaces with high adhesion depending on the gravitational fields, designing devices based on droplet evaporation, etc. The sharper the spatial transition from a hydrophilic to a hydrophobic surface, the more effects can be observed. From this point of view, the most attractive surfaces should have a micron-scale transition area from the superhydrophobic to the superhydrophilic region. These surfaces can be used to create microdevices in optics (e.g., plasmonic reflectors) and chemical sensors and biosensorics (substrates for surface-enhanced Raman spectroscopy, SERS).

The intensification of heat transfer, control of fluid movement, etc., were considered in some papers [48]. The evaporation of droplets has also received attention in the literature but there are still very few publications on this topic, e.g., [43,49]. Therefore, at present, it is not clear how evaporation from a material with high adhesion occurs. This study seeks to expand our understanding of the evaporation of a liquid droplet from a biphilic surface. Along with the case where a sessile droplet is located at the point of contact (point of attachment of the droplet to the surface/seat), we also consider the case where the droplet is inverted (suspended). It has been found that the special structure of the surface allows the evaporation of a suspended droplet simultaneously in the CCR and CCA modes. On this surface, a suspended droplet evaporates 30% faster than on a hydrophobic surface. The reasons for this, which are related to the heat and mass transfer processes, are discussed later in this paper.

2. Materials and Methods

2.1. Preparation of Materials

The creation of unique biphilic substrates with seat points was based on a combination of laser processing and hot-wire chemical vapor deposition (HW CVD) [50,51]. The superhydrophilicity effect was achieved on the surface of single-crystal silicon using laser irradiation. For this, an 8×18 mm^2 silicon wafer was treated with the fundamental harmonic of a Nd:YAG laser with a wavelength of 1064 nm, a pulse duration of 9 ns, and a Gaussian spatial profile. The laser beam scanned the substrate surface at a speed of 2 mm/s and the treated area was 10×10 mm^2. To achieve the superhydrophilicity effect, the silicon surface was irradiated with 40,000 laser pulses at a frequency of 5 Hz. Next, a fluoropolymer coating was applied to the obtained superhydrophobic surface by HW CVD. The method involved the activation of the precursor gas flow with a hot (670 °C) Nichrome wire catalyst, followed by the precipitation and polymerization of the free radicals formed on the substrate surface. Hexafluoropropylene oxide was used as the precursor gas. The initial surface contact angle of the fluoropolymer under the selected deposition conditions (i.e., without laser removal) was $155 \pm 3°$. The landing seats for droplet fixing were created in the last stage. For this, the resulting sample with fluoropolymer coating was locally (pointwise) irradiated with laser pulses. Round seats of various sizes were obtained by varying the number of laser pulses, radiation focusing, and beam energy (Figure 1a).

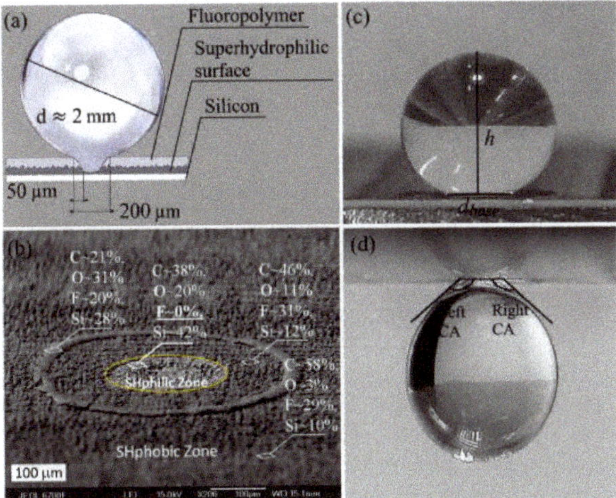

Figure 1. (**a**) Schematic of the substrate. (**b**) SEM image of the seat after the evaporation of colloidal solution droplets of SiO$_2$ particles in distilled water. The indicated compositions in the corresponding square regions were determined by the EDX method. (**c**) Optical image of a sessile droplet in the seat. (**d**) Optical image of a droplet in the seat. h, d_{base}, Left CA, and Right CA are the droplet height, base diameter, and left and right contact angles, respectively.

The layout of the seat is shown in Figure 1a. The thickness of the wafers was 0.5 mm and the thermal conductivity of the silicon wafer was 149 W/(m·K). The contribution of the Teflon layer can be neglected due to its very thin thickness (~50 nm). Figure 1b shows an SEM image with the seat and the annular coffee ring deposit obtained by the deposition of solid particles on the contact line during the evaporation of a 0.1 wt % SiO$_2$ nanofluid sessile droplet. It should be noted that the size of the "coffee ring" corresponds to the measured diameter of the base of the droplet. The obtained biphilic surfaces with seats allowed us to carry out experiments to study the heat and mass transfer processes during droplet evaporation by varying the orientation of the droplet relative to the gravitational forces (Figure 1c,d).

2.2. Experimental Setup

The study of the heat and mass transfer processes during the evaporation of the droplets was carried out on the experimental setup shown schematically in Figure 2. The experimental setup consisted of a copper plate to which the substrate was attached and a rotary mechanism. The substrate was attached to the copper surface using thermal paste, then, the required angle was set and the droplet was placed on the seat. The conditions around the droplet were measured with an AZ Instrument model 872 hygrometer to ensure the control of the ambient humidity and temperature with an uncertainty of ±4%.

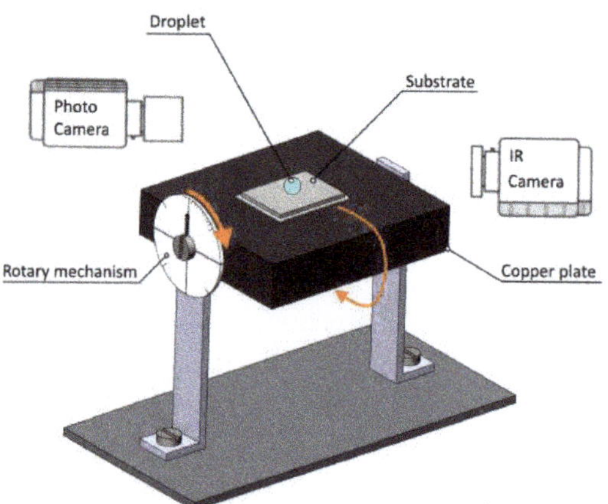

Figure 2. Schematic of the experimental setup.

We studied droplets of distilled water with an initial volume of about 6 µL. The ambient temperature and relative humidity corresponded to room-temperature conditions and were 25 °C and $\varphi = 21\%$, respectively. Droplets were formed on the substrate surface using a Thermo Scientific mechanical pipette with an accuracy of ~0.1 µL. The temperature distribution on the droplet surface was determined by infrared thermography (Figure 3a,b). The measurements were carried out with an NEC TH7102IR thermal imaging camera at wavelengths of $\lambda = 8$–14 µm using a TH 71-377 macro lens. Thermal images were averaged over the area near the central part of the thermal image of the droplet, as shown in Figure 3a,b. The obtained data were processed by the ThermoTracer program. The systematic and random errors of these measurements were estimated at 0.2 °C and 0.45 °C, respectively, resulting in a total error of 0.65 °C. During the experiment, the droplet shape was recorded with a Digi Scope II v3 digital microscope. The contact angle, contact line size, and droplet height were measured from the photographs as shown in Figure 1c,d. The contact angle was measured on the left and right sides (see Figure 1d), and the obtained values were then averaged. Each evaporation experiment was repeated 3 times and good reproducibility of all geometric parameters was observed. The error in the determination of the geometric parameters was the sum of the instrumental error (0.05 mm) and the statistical error (0.02 mm). The change in the shape of the evaporating droplets is shown in the photographs (Figure 3c,d).

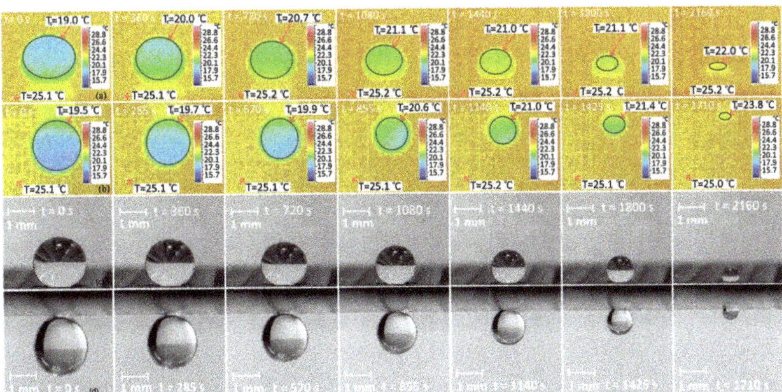

Figure 3. Thermal images of droplets: (**a**) sessile; (**b**) pendant. Change in the shape of evaporating droplets: (**c**) sessile; (**d**) pendant.

3. Results and Discussion

A comparison of the droplet shapes for the same relative time moment t/t_{evap}, where t_{evap} is the full evaporation time of 2400 s and 1800 s for the corresponding sessile and pendant droplets is shown in Figure 4a. The initial droplet diameters differed by ~3%. These data are presented for the case of the evaporation of the suspended and sessile droplets from the same seat so the surface preparation conditions could not affect the process. Usually, the evaporation of droplets on superhydrophobic substrates is accompanied by the instability of the contact line dynamics [52]. As can be seen in Figure 4a, this effect was not observed in the case of the suspended droplet and was absent in the initial stage of the evaporation of the sessile droplet during evaporation on biphilic surfaces. The sessile droplet had a large area of contact with the surface (base diameter). This effect may be due to the fact that the droplet moving under its own weight spread over a large area near the seat so the size of the coffee ring did not coincide with the size of the seat, as shown in Figure 1b.

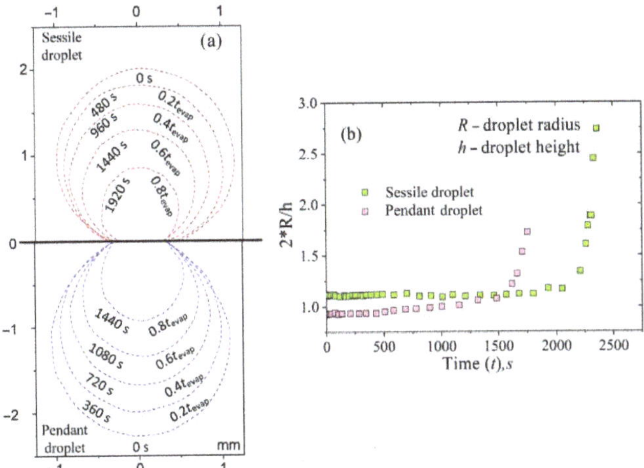

Figure 4. (**a**) Contours of sessile and suspended droplets at different times. (**b**) The ratio of the droplet diameter to its height in the process of evaporation of suspended and sessile droplets.

Figure 4b shows that the droplet height decreased almost linearly, regardless of the method of suspension. However, the height of the suspended droplet decreased much faster (Figure 4b). It should be noted that the droplet height is an important parameter that determines the degree of influence of certain forces acting on evaporating droplets.

A more detailed comparison of the evaporation dynamics of the suspended and sessile droplets is shown in Figure 5. The droplet height, base diameter, and contact angles were analyzed (Figure 5a,c). A comparison of the temperature dynamics of the sessile and suspended droplets is shown in Figure 5b.

For the sessile droplets, the changes in the geometric characteristics can be divided into several regions that are consistent with the temperature data (Figure 5a,b). In the first stage (time interval from 0 to 1440 s), the droplet base diameter changed linearly, the contact angle remained constant and close to the value of the fluoropolymer coating, and the droplet height decreased at a constant rate. The droplet base diameter in this stage was noticeably higher than the superhydrophilic area on the surface. For the size used in the work, the Bond number was quite large ~0.2 (boundary value) at the considered period of time. Thus, the sessile droplet was deformed under the influence of gravity. We need to point out that the fluoropolymer coating was superhydrophobic so the direct contact of the droplet with the fluoropolymer was excluded (we suggest that the superhydrophobic surface was wetted in the Wenzel state). As a result, the sessile droplet hung over the "true" contact area determined by the pinning to the area with superhydrophilic properties; in other words, the "true" contact line was hidden by the side wall of the droplet in the initial stage of evaporation. As the liquid mass decreased, this effect was leveled and the registered contact diameters for the sessile and suspended droplets became almost the same (Figure 5a,c after 1440 s and 1080 s).

In the initial stage of evaporation, a sharp decrease in the surface temperature of the water droplet to 19 °C was observed, after which the droplet temperature changed slightly within a couple of minutes (interval from 0 to 170 s). Further, a gradual increase in the temperature of the droplet in the interval up to 750 s was observed. In the second stage from 750 to 1440 s, the size of the contact line decreased, whereas the contact angle remained constant (depinning mode) and the droplet height decreased at the same rate as in the first stage. In the third time interval from 1440 to 1950 s, a decrease in the droplet height and a decrease in the contact angle and base diameter were observed; this evaporation mode is called the mixed mode. The interval from 750 to 1950 s corresponded to a plateau in the temperature vs. the time graph. In the last stage of evaporation after 1950 s, the contact angle decreased sharply and the height of the droplet changed at a higher rate than in the previous stages, whereas the size of the droplet base remained unchanged. In this time interval, a sharp increase in the droplet temperature to the ambient air temperature was recorded. Similar temperature dynamics were observed in [53].

In the case of the suspended droplet, a smaller number of characteristic regimes in the evaporation dynamics was observed. The diameter of the contact line remained almost unchanged during the entire evaporation process (Figure 5c). In the first stage from 0 to 1080 s, the droplet height decreased at a constant rate, whereas the diameter of the contact line and the contact angle remained constant over time.

As in the case of the sessile droplet, the temperature decreased sharply at the initial time and then gradually increased as the droplet decreased. In the second interval from 1080 to 1440 s, there was a slight change in the diameter of the droplet base and the contact angle, whereas the droplet height decreased at a constant rate. It should be noted that an increase in temperature was observed from 1080 s; the transition to exponential growth was more pronounced at 1440 s. In the last stage of evaporation from 1440 s to the complete drying of the droplet, the contact angle decreased significantly. In the same time interval, the height of the droplet decreased much faster, whereas its base remained constant. The evaporation rate was greater in the last stages than in the initial ones since the minimum film thickness ensured high thermal conductivity. The evaporation of the droplets from a surface with a contact angle of less than 90° was not uniform. That is, the evaporating material flow did not emanate from the droplet as from a sphere (uniformly in all directions) but diverged along the edges of the droplet [41]. This can also affect the rate of evaporation and deposition processes.

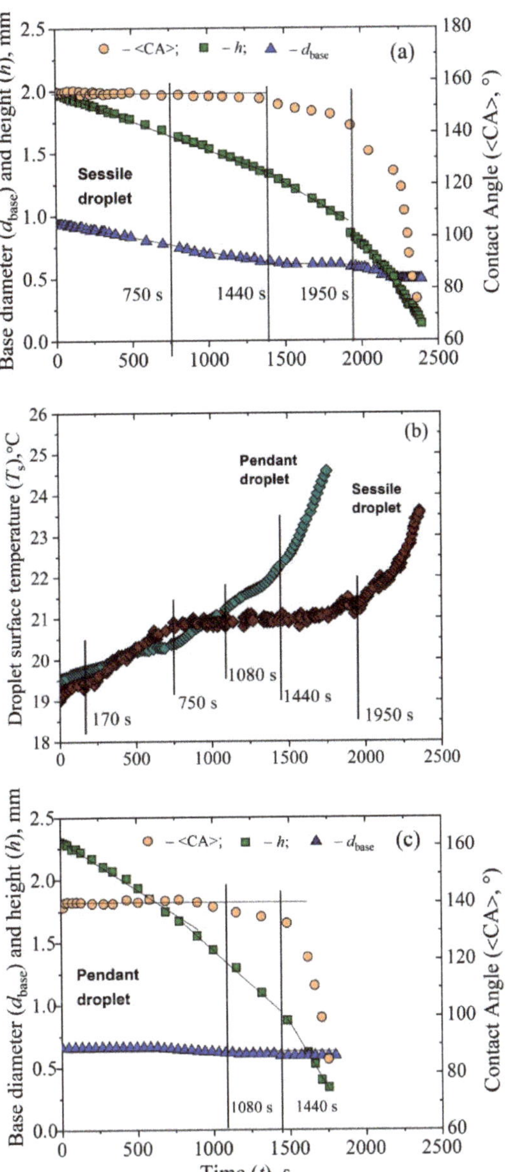

Figure 5. Change in the base diameter, height, and contact angle for sessile (**a**) and pendant (**c**) droplets. (**b**) Comparison of the surface temperature dynamics for sessile and suspended droplets.

One of the most interesting aspects of this work was the higher rate of evaporation of the pendant drop, which was accompanied by a higher temperature on its surface. It is well known that in the case of a free droplet, the more intense evaporation leads to a decrease in its surface temperature up to adiabatic temperatures. In this case, this was not observed. Possible explanations are (1) the intense heat flux from the side of the substrate and (2) the difference in the form of the droplets. As shown in Figure 4a, the main geometric characteristics of the droplets in dimensionless times had similar behavior, and differences were observed only in the surface area of the droplet and the diameter of the base (Figure 5). The surface areas of the droplets were quite close, however, it is unlikely that this significantly contributed to the differences in the evaporation dynamics. This is in agreement with the results of [54], which showed that pendant droplets evaporate slightly slower than sessile droplets in the case of CA = 90°, i.e., the effect of hiding the contact line discussed above did not occur.

The dynamics of the base diameter, unlike the other geometric parameters, were very different for the pendant and sessile droplets. Note that when reaching $0.6t_{evap}$ (these were ~1080 and ~1440 s for the pendant and sessile droplets, respectively), the base diameters became close. From these times onward, the temperature dependence (Figure 5b) and rapid heating of the droplets that were similar in dynamics began. At the earlier times, there was a dramatic difference in the base diameter (as noticed above, it exceeded the pinning region for the sessile droplet). It is known that the main evaporation occurs at the contact line [2]. It was expected that for the sessile droplet, the contact line would make a noticeably smaller contribution to evaporation since it was hidden by an overhanging droplet. Vapors were trapped between the droplet surfaces and the non-wetted wall. This made it possible to more efficiently distribute heat from the substrate over the volume of the droplet and, thereby, increase the evaporation flux from its surface, reducing the average temperature. For the pendant droplet, on the contrary, a lot of liquid evaporated from the contact line and the incoming heat was spent on this process, whereas the rest of the volume did not receive heat from the wall. As a result, the vapor flow from the rest of the droplet area decreased and its average temperature counterintuitively increased. In our experiments, we could not trace the convective flows inside the droplet and the diffusion flows of the vapor outside. A full explanation of the observed effect could be provided by detailed modeling of the processes.

A comparison of the evaporation dynamics of sessile and suspended liquid droplets has not been given much attention in the literature. The rate of evaporation of sessile and suspended droplets of methyl acetoacetate on a smooth surface was studied by Picknett and Bexon [27]. For the smooth surface, these authors used an equiconvex lens with a radius of curvature of 200 mm coated with a 50 μm thick PTFE sticky layer to keep the drop on the inverted surface and provide a sufficiently large contact angle. Due to its small size, the droplet had almost the same shape as the corresponding sessile droplet and the evaporation rate was almost the same. Due to the small initial droplet sizes, the study [27] did not observe the effect of hiding the contact line, which also agrees with our results at the later times of $>0.8\ t_{evap}$.

4. Conclusions

The shape and size of a droplet placed on a substrate depend on the contact angle, surface tension, substrate tilt, fixation (pinning or depinning) of the line of contact between the droplet and the substrate, etc. Droplet evaporation from biphilic surfaces is of great practical and scientific interest for a deeper understanding of heterophase processes such as wetting, adsorption, precipitation, coagulation, coating, etc. The paper presents a study of the heat and mass transfer during the evaporation of a water droplet from a biphilic surface with different orientations of the droplet relative to the gravitational forces. It should be noted that it is difficult to assess the effect of gravitational force on the rate of droplet evaporation. A feature of this study is the use of a unique approach to the placement and fixation of a droplet on a surface, which made it possible to evaluate the effect of droplet orientation on the evaporation dynamics. The results of the study show that the unique surface structure allows the evaporation of suspended droplets simultaneously in the CCR and CCA modes. An interesting effect was found when comparing the total evaporation time of sessile and suspended droplets: the suspended droplets evaporated 30% faster and, at the same time, had a higher temperature than the sessile droplets. The temperature evolution of sessile and suspended droplets has a stagewise character.

Author Contributions: Conceptualization, M.-K.L.; methodology, V.T. and Y.L.; validation, S.S.; investigation, N.M. and E.S.; resources, S.S. and A.S.; data curation, S.S. and A.S.; writing—original draft preparation, E.S.; writing—review and editing, S.S. and E.S.; visualization, N.M.; supervision, S.S.; project administration, E.S. All authors have read and agreed to the published version of the manuscript.

Funding: The investigation of droplet evaporation was supported by the Ministry of Science and Higher Education of the Russian Federation (mega-grant No. 075-15-2021-575). Creating biphilic surfaces was performed with the financial support of the RFBR and NSFC, project number No. 21-52-53025 GFEN_a.

Institutional Review Board Statement: Not applicable.

Informed Consent Statement: Not applicable.

Data Availability Statement: On request.

Conflicts of Interest: The authors declare no conflict of interest.

References

1. Lim, T.; Han, S.; Chung, J.; Chung, J.T.; Ko, S.; Grigoropoulos, C.P. Experimental Study on Spreading and Evaporation of Inkjet Printed Pico-Liter Droplet on a Heated Substrate. *Int. J. Heat Mass Transf.* **2009**, *52*, 431–441. [CrossRef]
2. Pan, Z.; Dash, S.; Weibel, J.A.; Garimella, S.V. Assessment of Water Droplet Evaporation Mechanisms on Hydrophobic and Superhydrophobic Substrates. *Langmuir* **2013**, *29*, 15831–15841. [CrossRef] [PubMed]
3. Yu, Y.; Zhu, H.; Frantz, J.M.; Reding, M.E.; Chan, K.C.; Ozkan, H.E. Evaporation and Coverage Area of Pesticide Droplets on Hairy and Waxy Leaves. *Biosyst. Eng.* **2009**, *104*, 324–334. [CrossRef]
4. Li, W.; Ji, W.; Sun, H.; Lan, D.; Wang, Y. Pattern Formation in Drying Sessile and Pendant Droplet: Interactions of Gravity Settling, Interface Shrinkage, and Capillary Flow. *Langmuir* **2019**, *35*, 113–119. [CrossRef]
5. Ghosh, J.; Mukhopadhyay, A.; Rao, G.V.; Sanyal, D. Analysis of Evaporation of Dense Cluster of Bicomponent Fuel Droplets in a Spray Using a Spherical Cell Model. *Int. J. Therm. Sci.* **2008**, *47*, 584–590. [CrossRef]
6. Eslamian, M. Spray-on Thin Film PV Solar Cells: Advances, Potentials and Challenges. *Coatings* **2014**, *4*, 60–84. [CrossRef]
7. Hsu, C.C.; Su, T.W.; Wu, C.H.; Kuo, L.S.; Chen, P.H. Influence of Surface Temperature and Wettability on Droplet Evaporation. *Appl. Phys. Lett.* **2015**, *106*, 141602. [CrossRef]
8. Shin, D.H.; Lee, S.H.; Jung, J.Y.; Yoo, J.Y. Evaporating Characteristics of Sessile Droplet on Hydrophobic and Hydrophilic Surfaces. *Microelectron. Eng.* **2009**, *86*, 1350–1353. [CrossRef]
9. Gao, M.; Kong, P.; Zhang, L.X.; Liu, J.N. An Experimental Investigation of Sessile Droplets Evaporation on Hydrophilic and Hydrophobic Heating Surface with Constant Heat Flux. *Int. Commun. Heat Mass Transf.* **2017**, *88*, 262–268. [CrossRef]
10. Hampton, M.A.; Nguyen, T.A.H.; Nguyen, A.V.; Xu, Z.P.; Huang, L.; Rudolph, V. Influence of Surface Orientation on the Organization of Nanoparticles in Drying Nanofluid Droplets. *J. Colloid Interface Sci.* **2012**, *377*, 456–462. [CrossRef]
11. Putnam, S.A.; Briones, A.M.; Ervin, J.S.; Hanchak, M.S.; Byrd, L.W.; Jones, J.G. Interfacial Heat Transfer during Microdroplet Evaporation on a Laser Heated Surface. *Int. J. Heat Mass Transf.* **2012**, *55*, 6307–6320. [CrossRef]
12. Deng, W.; Gomez, A. Electrospray Cooling for Microelectronics. *Int. J. Heat Mass Transf.* **2011**, *54*, 2270–2275. [CrossRef]
13. Dugas, V.; Broutin, J.; Souteyrand, E. Droplet Evaporation Study Applied to DNA Chip Manufacturing. *Langmuir* **2005**, *21*, 9130–9136. [CrossRef]
14. Li, G.; Flores, S.M.; Vavilala, C.; Schmittel, M.; Graf, K. Evaporation Dynamics of Microdroplets on Self-Assembled Monolayers of Dialkyl Disulfides. *Langmuir* **2009**, *25*, 13438–13447. [CrossRef]
15. Kim, J. Spray Cooling Heat Transfer: The State of the Art. *Int. J. Heat Fluid Flow* **2007**, *28*, 753–767. [CrossRef]
16. Putnam, S.A.; Briones, A.M.; Byrd, L.W.; Ervin, J.S.; Hanchak, M.S.; White, A.; Jones, J.G. Microdroplet Evaporation on Superheated Surfaces. *Int. J. Heat Mass Transf.* **2012**, *55*, 5793–5807. [CrossRef]
17. Chen, R.H.; Phuoc, T.X.; Martello, D. Effects of Nanoparticles on Nanofluid Droplet Evaporation. *Int. J. Heat Mass Transf.* **2010**, *53*, 3677–3682. [CrossRef]
18. Park, J.; Moon, J. Control of Colloidal Particle Deposit Patterns within Picoliter Droplets Ejected by Ink-Jet Printing. *Langmuir* **2006**, *22*, 3506–3513. [CrossRef]
19. Kumari, N.; Garimella, S.V. Characterization of the Heat Transfer Accompanying Electrowetting or Gravity-Induced Droplet Motion. *Int. J. Heat Mass Transf.* **2011**, *54*, 4037–4050. [CrossRef]
20. Cheng, W.L.; Han, F.Y.; Liu, Q.N.; Zhao, R.; Fan, H.-l. Experimental and Theoretical Investigation of Surface Temperature Non-Uniformity of Spray Cooling. *Energy* **2011**, *36*, 249–257. [CrossRef]

21. Zhang, Z.; Jiang, P.X.; Hu, Y.T.; Li, J. Experimental Investigation of Continual-and Intermittent-Spray Cooling. *Exp. Heat Transf.* **2013**, *26*, 453–469. [CrossRef]

22. Chen, P.; Harmand, S.; Szunerits, S.; Boukherroub, R. Evaporation Behavior of PEGylated Graphene Oxide Nanofluid Droplets on Heated Substrate. *Int. J. Therm. Sci.* **2019**, *135*, 445–458. [CrossRef]

23. Kelly-Zion, P.L.; Pursell, C.J.; Vaidya, S.; Batra, J. Evaporation of Sessile Drops under Combined Diffusion and Natural Convection. *Colloids Surf. A Physicochem. Eng. Asp.* **2011**, *381*, 31–36. [CrossRef]

24. Lopes, M.C.; Bonaccurso, E.; Gambaryan-Roisman, T.; Stephan, P. Influence of the Substrate Thermal Properties on Sessile Droplet Evaporation: Effect of Transient Heat Transport. *Colloids Surf. A Physicochem. Eng. Asp.* **2013**, *432*, 64–70. [CrossRef]

25. Crafton, E.F.; Black, W.Z. Heat Transfer and Evaporation Rates of Small Liquid Droplets on Heated Horizontal Surfaces. *Int. J. Heat Mass Transf.* **2004**, *47*, 1187–1200. [CrossRef]

26. Tran, T.; Staat, H.J.J.; Prosperetti, A.; Sun, C.; Lohse, D. Drop Impact on Superheated Surfaces. *Phys. Rev. Lett.* **2012**, *108*, 036101. [CrossRef]

27. Picknett, R.G.; Bexon, R. The Evaporation of Sessile or Pendant Drops in Still Air. *J. Colloid Interface Sci.* **1977**, *61*, 336–350. [CrossRef]

28. Yu, H.-Z.; Soolaman, D.M.; Rowe, A.W.; Banks, J.T. Evaporation of Water Microdroplets on Self-Assembled Monolayers: From Pinning to Shrinking. *ChemPhysChem* **2004**, *5*, 1035–1038. [CrossRef]

29. Fang, X.; Pimentel, M.; Sokolov, J.; Rafailovich, M. Dewetting of the Three-Phase Contact Line on Solids. *Langmuir* **2010**, *26*, 7682–7685. [CrossRef]

30. Uno, K.; Hayashi, K.; Hayashi, T.; Ito, K.; Kitano, H. Particle Adsorption in Evaporating Droplets of Polymer Latex Dispersions on Hydrophilic and Hydrophobic Surfaces. *Colloid Polym. Sci.* **1998**, *276*, 810–815. [CrossRef]

31. Nguyen, T.A.H.; Nguyen, A.V.; Hampton, M.A.; Xu, Z.P.; Huang, L.; Rudolph, V. Theoretical and Experimental Analysis of Droplet Evaporation on Solid Surfaces. *Chem. Eng. Sci.* **2012**, *69*, 522–529. [CrossRef]

32. Hu, H.; Larson, R.G. Evaporation of a Sessile Droplet on a Substrate. *J. Phys. Chem. B* **2002**, *106*, 1334–1344. [CrossRef]

33. Lee, K.S.; Cheah, C.Y.; Copleston, R.J.; Starov, V.M.; Sefiane, K. Spreading and Evaporation of Sessile Droplets: Universal Behaviour in the Case of Complete Wetting. *Colloids Surf. A Physicochem. Eng. Asp.* **2008**, *323*, 63–72. [CrossRef]

34. Furuta, T.; Isobe, T.; Sakai, M.; Matsushita, S.; Nakajima, A. Wetting Mode Transition of Nanoliter Scale Water Droplets during Evaporation on Superhydrophobic Surfaces with Random Roughness Structure. *Appl. Surf. Sci.* **2012**, *258*, 2378–2383. [CrossRef]

35. Deegan, R.D.; Bakajin, O.; Dupont, T.F.; Huber, G.; Nagel, S.R.; Witten, T.A. Capillary Flow as the Cause of Ring Stains from Dried Liquid Drops. *Nature* **1997**, *389*, 827–829. [CrossRef]

36. Gao, L.; McCarthy, T.J. Wetting 101°. *Langmuir* **2009**, *24*, 14105–14115. [CrossRef] [PubMed]

37. Zheng, Z.; Zhou, L.; Du, X.; Yang, Y. Numerical Investigation on Conjugate Heat Transfer of Evaporating Thin Film in a Sessile Droplet. *Int. J. Heat Mass Transf.* **2016**, *101*, 10–19. [CrossRef]

38. Gibbons, M.J.; Di Marco, P.; Robinson, A.J. Local Heat Transfer to an Evaporating Superhydrophobic Droplet. *Int. J. Heat Mass Transf.* **2018**, *121*, 641–652. [CrossRef]

39. Rahimzadeh, A.; Eslamian, M. Experimental Study on the Evaporation of Sessile Droplets Excited by Vertical and Horizontal Ultrasonic Vibration. *Int. J. Heat Mass Transf.* **2017**, *114*, 786–795. [CrossRef]

40. Yu, Y.S.; Wang, Z.; Zhao, Y.P. Experimental and Theoretical Investigations of Evaporation of Sessile Water Droplet on Hydrophobic Surfaces. *J. Colloid Interface Sci.* **2012**, *365*, 254–259. [CrossRef]

41. Deegan, R.D.; Bakajin, O.; Dupont, T.F.; Huber, G.; Nagel, S.R.; Witten, T.A. Contact Line Deposits in an Evaporating Drop. *Phys. Rev. E Stat. Phys. Plasmas Fluids Relat. Interdiscip. Top.* **2000**, *62*, 756–765. [CrossRef] [PubMed]

42. Birdi, K.S.; Vu, D.T. Wettability and the Evaporation Rates of Fluids from Solid Surfaces. *J. Adhes. Sci. Technol.* **1993**, *7*, 485–493. [CrossRef]

43. Qi, W.; Li, J.; Weisensee, P.B. Evaporation of Sessile Water Droplets on Horizontal and Vertical Biphobic Patterned Surfaces. *Langmuir* **2019**, *35*, 17185–17192. [CrossRef]

44. Raj, R.; Enright, R.; Zhu, Y.; Adera, S.; Wang, E.N. Unified Model for Contact Angle Hysteresis on Heterogeneous and Superhydrophobic Surfaces. *Langmuir* **2012**, *28*, 15777–15788. [CrossRef] [PubMed]

45. Serdyukov, V.; Starinskiy, S.; Malakhov, I.; Safonov, A.; Surtaev, A. Laser Texturing of Silicon Surface to Enhance Nucleate Pool Boiling Heat Transfer. *Appl. Therm. Eng.* **2021**, *194*, 117102. [CrossRef]

46. Khanikar, V.; Mudawar, I.; Fisher, T. Effects of Carbon Nanotube Coating on Flow Boiling in a Micro-Channel. *Int. J. Heat Mass Transf.* **2009**, *52*, 3805–3817. [CrossRef]

47. Sathyamurthi, V.; Ahn, H.S.; Banerjee, D.; Lau, S.C. Subcooled Pool Boiling Experiments on Horizontal Heaters Coated with Carbon Nanotubes. *J. Heat Transf.* **2009**, *131*, 071501. [CrossRef]

48. Motezakker, A.R.; Sadaghiani, A.K.; Çelik, S.; Larsen, T.; Villanueva, L.G.; Koşar, A. Optimum Ratio of Hydrophobic to Hydrophilic Areas of Biphilic Surfaces in Thermal Fluid Systems Involving Boiling. *Int. J. Heat Mass Transf.* **2019**, *135*, 164–174. [CrossRef]

49. Pradhan, T.K.; Panigrahi, P.K. Evaporation Induced Natural Convection inside a Droplet of Aqueous Solution Placed on a Superhydrophobic Surface. *Colloids Surf. A Physicochem. Eng. Asp.* **2017**, *530*, 1–12. [CrossRef]

50. Starinskiy, S.V.; Rodionov, A.A.; Shukhov, Y.G.; Safonov, A.I.; Maximovskiy, E.A.; Sulyaeva, V.S.; Bulgakov, A. V Formation of Periodic Superhydrophilic Microstructures by Infrared Nanosecond Laser Processing of Single-Crystal Silicon. *Appl. Surf. Sci.* **2020**, *512*, 145753. [CrossRef]

51. Starinskiy, S.V.; Bulgakov, A.V.; Gatapova, E.Y.; Shukhov, Y.G.; Sulyaeva, V.S.; Timoshenko, N.I.; Safonov, A.I. Transition from Superhydrophilic to Superhydrophobic of Silicon Wafer by a Combination of Laser Treatment and Fluoropolymer Deposition. *J. Phys. D Appl. Phys.* **2018**, *51*, 255307. [CrossRef]

52. Tsai, P.; Lammertink, R.G.H.; Wessling, M.; Lohse, D. Evaporation-Triggered Wetting Transition for Water Droplets upon Hydrophobic Microstructures. *Phys. Rev. Lett.* **2010**, *104*, 2–3. [CrossRef] [PubMed]

53. Borodulin, V.Y.; Letushko, V.N.; Nizovtsev, M.I.; Sterlyagov, A.N. The Experimental Study of Evaporation of Water–Alcohol Solution Droplets. *Colloid J.* **2019**, *81*, 219–225. [CrossRef]

54. Kim, J.Y.; Hwang, I.G.; Weon, B.M. Evaporation of inclined water droplets. *Sci. Rep.* **2017**, *7*, 42848. [CrossRef]

Article

RANS Modeling of Turbulent Flow and Heat Transfer in a Droplet-Laden Mist Flow through a Ribbed Duct

Maksim A. Pakhomov * and Viktor I. Terekhov

Laboratory of Thermal and Gas Dynamics, Kutateladze Institute of Thermophysics, Siberian Branch of Russian Academy of Sciences, Acad. Lavrent'ev Avenue, 1, 630090 Novosibirsk, Russia
* Correspondence: pakhomov@ngs.ru

Abstract: The local structure, turbulence, and heat transfer in a flat ribbed duct during the evaporation of water droplets in a gas flow were studied numerically using the Eulerian approach. The structure of a turbulent two-phase flow underwent significant changes in comparison with a two-phase flow in a flat duct without ribs. The maximum value of gas-phase turbulence was obtained in the region of the downstream rib, and it was almost twice as high as the value of the kinetic energy of the turbulence between the ribs. Finely dispersed droplets with small Stokes numbers penetrated well into the region of flow separation and were observed over the duct cross section; they could leave the region between the ribs due to their low inertia. Large inertial droplets with large Stokes numbers were present only in the mixing layer and the flow core, and they accumulated close to the duct ribbed wall in the flow towards the downstream rib. An addition of evaporating water droplets caused a significant enhancement in the heat transfer (up to 2.5 times) in comparison with a single-phase flow in a ribbed channel.

Keywords: droplet-laden flow; ribbed duct; heat transfer; evaporation; RANS; RSM

Citation: Pakhomov, M.A.; Terekhov, V.I. RANS Modeling of Turbulent Flow and Heat Transfer in a Droplet-Laden Mist Flow through a Ribbed Duct. *Water* **2022**, *14*, 3829. https://doi.org/10.3390/w14233829

Academic Editor: Giuseppe Pezzinga

Received: 26 October 2022
Accepted: 20 November 2022
Published: 24 November 2022

Publisher's Note: MDPI stays neutral with regard to jurisdictional claims in published maps and institutional affiliations.

1. Introduction

The intensification of heat transfer in the internal cooling channels of gas turbine (GT) blades remains one of the key problems due to the constant growth in the inlet gas temperature of the GT. This temperature already reaches 2000 K and significantly exceeds the allowable temperatures for the long-term operation of the blades and power equipment of gas turbines [1–4]. Therefore, cooling the working surfaces of heat-loaded elements is an important and urgent problem of heat transfer. Various cooling methods (film cooling, jet impingement cooling, internal convective cooling, thermal barrier coatings, and spray cooling by the evaporation of various atomized droplets) have been developed for the effective thermal protection of working surfaces and increasing the operating times of power equipment elements. Internal convective cooling is a reliable and simple method for efficient cooling and heat removal from the GT heat-loaded elements.

One of the most effective methods for increasing heat transfer is the use of passive heat transfer intensifiers with various surface shapes. The use of various ribs or obstacles installed on a duct wall is one of the most effective ways to increase heat transfer (see monographs [5–7]). The rib height, h; duct height, H; rib pitch, p; obstacle shape; rib-to-channel height expansion ratio, $ER = h/H$; pitch-to-height ratio, p/h; and some other factors have a great effect on the formation and development of the recirculation region and heat transfer in such flows.

The heat transfer enhancement (HTE) of ribbed ducts (by 2–5 times) is accompanied by a significant increase in the pressure drop (of more than ten times) for most of these surfaces [1,3,4]. Two-dimensional obstacles most often have the form of ribs and protrusions of various configurations located at different angles to the flow on the duct walls [4–7]. They deflect and mix the flow, give rise to multiscale separated flows, and generate vorticity and three-dimensional velocity gradients [4–7]. A two-phase flow around two-dimensional obstacles is one of the most common cases of shear separated flow for flows with both solid particles [8–12] and gas bubbles [13,14]. We should note that all of the abovementioned works were performed without taking into account the interfacial heat transfer between a dispersed phase (solid particles, droplets, and gas bubbles) and carrier fluid flow or in gas–liquid flows. The state-of-the-art research studying the movements and interactions of two-phase gas-dispersed flows with various obstacles was reviewed in [12,15].

The use of the latent heat of phase transition during the evaporation of droplets leads to a significant increase in HTE (up to several times) in comparison with conventional single-phase forced convection. Studies of the flow structure, friction, and heat transfer in a flow around ribs of various shapes with two-phase mist/steam and gas-droplet flows were carried out in several experimental [16–18] and numerical works [17,19–21].

The heat transfer in the case of a gas-droplet flow between two ribs was studied experimentally in [17]. The study was carried out with an initial mass fraction of water droplets of $M_{L1} = 15\%$. Their initial diameter was $d_1 = 50$–60 μm, the flow Reynolds number based on the mean mass flow velocity at the inlet and the hydraulic diameter was $\mathrm{Re} = U_m D_h / \nu = (0.8$–$2.4) \times 10^4$, the heat flux density was $q_W = \mathrm{const} = 2.6\ \mathrm{kW/m^2}$, and $p/h = 10$ and 20 for the ribs installed at an angle to the free-stream flow of $\varphi = 90°$. Heat transfer measurements in the case of a gas-droplet flow between two ribs in a system of continuous V-shaped ribs and broken V-shaped ribs were carried out in [18]. The study was carried out at $M_{L1} = 10\%$, $d_1 = 50$–60 μm, $\mathrm{Re} = (0.8$–$2.4) \times 10^4$, $q_W = \mathrm{const} = 1$–$10\ \mathrm{kW/m^2}$, $p/h = 10$ and 20, and $\varphi = 45°$.

Numerical simulations [17,19–21] were performed using the commercial CFD package ANSYS Fluent using isotropic k–ε (in [17,19]) and k–ω SST (in [20,21]) turbulence models. The effect of ribs installed at an angle ($\varphi = 45°$) to the flow on the heat transfer in a two-phase flow was studied numerically in [22–24]. The predictions [17,20] were carried out for a two-phase flow in a smooth duct, and in [19,21] they were made in a U-shaped duct. Computations [19] were carried out for the following range of initial parameters: $M_{L1} = 2\%$, $d_1 = 5$ μm, $\mathrm{Re} = (0.5$–$4) \times 10^4$, $q_W = \mathrm{const} = 10\ \mathrm{kW/m^2}$, and $p/h = 10$ in a gas-droplet flow. The range of variation in the initial parameters in [20] was as follows: $M_{L1} = 1\%$, $d_1 = 10$ μm, $\mathrm{Re} = (1$–$6) \times 10^4$, $q_W = \mathrm{const} = 4.8\ \mathrm{kW/m^2}$, and $p/h = 10$ in the flow of steam water droplets.

An analysis of previous works allowed us to draw the following conclusions. Note that the studies in [16,21] were performed for a single-component mist/steam flow. In this area, the first steps have been taken, which revealed the great potential of such a cooling method. There are an extremely limited number of papers in the literature concerned with the study of heat transfer in a turbulent droplet-laden flow in a ribbed duct. In these works, a significant HTE up to three times was experimentally and numerically shown in comparison with a single-phase flow in a smooth duct with a fixed Reynolds number in the flow. All numerical works [17,19–21] used the RANS approach and isotropic turbulent models (ITM). The use of such models for the simulation of a complicated vortex turbulent flow has a number of limitations [22–24], even for a simpler case of flow in a backward-facing step. The Euler–Lagrangian approach was used to model the dynamics and heat transfer in the two-phase flow. Research in this direction should be deepened and detailed. In addition to the flow and turbulent characteristics, heat transfer should also be studied.

In the present study, the authors used the Euler–Euler approach for flow and heat transfer simulation in the dispersed phase [25]. The turbulent characteristics of the carrier phase were predicted using the elliptical blending Reynolds stress model (RSM). This approach allows the partial elimination of the problems associated with the significant anisotropy of turbulent velocity fluctuations for the flows with recirculating regions [22–24]. This work is aimed at the numerical study of the effect of droplet evaporation on the flow, turbulence, and heat transfer in a ribbed duct in comparison with a smooth one.

2. Mathematical Models

The paper considers the flow dynamics and heat transfer in 2D two-phase gas-droplet turbulent flow in the presence of interfacial heat transfer between the ribs. The two-fluid Euler approach is used to describe the flow dynamics and heat and mass transfer in the gaseous and dispersed phases [26,27]. The carrier phase turbulence is predicted using the elliptical Reynolds stress model [28], taking into account the effect of droplets [29,30]. The dispersed phase (water droplets) is described using steady-state continuity equations, two momentum equations, and energy equations. The authors used their own in-house code for all numerical simulations presented in this paper.

2.1. Governing Equations for the Two-Phase Turbulent Mist Phase

The set of incompressible steady-state 2D RANS equations of the carrier phase includes continuity equations, two momentum equations (in streamwise and transverse directions), energy equations, and steam diffusion into the binary air–steam medium [25]. The effect of evaporating water droplets on the motion and heat transfer in the carrier phase (air) is considered using the sink or source terms.

$$
\rho \frac{\partial U_j}{\partial x_j} = \frac{6J}{d} \Phi
$$
$$
\frac{\partial (U_i U_j)}{\partial x_i} = -\frac{\partial P}{\rho \partial x_i} + \frac{\partial}{\partial x_j}\left(\nu \frac{\partial U_i}{\partial x_j} - \langle u_i' u_j' \rangle\right) - (U_i - U_{Li})\frac{M_L}{\tau}
$$
$$
\frac{\partial (U_i T)}{\partial x_i} = \frac{\partial}{\partial x_i}\left(\frac{\nu}{\mathrm{Pr}}\frac{\partial T}{\partial x_i} - \langle u_j' t \rangle\right) + D_T \frac{(C_{PV} - C_{PA})}{C_P}\left(\frac{\partial K_V}{\partial x_i}\frac{\partial T}{\partial x_i}\right) - \frac{6\Phi}{\rho C_{Pd}}[\alpha(T - T_L) + JL] \qquad (1)
$$
$$
\frac{\partial (U_i K_V)}{\partial x_i} = \frac{\partial}{\partial x_i}\left(\frac{\nu}{\mathrm{Sc}}\frac{\partial K_V}{\partial x_i} - \langle u_j' k_V \rangle\right) + \frac{6J\Phi}{d}
$$
$$
\rho = P/(R_g T)
$$

Here, U_i ($U_x \equiv U$, $U_y \equiv V$) and u_i' ($u_x' \equiv u'$, $u_y' \equiv v'$) are components of mean gas velocities and their pulsations; x_i are projections on the coordinate axis; $2k = \langle u_i u_i \rangle = u'^2 + v'^2 + w'^2 \approx u'^2 + v'^2 + 0.5(u'^2 + v'^2) \approx 1.5(u'^2 + v'^2)$ is the kinetic energy of gas-phase turbulence; $\tau = \rho_L d^2/(18\rho\nu W)$; $W = 1 + \mathrm{Re}_L^{2/3}/6$ is the particle relaxation time, taking into account the deviation from the Stokes power law; and $\mathrm{Re}_L = |\mathbf{U} - \mathbf{U}_L|d/\nu$ is the Reynolds number of the dispersed phase.

The turbulent heat $\langle u_j' t \rangle = -\frac{\nu_T}{\mathrm{Pr}_T}\frac{\partial T}{\partial x_j}$, and the mass $\langle u_j' k_V \rangle = -\frac{\nu_T}{\mathrm{Sc}_T}\frac{\partial K_V}{\partial x_j}$ fluxes in the gas phase are predicted using simple eddy diffusivity (Boussinesq hypothesis). The constant value of the turbulent Prandtl and Schmidt numbers, Pr_T and Sc_T equal to 0.9, is used in this work.

2.2. Evaporation Model

The set of Eulerian Equation (1) of two-phase flow is supplemented by the equations of heat transfer on the droplet surface and the conservation equation of steam on the surface of the evaporating droplet [31]. It is assumed that the temperature over the droplet radius is constant [31].

$$
\lambda_L \left(\frac{\partial T_L}{\partial y}\right)_L = \alpha(T - T_L) - JL, \; \alpha = \frac{\alpha_P}{1 + C_P(T - T_L)/L} = \frac{\alpha_P}{1 + \mathrm{Ja}} \qquad (2)
$$

$$
J = J K_V^* - \rho D \left(\frac{\partial K_V}{\partial y}\right)_L \qquad (3)
$$

Here, λ_L is the coefficient of heat conductivity of the droplet; α and α_P are the heat transfer coefficient for the evaporating droplet and non-evaporating particle, respectively; T_L is the temperature of the droplet; J is the mass flux of steam from the surface of the evaporating droplet; L is the latent heat of evaporation; ρ is the density of the gas–steam mixture; D is the diffusion coefficient; and K_V^* is the steam mass fraction at the "steam-gas mixture–droplet" interface, corresponding to the saturation parameters at droplet temperature T_L. Subscript "L" corresponds to the parameter on the droplet surface. The

Jacob number, $Ja = C_P(T - T_L)/L$, is the ratio of sensible heat to latent heat during droplet evaporation. It characterizes the rate of the evaporation process and is the reciprocal of the Kutateladze number, Ku. For our conditions, the Jakob number is $Ja \leq 0.01$.

The expression for the diffusional Stanton number has the form

$$St_D = -\rho D \left(\frac{\partial K_V}{\partial y}\right)_L / [\rho \mathbf{U}(K_V^* - K_V)] \tag{4}$$

We can insert Equation (4) into Equation (3). Equation (3) can be written in the form

$$J = St_D \rho \mathbf{U} b_{1D}, \tag{5}$$

where $b_{1D} = (K_V^* - K_V)/(1 - K_V^*)$ is the diffusion parameter of vapor (steam) blowing, determined with the use of a saturation curve.

A droplet evaporates at the saturation temperature, and the temperature distribution inside a droplet is uniform. The droplet temperature along the droplet radius remains constant because the Biot number is $Bi = \alpha_L d_1 / \lambda_L << 1$ and the Fourier number is $Fo = \tau_{eq}/\tau_{evap} << 1$. Here, τ_{eq} is the period when an internal temperature gradient inside a droplet exists, and τ_{evap} is the droplet's lifetime. In this case, a droplet evaporates at the saturation temperature, and the temperature distribution inside a droplet is uniform.

2.3. The Elliptic Blending Reynolds Stress Model (RSM) for the Gas Phase

In the present study, the low-Reynolds-number elliptic blending RSM of [28] is employed. The transport equations for $\langle u_i' u_j' \rangle$ and the kinetic energy dissipation rate, ε, are written in the following general form:

$$\frac{\partial \left(U_j \langle u_i' u_j' \rangle \right)}{\partial x_j} = P_{ij} + \phi_{ij} - \varepsilon_{ij} + \frac{\partial}{\partial x_l}\left(\nu \delta_{lm} + \frac{C_\mu T_T}{\sigma_k}\langle u_l' u_m' \rangle\right)\frac{\partial \langle u_i' u_j' \rangle}{\partial x_m} - A_L, \tag{6}$$

$$\frac{\partial (U_j \varepsilon)}{\partial x_j} = \frac{1}{T_T}\left(C_{\varepsilon 1}' P_2 - C_{\varepsilon 2}\varepsilon\right) + \frac{\partial}{\partial x_l}\left(\nu \delta_{lm} + \frac{C_\mu T_T}{\sigma_\varepsilon}\frac{\partial \varepsilon}{\partial x_m}\right) - \varepsilon_L, \tag{7}$$

$$\beta - L_T^2 \nabla^2 \beta = 1, \quad \phi_{ij} = \left(1 - \beta^2\right)\phi_{ij}^W + \beta^2 \phi_{ij}^H \tag{8}$$

Here, P_{ij} is the stress-production term, T_T and L_T are the turbulent time and geometrical macroscales, and ϕ_{ij} is the velocity–pressure–gradient correlation, well-known as the pressure term. The blending model (8) presented in [28] is used to predict ϕ_{ij} in Equations (6) and (7), where β is the blending coefficient, which goes from zero at the wall to unity far from the wall; ϕ_{ij}^H is the "homogeneous" part (valid away from the wall) of the model; and ϕ_{ij}^W is the "inhomogeneous" part (valid in the wall region).

The other constants and functions of the turbulence model are presented in detail in [28]. The last terms of the system of Equations (6) and (7), A_L and ε_L, represent the effects of particles on carrier phase turbulence [29,30].

2.4. Governing Equations for the Dispersed Phase

The set of incompressible steady-state 2D governing mean equations for the dispersed phase consists of continuity equation, two momentum equations (in streamwise and transverse directions), energy equations.

$$\frac{\partial (\rho_L \Phi U_{Lj})}{\partial x_j} = -\frac{6J\Phi}{d},$$

$$\frac{\partial (\rho_L \Phi U_{Lj} U_{Li})}{\partial x_j} + \frac{\partial (\rho_L \Phi \langle u_{Li} u_{Lj} \rangle)}{\partial x_j} = \Phi(U_i - U_{Li})\frac{\rho_L}{\tau} + \Phi \rho_L g - \frac{1}{\tau}\frac{\partial (\rho_L D_{Lij}\Phi)}{\partial x_j} - \frac{\partial (\Phi P)}{\partial x_i}, \tag{9}$$

$$\frac{\partial\left(\rho_L \Phi U_{Lj} T_{Li}\right)}{\partial x_j} + \frac{\partial}{\partial x_j}\left(\rho_L \Phi \langle \theta u_{Lj} \rangle\right) = \Phi(T_i - T_{Li})\frac{\rho_L}{\tau_\Theta} - \frac{1}{\tau_\Theta}\frac{\partial\left(\rho_L D^\Theta_{Lij} \Phi\right)}{\partial x_j}. \tag{10}$$

where D_{Lij} and D^Θ_{Lij} are the turbulent diffusivity tensor and the particle turbulent heat transport tensor [29,30], $\tau_\Theta = C_{PL}\rho_L d^2/(12\lambda Y)$ is the thermal relaxation time, and $Y = \left(1 + 0.3\mathrm{Re}_L^{1/2}\mathrm{Pr}^{1/3}\right)$.

The set of governing mean equations for the dispersed phase (8–10) is completed by the kinetic stress equations, temperature fluctuations, and turbulent heat flux in the dispersed phase, which are in the form presented in [29,30].

The volume fraction of the dispersed phase is lower ($\Phi_1 < 10^{-4}$), and the droplets are finely dispersed ($d_1 < 100$ μm); therefore, the effects of interparticle collisions and break-up are neglected [25,32,33]. Droplet bag break-up is observed at We $= \rho(\mathbf{U}_S - \mathbf{U}_L)^2 d/\sigma \geq \mathrm{We}_{cr}$ $= 7$ [33]. Here, $U_S = U + \langle u'_S \rangle$ and $\mathbf{U}_L$ are the gas velocity seen by the droplet [34] and the mean droplet velocity, respectively, U is the mean gas velocity (derived directly from the RANS predictions), $\langle u'_S \rangle$ is the drift velocity between the fluid and the particles [34], and ρ and ρ_L are the densities of the gas and dispersed phases. For all droplet sizes investigated in the present paper, the Weber number is very small (We $<< 1$). Droplet fragmentation at its contact with a duct wall also is not considered. The effect of break-up and coalescence in the two-phase mist flow can be neglected due to a low droplet volume fraction at the inlet ($\Phi_1 = M_{L1}\rho/\rho_L < 2 \times 10^{-4}$). Here, M_{L1} is the initial droplet mass fraction, and ρ_L is the density of the dispersed phase.

A scheme of the flow is shown in Figure 1. A similar Euler approach was used by the authors to describe gas-droplet axisymmetric flows behind a sudden pipe expansion [25] and behind a backward-facing step in a flat duct [35].

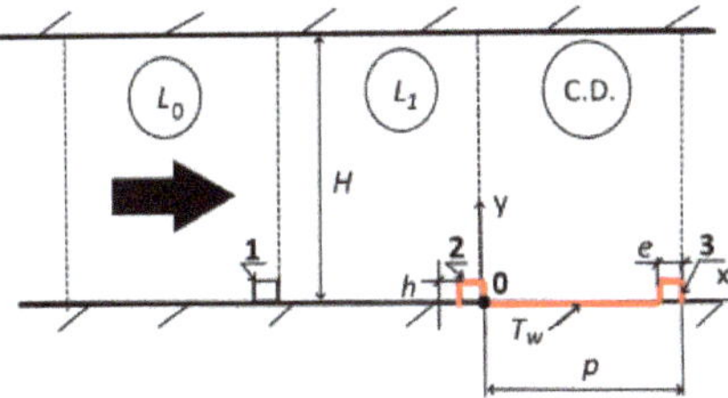

Figure 1. Scheme of flow in a two-phase turbulent flow in a ribbed flat duct (not to scale). Abbreviations L_0 and L_1 are the computational domains for the preliminary simulations; C.D. is the computational domain; and 1, 2, and 3 are the 1st, 2nd, and 3rd ribs.

3. Numerical Solution and Model Validation

3.1. Numerical Solution

The solution was obtained using the finite volume method on staggered grids. The QUICK procedure of the third order of accuracy was used for the convective terms. Central differences of the second order of accuracy were used for diffusion fluxes. The pressure field was corrected according to the agreed finite volume SIMPLEC procedure. The components of the Reynolds stress of the carrier fluid phase were simulated according to the method proposed in [36]. The components of the Reynolds stress were determined at the same points along the control volume faces as the corresponding components of the average velocity of the carrier phase. The computational grid consisted of rectangular cells. It was inhomogeneous and thickened towards all solid walls, which was necessary to resolve the details of the turbulent flow in the near-wall zone (see Figure 2). In the viscous sublayer, at least 10 computational volumes (CVs) were set. The correct simulation of sharp gradients of two-phase flow parameters was necessary. The coordinate transformation given in [37] was suitable for such a two-dimensional boundary layer problem.

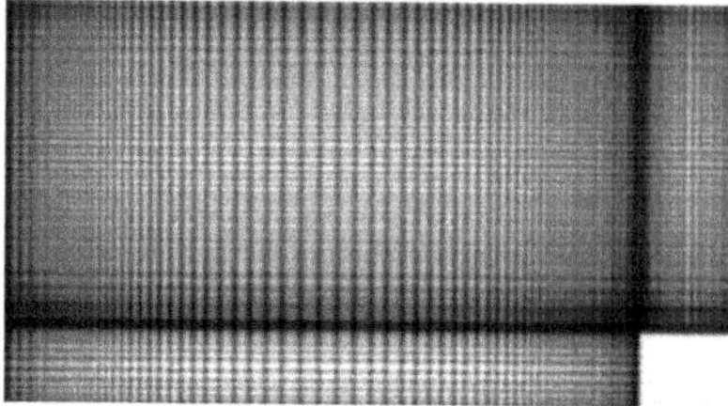

Figure 2. Computational mesh "medium" (not to scale).

All predictions were carried out on a "medium" grid containing 256×120 control volumes (CVs). The first computational cell was located at a distance from the wall of $y_+ = u_*y/\nu \approx 0.5$ (the friction velocity u_* was determined for a single-phase air flow with other identical parameters). Additionally, simulations were carried out on grids containing "coarse" 128×60 and "fine" 512×200 CVs. The difference in the results of the calculations of the wall friction coefficient (a) and the Nusselt number (b) for two-phase flow did not exceed 0.1% (see Figure 3). The Nusselt number at $T_W = $ const was determined by the formula:

$$\mathrm{Nu} = -(\partial T/\partial y)_W H/(T_W - T_m),$$

where T_W and T_m are the wall and the mass-averaged temperatures of the gas in the corresponding cross section.

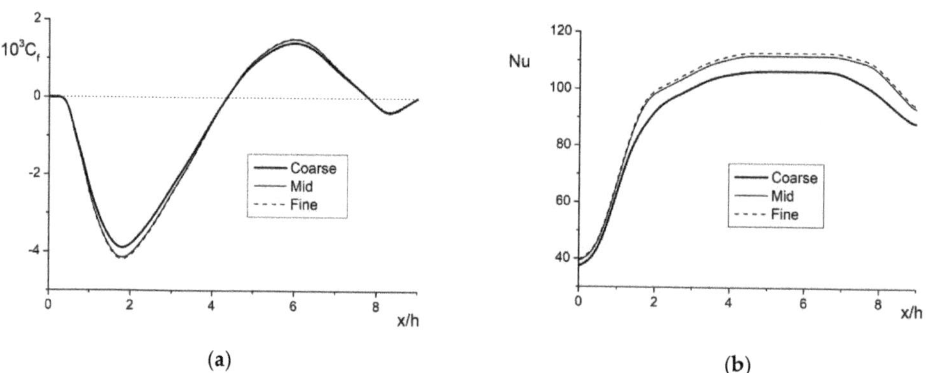

(a)

(b)

Figure 3. Grid independence test for the wall friction (a) and heat transfer (b) distributions along the duct length. $M_{L1} = 5\%$, $d_1 = 15$ μm, Re $= HU_{m1}/\nu = 1.6 \times 10^4$.

Periodic boundary conditions were set at the inlet of the computational domain. Initially, a single-phase fully hydrodynamically developed air flow was supplied to the inlet to the computational domain $L_0 = 10p$, where p is the rib pitch (the spacing between upstream and downstream ribs). The 1st rib was installed at the end of this domain. The output parameters from section L_0 were the input values for section $L_1 = 10p$, located between the 1st and 2nd ribs (see Figure 1). All simulations were performed for the two-dimensional case of a gas-droplet flow for the 2nd and 3rd obstacles. Drops were fed into a single-phase turbulent air flow along the entire cross section of the duct in the inlet cross section behind the 2nd rib. The initial temperatures of the gas and dispersed phases at the inlet to the computational domain were $T_1 = T_{L1} = 293$ K. The boundary condition $T_W = $ const $= 373$ K was set on the ribbed wall; the opposite smooth (without obstacles) wall of the flat duct was adiabatic. The entire ribbed duct surface and all the ribs were heated to eliminate the influence of the possible formation of liquid spots during the deposition of droplets on the wall from a two-phase mist flow. The impermeability and no-slip conditions for the gas phase were imposed on the duct walls. For the dispersed phase on the duct wall, the boundary condition of the "absorbing wall" [30] was used when a droplet did not return to the flow after contact with the wall surface. All droplets deposited from two-phase flow onto the wall momentarily evaporated. Thus, the pipe surface was always dry, and there was no liquid film or spots of deposited droplets formed on the wall [25,31,35]. This assumption for the heated surface is valid (see, for example, papers [25,35]). Furthermore, this condition is valid if the temperature difference between the wall and the droplet is greater than $T_W - T_L \geq 40\ K$ [38]. In the outlet cross section, the conditions for the equality to zero of the derivatives of all variables in the streamwise direction were set.

3.2. Model Validation

At the first stage, a comparison with the data of recent LDA measurements [39] for a single-phase air flow in the presence of ribs was performed. The results of the experiments and predictions are shown in Figure 4. This figure shows comparisons of measured and predicted data in the form of transverse profiles of mean longitudinal velocity, U/U_{m1} (a), and the velocity of its fluctuations, u'/U_{m1} (b), along the duct length. The averaged and fluctuating components of the streamwise velocity were normalized by the value of the average mass velocity of a single-phase flow at the duct inlet U_{m1}. Comparisons with the data of [39] were made for the 17th and 18th obstacles. The height of the duct with a square cross section was $H = 60$ mm. The profiles of the mean longitudinal velocity component agreed well with the experimental data (the difference did not exceed 5–7%). The agreement between the measurements and numerical predictions for longitudinal velocity pulsations was also quite good (the difference did not exceed 10%) except for the near-wall region.

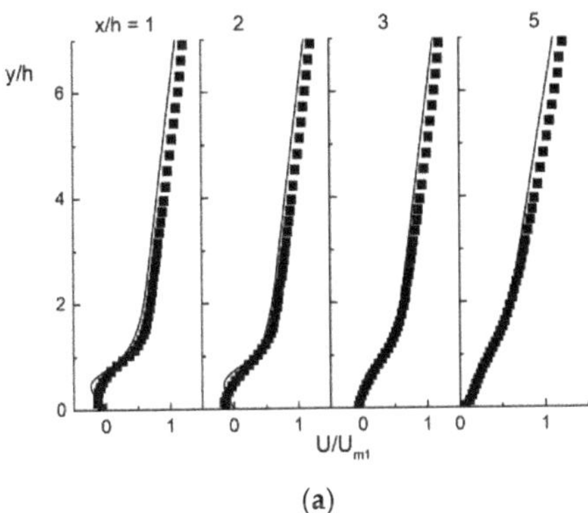
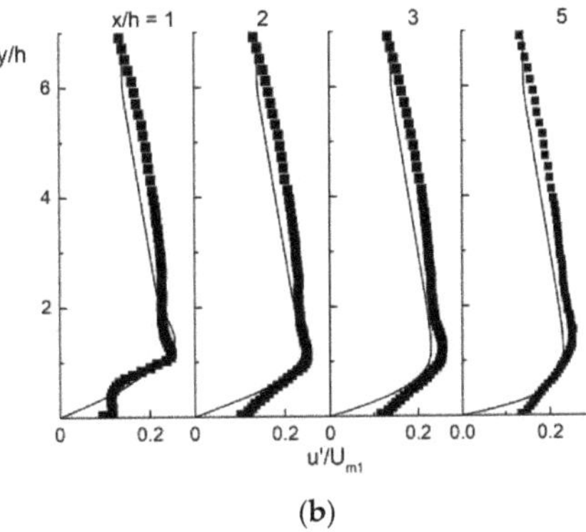

(a)   (b)

Figure 4. Profiles of the mean longitudinal velocity (**a**) and its fluctuations (**b**) in the flow around a two-dimensional square obstacle. (**a**): $h/H = 0.067$, $Re = U_{m1}H/\nu = 5 \times 10^4$, $p/h = 9$, $H = 60$ mm, $h = e = 4$ mm, $p = 36$ mm, $U_{m1} = 12.5$ m/s. The symbols are the measurements of [39]; the lines are the authors' simulations.

The results of measurements [40] and RANS numerical simulations with various isotropic turbulence models (k–ε, $v2f$, and k–ω shear stress tensor (SST)) [41] for the flow in the ribbed duct were used for heat transfer comparisons. Satisfactory agreement with the data of other authors for a single-phase flow around a two-dimensional obstacle was obtained (the maximum differences did not exceed 15%), except for the duct cross section near the upstream obstacle at $x/h < 2$ (see Figure 5). Here, Nu is the Nusselt number in a ribbed duct and Nu_0 is the Nusselt number in a smooth duct for a single-phase flow. The Nusselt number at a constant value of heat flux density ($q_W = const$) is determined by the formula:

$$Nu = q_W H / [\lambda (T_W - T_m)].$$

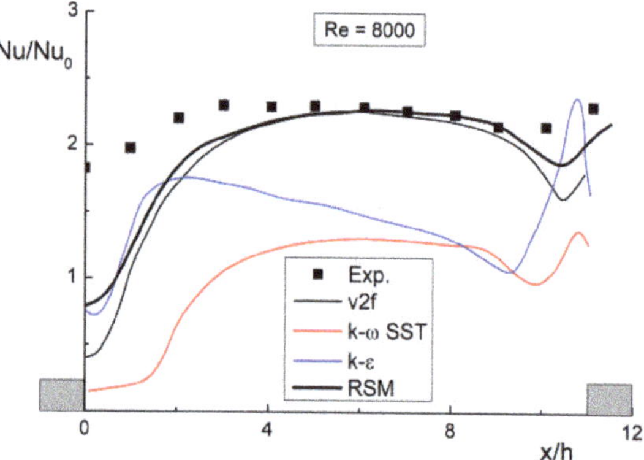

Figure 5. Distribution of heat transfer enhancement ratio in the flow around a two-dimensional square obstacle. $\mathrm{Re}_D = 0.8 \times 10^4$, $h = e = 4$ mm, $p = 40$ mm, $H = 30$ mm, $H/h = 7.5$, $T_1 = 300$ K, $q_W = 1$ kW/m^2, Tu$_1 = 5\%$. The symbols are the experiments of [40]; the lines are predictions: *v2f*, *k–ω* SST, and *k–ε* are predictions of [41], and RSM is the authors' simulations. Reprinted with permission from (Liu, J. et al.).

Comparisons with the data [40,41] were made for the 7th and 8th obstacles. All predictions were carried out for a flat duct with a square cross section and a height of $H = 30$ mm.

4. The RANS Results and Discussion

All 2D numerical simulations were carried out for a mixture of air with water drops at the duct inlet for the case of a downward two-phase flow at atmospheric pressure. Ribs were installed on the "bottom" wall of the flat duct. All simulations were performed for the flow around the system of the 2nd and 3rd obstacles. The computational domain included two square ribs with a height of $h = 4$ mm. The height of a smooth duct was $H = 40$ mm ($H/h = 10$), and the distance between two ribs was $p/h = 5$–12. The mass-average gas velocity in the inlet cross section in the computational domain varied within $U_{m1} = 5$–20 m/s, and the Reynolds number for the gas phase, constructed from the mass-average gas velocity at the inlet and the duct height, was $\mathrm{Re}_H = HU_{m1}/\nu \approx (0.6$–5$) \times 10^4$. The initial average droplet diameter was $d_1 = 5$–50 μm, and their mass concentration was $M_{L1} = 0$–10%. The initial temperature of the gaseous and dispersed phases was $T_1 = T_{L1} = 293$ K.

A turbulent flow is 3D in nature. Nevertheless, there are many cases when it is possible to use a 2D approach to describe a quasi-two-dimensional turbulent flow, for example, if the duct width, Z, is much greater than its height, H ($Z/H > 10$). The authors of [42] recommended the consideration of the turbulent solid particle-laden flow in a backward-facing step in a flat channel as two-dimensional due to the large aspect ratio of Z/H.

4.1. Flow Structure

The streamlines for a gas-droplet flow around the system of two ribs are shown in Figure 6. The complex vortex structures of the averaged flow between two ribs are clearly visible. The formation of two regions of the flow recirculation is shown. The first large recirculation region formed behind the upstream rib due to the separation of the two-phase flow at the backward-facing step (BFS). A small corner vortex was located at the end of the reverse step. The second one formed due to the droplet-laden flow separation before the downstream rib (forward-facing step (FFS)) when the fluid flow left the cell between the two ribs. It was much shorter than the previous one.

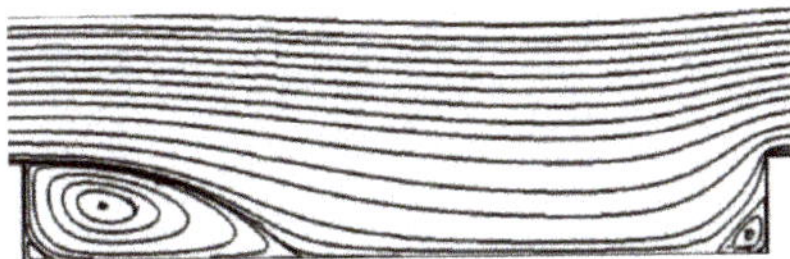

Figure 6. The streamlines for a gas-droplet flow between two ribs. Re = $U_{m1}H/\nu = 1.6 \times 10^4$, $h/H = 0.1$, $p/h = 10$, $H = 40$ mm, $h = 4$ mm, $p = 40$ mm, $U_{m1} = 6$ m/s, $T_1 = 293$ K, $T_W = 373$ K, $d_1 = 15$ μm, $M_{L1} = 0.05$.

Figure 7 shows the profiles of the average longitudinal velocities, U/U_{m1} (a); turbulent kinetic energy (TKE), k/U_{m1}^2 (b); and gas-phase temperature, $\Theta = (T_W - T)/(T_W - T_{1,m})$, in a single-phase flow ($M_{L1} = 0$), a gas-droplet flow at $M_{L1} = 0.05$, and liquid drops $\Theta_L = (T_{L,max} - T_L)/(T_{L,max} - T_{L1})$ (c) as well as vorticity, $\Omega_z = \omega_z h/U_{m1}$ (d). The solid lines are the single-phase flow at $M_{L1} = 0$, the dotted line is the gas phase at $M_{L1} = 0.05$, and the dash-dot line is the dispersed phase behind the two-dimensional obstacle. Here, $T_{L,max}$ $T_{L1,m}$ are the droplet temperatures, which were highest in the corresponding cross section and at the inlet.

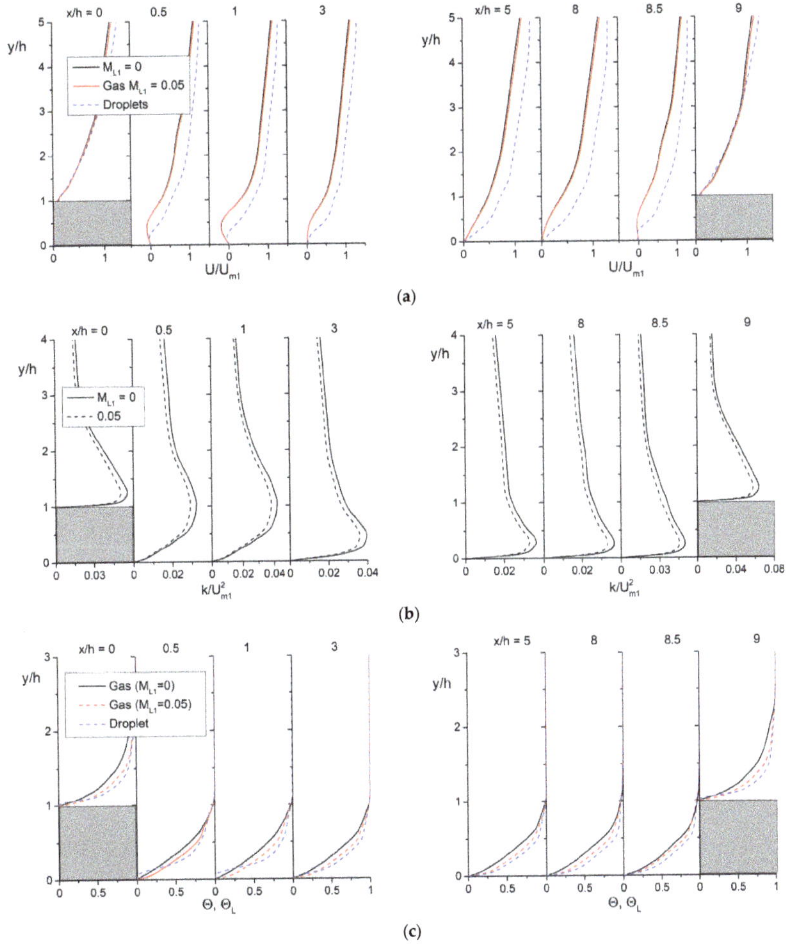

Figure 7. *Cont.*

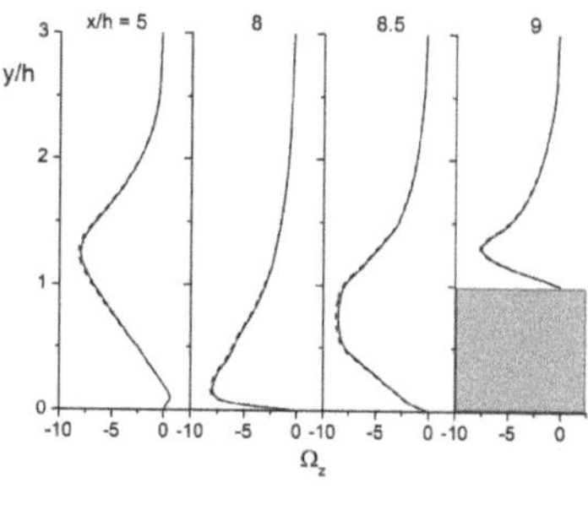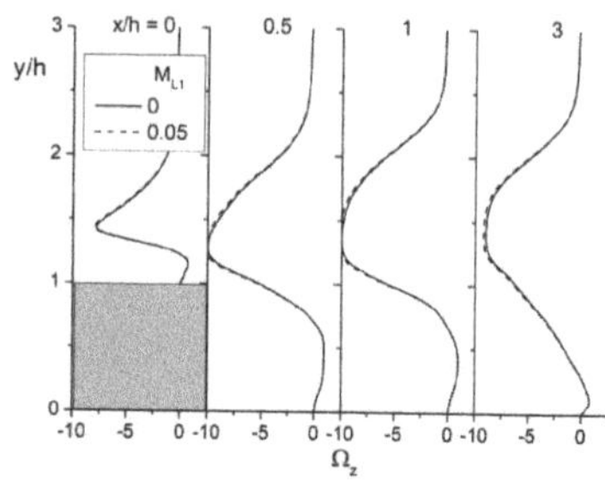

(d)

Figure 7. Transverse profiles of averaged streamwise velocities (**a**), turbulent kinetic energy (**b**), and gas-phase temperature in a single-phase flow ($M_{L1} = 0$), a gas-droplet flow at $M_{L1} = 0.05$, and liquid drops (**c**) as well as vorticity (**d**). Re = 1.6×10^4, $h/H = 0.1$, $p/h = 10$, $d_1 = 15$ μm, $M_{L1} = 0.05$.

The structure of a turbulent two-phase flow showed significant changes when flowing around a system of obstacles installed on one of the duct walls. The profiles of the averaged streamwise velocity components of the gaseous and dispersed phases were similar to those for the single-phase flow regime (see Figure 7a). The gas velocity in the gas-droplet flow was slightly ($\leq$3%) ahead of the single-phase flow velocity. The drop velocity had the greatest value for the downward flow due to their inertia. Two regions with negative values for the longitudinal velocity of the gas-droplet flow are shown, which were confirmed by the data in Figure 6. The length of the main recirculation zone of the flow was $x_{R1} \approx 4.1h$, and the length of the second recirculation region in front of the step ahead was $x_{R2} \approx 1.1h$. The lengths of the recirculation zones were determined from the zero value of the flow velocity.

Figure 7b shows the transverse distributions of the kinetic energy (TKE) of carrier phase turbulence for a 2D flow. The TKE was calculated by the formula for a two-dimensional flow:

$$2k = \langle u_i' u_i' \rangle = u'^2 + v'^2 + w'^2 \approx u'^2 + v'^2 + 0.5\left(u'^2 + v'^2\right) \approx 1.5\left(u'^2 + v'^2\right)$$

The highest turbulence values were obtained for the mixing layer. The level of kinetic energy of turbulence increased as the downstream obstacle approached. The maximum value of gas-phase turbulence was obtained at $x/h = 9$ (the upper corner of the downstream rib), and it was almost twice as high as the values for the TKE between the ribs. The turbulence of the flow was associated with the flow around the obstacle.

The dimensionless temperature distributions of the single-phase flow and the gas and dispersed phases are shown in Figure 7c. All profiles in Figure 7c are qualitatively similar to each other. The gas temperature in the gas-droplet flow was lower than the corresponding value for a single-phase flow due to droplet evaporation. Let us note that the droplet temperature profile for the first two sections, $x/h =$ and 3, did not start from the wall ($y/h = 0$) as for the gas phase, but it is shifted from the wall by a small distance towards the flow core. This is explained by the absence of droplets in the near-wall zone in the area of flow separation due to their evaporation close to the wall between the ribs. The non-dimensional vorticity profiles are given in Figure 7d. They were calculated using the well-known formula:

$$\omega_z = \frac{\partial V}{\partial x} - \frac{\partial U}{\partial y}. \tag{11}$$

The magnitudes of vorticity were mainly negative values (because $\frac{\partial V}{\partial x} \ll \frac{\partial U}{\partial y}$), except in the near-wall region inside the flow recirculation zone (see Figure 7d). The minimal values are shown in the outer shear layer of the separation zone and on the top wall of the downstream rib. The maximal positive value was observed close to the wall of the ribbed wall. In the case of two-phase mist flow, the magnitude of vorticity was slightly higher than that of the single-phase flow (up to 4%).

Figure 8 shows the profiles of the dispersed-phase mass concentration, M_L/M_{L1}, for various droplet mass fractions (a) and their initial diameters (b). Obviously, due to the evaporation of droplets, their mass fraction decreased continuously, both streamwise and in traverse directions, when approaching the wall of the heated duct between the ribs. This was typical of the numerical data given in Figure 8a,b. The distributions of the mass fraction of droplets with changes in their initial amounts had qualitatively similar forms (see Figure 8a).

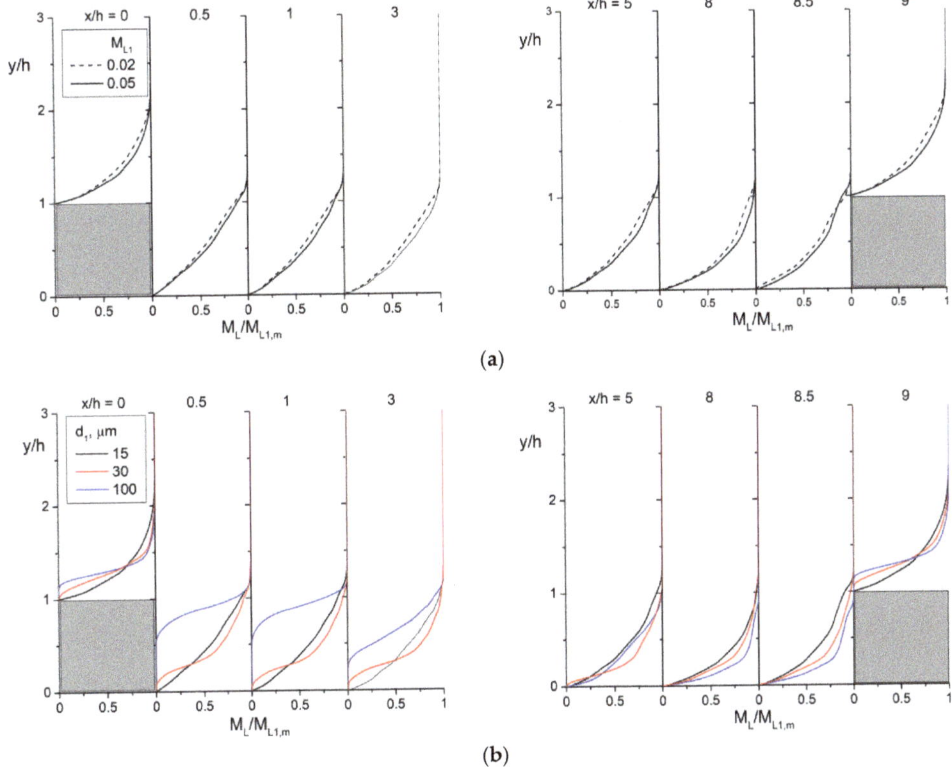

Figure 8. Transverse profiles of the water droplet mass fraction vs. the initial mass fraction (a) and the droplet diameter (b). Re = 1.6 × 10⁴. (a) d_1 = 15 μm; (b) M_{L1} = 0.05.

A change in the initial diameter of the liquid droplets had a more complex effect on the course of the evaporation processes (see Figure 8b). In the flow core, this value trended toward the corresponding value at the inlet to the computational domain, and $M_L/M_{L1} \rightarrow 1$. This is explained by the almost complete absence of droplet evaporation. Fine particles at Stk < 1 penetrated into the region of flow separation and were observed over the entire cross section of the duct. Large inertial droplets (d_1 = 100 μm, Stk > 1) almost did not penetrate into the flow recirculation zone, and they were present in the mixing layer and the flow core. In the near-wall zone, large drops were observed only behind the reattachment point. The largest and inertial droplets (d_1 = 100 μm) accumulated in the near-wall region towards the downstream obstacle. Finely dispersed low-inertia droplets could leave the region between the two ribs due to their low inertia, while large drops could not leave this region. This led to an increase in the droplet mass fraction in this flow region and towards the downstream obstacle.

In order to clearly display the flow structure in the inter-rib cavity, the contours of the nondimensional mean streamwise velocity, U/U_{m1} (a), and the temperature, $\Theta = (T_W - T)/(T_W - T_{1,m})$ (b), in two-phase mist flow are shown in Figure 9. Large-scale and small-scale flow recirculation zones behind the upwind rib (BFS) and before the downstream rib (FFS) can be found in Figure 9a. The small corner vortex directly behind the upstream rib was observed. The length of the main recirculation zone of the flow was $x_{R1} \approx 4.1h$, and the length of the second recirculation region in front of the step ahead was $x_{R2} \approx 1.1h$. The lengths of the recirculation zones were determined from the zero value of the mean streamwise flow velocity ($U = 0$). In this region, the gas temperature increased, and it led to the suppression of heat transfer (see Figure 9b). These conclusions agree with the data of Figures 6 and 7a,c.

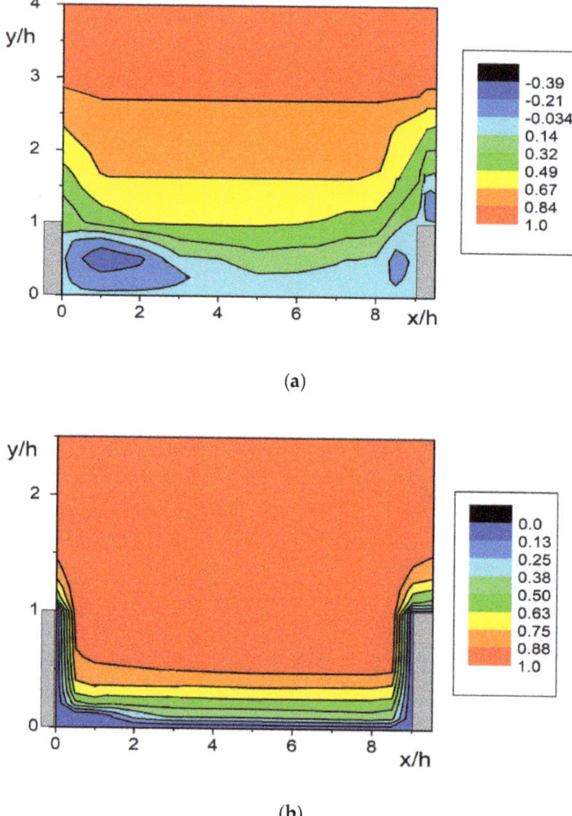

(a)

(b)

Figure 9. The contour plots of the mean streamwise velocity (**a**) and temperature (**b**). Re = 1.6×10^4, $d_1 = 15\ \mu m$, $M_{L1} = 0.05$.

4.2. Heat Transfer

The influence of the initial mass fraction (a) and droplet diameter (b) of the dispersed phase on the Nusselt number distribution in a two-phase flow along the duct length is shown in Figure 10. A significant HTE in the two-phase mist flow (up to 2.5 times) compared to a single-phase flow in a ribbed channel was obtained with the addition of evaporating water drops into a single-phase gas flow (see Figure 10a). Droplets of the minimum diameter (d_1 = 5 µm) evaporated most intensely, and the largest ones evaporated least intensely (d_1 = 100 µm) (see Figure 10b). The sizes of the zone of two-phase flow and the zone of HTE also decreased. This was an obvious fact for the evaporation of droplets in the two-phase mist flows, which was associated with a significant interface reduction; it was first shown by the authors of this work for a gas-droplet flow in a system of two-dimensional obstacles. Heat transfer was attenuated and trended toward the corresponding value for the single-phase flow in the region of flow separation for the most inertial droplets. These drops did not penetrate into the flow separation region behind the upstream rib (BFS). An increase in heat transfer was obtained in the region behind the point of flow reattachment. A decrease in heat transfer was shown in the section of flow separation towards the downstream rib (FFS). The most inertial droplets also did not leave the region between the two ribs and accumulated in front of the downstream obstacle.

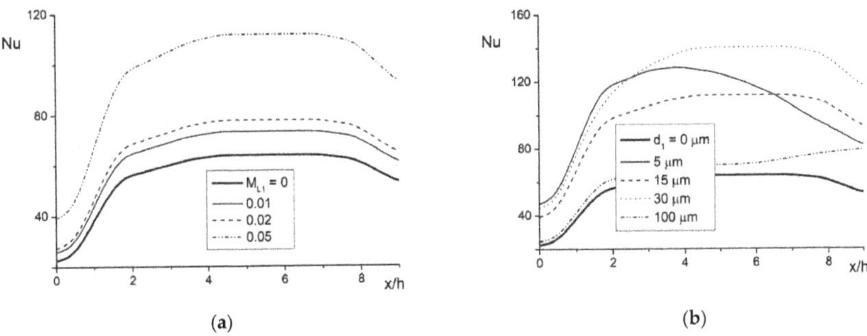

(a)   (b)

Figure 10. The effect on the heat transfer rate of the droplet mass fraction at the inlet (**a**) and their diameter (**b**). Re = 1.6 × 10^4. (**a**) d_1 = 15 µm; (**b**) M_{L1} = 0.05.

The effect of the gas Reynolds number, Re, and the initial mass fraction of the dispersed phase, M_{L1}, on the thermal hydraulic performance parameter is shown in Figure 11. The wall friction coefficient, C_f, was calculated using the formula $C_f/2 = \tau_W / (\rho U_{m1}^2)$. Here, Nu_0 and C_{f0} are the maximal Nusselt number and wall friction coefficient in the two-phase mist flow of a fully developed smooth duct, other conditions being equal. $Nu/Nu_0/(C_f/C_{f0})$ is the thermal hydraulic performance parameter. This is the ratio of the maximal Nusselt numbers divided by the maximal wall friction coefficient ratio. The ribbed surface provided a much better thermohydraulic performance than a smooth duct in the case of a droplet-laden turbulent mist flow, with other conditions being identical. This effect was quite pronounced at small Reynolds number values of Re < 10^4. It should be noted that the wall friction coefficient ratio, C_f/C_{f0}, was taken to the first power.

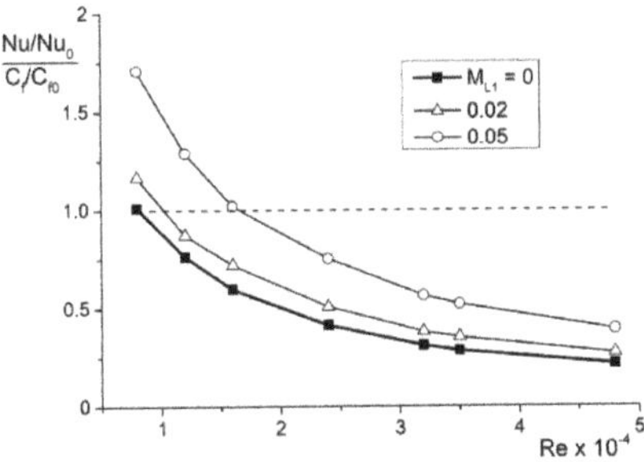

Figure 11. The effect on the thermal hydraulic performance parameter of the Reynolds number of the gas flow and the mass fraction of the dispersed phase at the inlet. d_1 = 15 µm.

5. Comparison with Results of Other Authors

Comparisons with the data of LES simulations of a solid particle-laden flow around a two-dimensional obstacle were made according to the conditions of [10]. The following data were used for the comparative analysis: $h = H/7$, $V_{P1} = U_{m1}/25$, the Reynolds number was plotted from the obstacle height, $Re = HU_{m1}/\nu = 740$, $\rho_P/\rho = 769.2$, and ρ_P was the particle material density. The height of the boundary layer for a single-phase flow in the inlet section of the computational domain was $\delta = 7h$, and the carrier phase was atmospheric air at $T = 293$ K (see Figure 12). Here, $h = 7$ mm was the obstacle height, $H = 1$ mm, $U_{m1} = 1.59$ m/s was the free flow velocity, and $V_{P1} = 0.06$ m/s. The two-dimensional obstacle was square in cross section and was mounted on the bottom wall. The flow of solid particles was blown vertically through a flat slot along the normal surface at distance h from the trailing edge of the obstacle. The number of solid particles during the LES calculation was 2×10^5. The calculations were performed for three Stokes numbers, $St^+ = \tau u_*^2/\nu = 0.25$, 1, 5, and 25, where $\tau = \rho_P d^2/(18\mu)$ was the particle relaxation time and $u_* = 0.5$ m/s was the friction velocity for a single-phase flow without particles, other things being equal. This corresponded to the solid particle diameters $d = 8, 15, 34$, and 76 μm. The calculations were carried out in a two-dimensional formulation for an isothermal two-phase flow around a single obstacle.

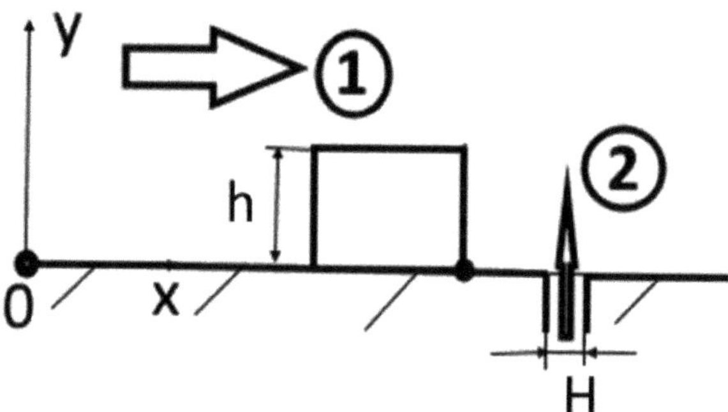

Figure 12. Scheme of two-phase flow behind a 2D square obstacle. The 1 is the single-phase air flow with the mean mass velocity U_{m1}, and the 2 is the dispersed phase stream U_{P1}.

The profiles of the dispersed phase concentration in the near-wall zone of the plate at $y = 0.02h$ are shown in Figure 13. Here, C_b is the mean concentration of particles over the hole (slot) width at the inlet to the computational domain. As the Stokes number, St_+, increased, heavier particles stopped penetrating into the recirculation region, resulting in lower concentrations along the obstacle wall. The local maximum concentration at $x/h \approx 1$ for all studied Stokes numbers (particle diameters) is explained by the injection of the particle flow. A characteristic feature of the low-inertia particles was a significant increase in the concentration of particles near the obstacle wall, according to the LES data ($C/C_b \approx 10$–20). Most likely, such an accumulation of particles in the corner near the wall of the obstacle can be explained by the effect of the accumulation of particles in [42]. For our Eulerian simulations, an increase in concentration was also obtained, but the values were much smaller (by a factor of approximately 8–10). For inertial particles at $St_+ = 5$, the region turned out to be almost completely free of solid particles. This was typical for both the data of the LES calculations [10] and our numerical calculations. Behind the obstacle, a decrease in the particle concentration in the near-wall region was observed, and here our numerical predictions agreed satisfactorily with the LES data (the difference did not exceed 20% at $St_+ = 1$ and 5 and did not exceed 100% at $St_+ = 0.25$).

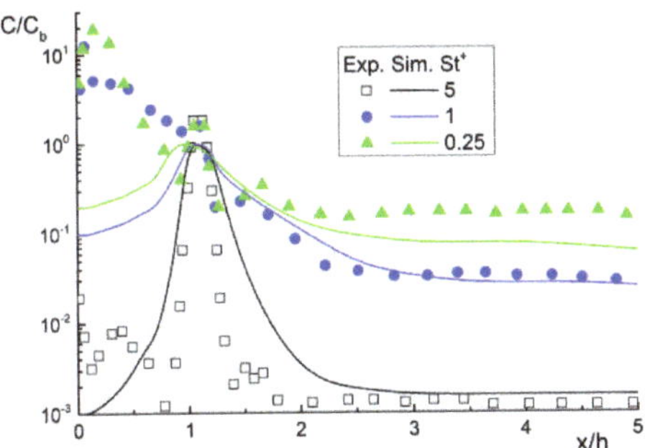

Figure 13. Concentration profiles of the dispersed phase at $y = 0.02h$ vs. the Stokes number St_+ along the length of the channel behind a 2D obstacle. The points are LES calculations [10]; the lines are the authors' predictions. $h = H/7$, $V_{P1} = U_{m1}/25$, $Re = HU_{m1}/\nu = 740$, $\rho_P/\rho = 769.2$. Reprinted with permission from (Grigoriadis, D.G.E. et al.)

Figure 13 shows the concentration profiles of the dispersed phase when the Stokes number, St_+, is varied along the length of the channel behind a two-dimensional obstacle. Particles at $St_+ = 0.25$ accumulated in the near-wall region near the bottom wall. Further downstream, heavier particles gradually left the recirculation region, and at $St_+ > 1$ the decrease in their distribution profile was similar to a Gaussian distribution. For the largest particles at $St_+ = 25$, according to the results of our numerical predictions, an underestimation of the position of the concentration maximum was observed, and in general the particles rose lower than according to the LES results [10].

Figure 14 shows the profiles of the dispersed phase concentration when the Stokes number, St_+, was varied along the length of the duct behind a two-dimensional obstacle. Particles at $St_+ = 0.25$ accumulated in the near-wall region near the bottom wall. Further downstream, heavier particles gradually left the recirculation region, and at $St_+ > 1$ the decrease in their distribution profile was similar to a Gaussian distribution. An underestimation of the position of the concentration maximum was observed, according to the results of our numerical predictions for the largest particles at $St_+ = 25$. The maximal penetration coordinate in the transverse directions in our RANS predictions was smaller than that in the LES results [10].

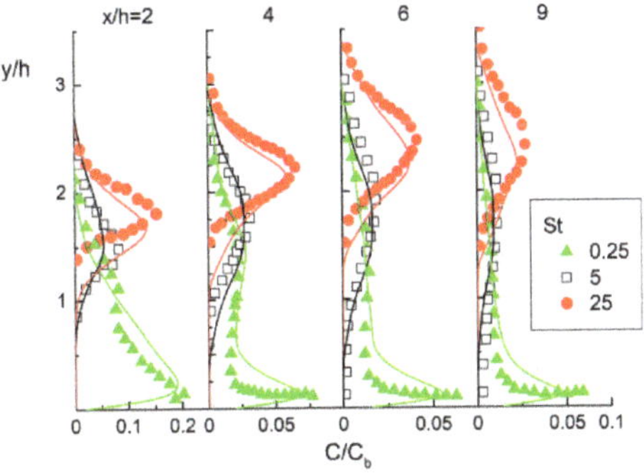

Figure 14. The transverse profiles of particle concentrations for various Stokes numbers, St_+, after a 2D obstacle. The points are LES calculations [10]; the lines are the authors' predictions.

6. Conclusions

Two-dimensional numerical simulations of the local flow structure, turbulence, and heat transfer in a ribbed flat duct during the evaporation of water droplets in a gas flow were carried out. The set of steady-state RANS equations written with consideration of the influence of droplet evaporation on the transport processes in gas is used. The two-fluid Eulerian approach was used to describe the flow dynamics and heat and mass transfer in the dispersed phase. To describe the turbulence of the gas phase, an elliptical blending RSM model was employed.

It was shown that the transverse profiles of the averaged longitudinal velocity components of the gaseous and dispersed phases were similar to those of the single-phase flow regime. The gas velocity in the gas-droplet flow was slightly ($\leq 3\%$) higher than those in the single-phase flow. The droplet velocity is higher than the gas-phase velocity in the two-phase flow. Finely dispersed droplets (Stk < 1) penetrated well into the region of flow recirculation and were observed over the entire cross section of the duct. They could leave the region between the two ribs due to their low inertia. Large inertial droplets (Stk > 1) were present only in the mixing layer and the flow core and accumulated in the near-wall region close to the downstream wall of the rib. A significant increase in heat transfer (up to 2.5 times) in comparison with a single-phase flow in a ribbed duct was shown when evaporating water drops were added to a single-phase gas turbulent flow. For the most inertial droplets, which were not involved in the separation motion in the region of the main recirculation zone behind the BFS (upstream rib), the heat transfer intensification decreased and trended toward the corresponding value for the single-phase flow regime in the recirculation zone. An increase in heat transfer was obtained behind the reattachment point. A decrease in heat transfer was shown in the zone close to the FFS (downstream rib).

Author Contributions: Conceptualization, M.A.P. and V.I.T.; methodology, M.A.P. and V.I.T.; investigation, M.A.P.; data curation, M.A.P. and V.I.T.; formal analysis, M.A.P. and V.I.T.; writing—original draft preparation, M.A.P. and V.I.T.; writing—review and editing, M.A.P. and V.I.T.; resources, M.A.P. and V.I.T.; project administration, V.I.T.; All authors have read and agreed to the published version of the manuscript.

Funding: This work was partially supported by the Ministry of Science and Higher Education of the Russian Federation (mega-grant 075-15-2021-575). The RANS and turbulence models for the single-phase turbulent flow were developed under a state contract with IT SB RAS (121031800217-8).

Acknowledgments: The authors thank Sebastian Ruck (Karlsruhe Institute of Technology, Karlsruhe, Germany) for providing the experimental database in an electronic form.

Conflicts of Interest: The authors declare no conflict of interest.

Nomenclature

$C_f/2 = \tau_W/(\rho U_{m1}^2)$	wall friction coefficient
D	droplet diameter
H	rib height
$2k = \langle u_i' u_i' \rangle$	turbulent kinetic energy
M_L	mass fraction
$\mathrm{Nu} = -(\partial T/\partial y)_W H/(T_W - T_m)$	Nusselt number
p	rib pitch
q_W	heat flux density
$\mathrm{Re}_D = U_m D_h/\nu$	Reynolds number, based on hydraulic diameter
$\mathrm{Re} = U_{m1} H/\nu$	Reynolds number, based on the duct height
$\mathrm{Stk} = \tau/\tau_f$	the mean Stokes number
T	temperature
$\mathbf{U}_L$	the mean droplet velocity
U_{m1}	mean mass flow velocity
$\mathbf{U}_S$	the fluid (gas) velocity seen by the droplet
u_*	wall friction velocity
x	streamwise coordinate
x_R	position of the flow reattachment point
y	distance normal from the wall

Subscripts	
0	two-phase mist flow in a smooth duct
1	initial condition
W	wall
L	liquid
M	mean mass
Greek	
Φ	volume fraction
ρ	density
μ	the dynamic viscosity
ν	kinematic viscosity
τ	the droplet relaxation time
τ_W	wall shear stress
Acronym	
BFS	backward-facing step
CV	control volume
FFS	forward-facing step
THE	heat transfer enhancement
RANS	Reynolds-averaged Navier–Stokes
SMC	second-moment closure
TKE	turbulent kinetic energy

References

1. Webb, R.L.; Kim, N.N. *Principles of Enhanced Heat Transfer*, 2nd ed.; Taylor & Francis: New York, NY, USA, 2005.
2. Han, J.C.; Dutta, S.; Ekkad, S. *Gas Turbine Heat Transfer and Cooling Technology*, 2nd ed.; CRC Press: New York, NY, USA, 2012; pp. 363–441.
3. Kalinin, E.K.; Dreitser, G.A.; Yarkho, S.A. *Enhancement of Heat Transfer in Channels*; Mashinostroenie: Moscow, Russia, 1990. (In Russian)
4. Terekhov, V.; Dyachenko, A.; Smulsky, Y.; Bogatko, T.; Yarygina, N. *Heat Transfer in Subsonic Separated Flows*; Springer: Cham, Switzerland, 2022.
5. Han, J.C.; Glicksman, L.R.; Rohsenow, W.M. An investigation of heat transfer and friction for rib-roughened surfaces. *Int. J. Heat Mass Transf.* **1978**, *21*, 1143–1156. [CrossRef]
6. Liou, T.M.; Hwang, J.J.; Chen, S.H. Simulation and measurement of enhanced turbulent heat transfer in a channel with periodic ribs on one principal wall. *Int. J. Heat Mass Transf.* **1993**, *36*, 507–517. [CrossRef]
7. Leontiev, A.I.; Olimpiev, V.V. Thermal physics and heat engineering of promising heat transfer intensifiers. *Rep. Russ. Acad. Sci. Energetics* **2011**, 7–31. (In Russian)
8. Fitzpatrick, X.A.; Lambert, B.; Murray, D.B. Measurements in the separation region of a gas-particle cross flow. *Exp. Fluids* **1992**, *12*, 329–341. [CrossRef]
9. Marchioli, C.; Armenio, V.; Salvetti, M.V.; Soldati, A. Mechanisms for deposition and resuspension of heavy particles in turbulent flow over wavy interfaces. *Phys. Fluids* **2006**, *18*, 025102. [CrossRef]
10. Grigoriadis, D.G.E.; Kassinos, S.C. Lagrangian particle dispersion in turbulent flow over a wall mounted obstacle. *Int. J. Heat Fluid Flow* **2009**, *30*, 462–470. [CrossRef]
11. Lin, S.; Liu, J.; Xia, H.; Zhang, Z.; Ao, X. A numerical study of particle-laden flow around an obstacle: Flow evolution and Stokes number effects. *Appl. Math. Model.* **2022**, *103*, 287–307. [CrossRef]
12. Varaksin, A.Y. Gas-solid flows past bodies. *High Temp.* **2018**, *56*, 275–295. [CrossRef]
13. Krepper, E.; Beyer, M.; Frank, T.; Lucas, D.; Prasser, H.-M. CFD modelling of polydispersed bubbly two-phase flow around an obstacle. *Nucl. Eng. Des.* **2009**, *239*, 2372–2381. [CrossRef]
14. Tas-Koehler, S.; Neumann-Kipping, M.; Liao, Y.; Krepper, E.; Hampel, U. CFD simulation of bubbly flow around an obstacle in a vertical pipe with a focus on breakup and coalescence modelling. *Int. J. Multiph. Flow* **2021**, *135*, 10352. [CrossRef]
15. Lakehal, D. On the modelling of multiphase turbulent flows for environmental and hydrodynamic applications. *Int. J. Multiph. Flow* **2002**, *28*, 823–863. [CrossRef]
16. Huang, Y.-H.; Chen, C.-H.; Liu, Y.-H. Nonboiling heat transfer and friction of air/water mist flow in a square duct with orthogonal ribs. *ASME J. Therm. Sci. Eng. Appl.* **2017**, *9*, 041014. [CrossRef]
17. Jiang, G.; Shi, X.; Chen, G.; Gao, J. Study on flow and heat transfer characteristics of the mist/steam two-phase flow in rectangular channels with 60 deg. ribs. *Int. J. Heat Mass Transf.* **2018**, *120*, 1101–1117. [CrossRef]
18. Huang, K.-T.; Liu, Y.-H. Enhancement of mist flow cooling by using V-shaped broken ribs. *Energies* **2019**, *12*, 3785. [CrossRef]
19. Dhanasekaran, T.S.; Wang, T. Computational analysis of mist/air cooling in a two-pass rectangular rotating channel with 45-deg angled rib turbulators. *Int. J. Heat Mass Transf.* **2013**, *61*, 554–564. [CrossRef]

20. Elwekeel, F.N.M.; Zheng, Q.; Abdala, A.M.M. Air/mist cooling in a rectangular duct with varying shapes of ribs. *Proc. IMechE Part C J. Mech. Eng. Sci.* **2014**, *228*, 1925–1935. [CrossRef]

21. Gong, J.; Ma, C.; Lu, J.; Gao, T. Effect of rib orientation on heat transfer and flow characteristics of mist/steam in square channels. *Int. Comm. Heat Mass Transf.* **2020**, *118*, 104900. [CrossRef]

22. Thangam, S.; Speziale, C.S. Turbulent flow past a backward-facing step: A critical evaluation of two-equation models. *AIAA J.* **1992**, *30*, 1314–1320. [CrossRef]

23. Zaikov, L.A.; Strelets, M.K.; Shur, M.L. A comparison between one- and two-equation differential models of turbulence in application to separated and attached flows: Flow in channels with counterstep. *High Temp.* **1996**, *34*, 713–725.

24. Heyerichs, K.; Pollard, A. Heat transfer in separated and impinging turbulent flows. *Int. J. Heat Mass Transf.* **1996**, *39*, 2385–2400. [CrossRef]

25. Pakhomov, M.A.; Terekhov, V.I. Second moment closure modelling of flow, turbulence and heat transfer in droplet-laden mist flow in a vertical pipe with sudden expansion. *Int. J. Heat Mass Transf.* **2013**, *66*, 210–222. [CrossRef]

26. Derevich, I.V.; Zaichik, L.I. Particle deposition from a turbulent flow. *Fluid Dyn.* **1988**, *23*, 722–729. [CrossRef]

27. Reeks, M.W. On a kinetic equation for the transport of particles in turbulent flows. *Phys. Fluids A* **1991**, *3*, 446–456. [CrossRef]

28. Fadai-Ghotbi, A.; Manceau, R.; Boree, J. Revisiting URANS computations of the backward-facing step flow using second moment closures. Influence of the numerics. *Flow Turbul. Combust.* **2008**, *81*, 395–410. [CrossRef]

29. Zaichik, L.I. A statistical model of particle transport and heat transfer in turbulent shear flows. *Phys. Fluids* **1999**, *11*, 1521–1534. [CrossRef]

30. Derevich, I.V. Statistical modelling of mass transfer in turbulent two-phase dispersed flows. 1. Model development. *Int. J. Heat Mass Transf.* **2000**, *43*, 3709–3723. [CrossRef]

31. Terekhov, V.I.; Pakhomov, M.A. Numerical study of heat transfer in a laminar mist flow over an isothermal flat plate. *Int. J. Heat Mass Transfer* **2002**, *45*, 2077–2085. [CrossRef]

32. Elgobashi, S. An updated classification map of particle-laden turbulent flows. In *IUTAM Symposium on Computational Approaches to Multiphase Flow, Fluid Mechanics and Its Applications*; Balachandar, S., Prosperetti, A., Eds.; Springer: Dordrecht, The Netherlands, 2006; Volume 81, pp. 3–10.

33. Lin, S.P.; Reitz, R.D. Drop and spray formation from a liquid jet. *Ann. Rev. Fluid Mech.* **1998**, *30*, 85–105. [CrossRef]

34. Mukin, R.V.; Zaichik, L.I. Nonlinear algebraic Reynolds stress model for two-phase turbulent flows laden with small heavy particles. *Int. J. Heat Fluid Flow* **2012**, *33*, 81–91. [CrossRef]

35. Pakhomov, M.A.; Terekhov, V.I. let evaporation in a gas-droplet mist dilute turbulent flow behind a backward-facing step. *Water* **2021**, *13*, 2333. [CrossRef]

36. Hanjalic, K.; Jakirlic, S. Contribution towards the second-moment closure modelling of separating turbulent flows. *Comput. Fluids* **1998**, *27*, 137–156. [CrossRef]

37. Anderson, D.A.; Tannehill, J.C.; Pletcher, R.H. *Computational Fluid Mechanics and Heat Transfer*, 2nd ed.; Taylor & Francis: New York, NY, USA, 1997.

38. Mastanaiah, K.; Ganic, E.N. Heat transfer in two-component dispersed flow. *ASME J. Heat Transf.* **1981**, *103*, 300–306. [CrossRef]

39. Ruck, S.; Arbeiter, F. LDA measurements in a one-sided ribbed square channel at Reynolds numbers of 50 000 and 100 000. *Exp. Fluids* **2021**, *62*, 232. [CrossRef]

40. Wang, L.; Sunden, B. Experimental investigation of local heat transfer in a square duct with continuous and truncated ribs. *Exp. Heat Transf.* **2005**, *18*, 179–197. [CrossRef]

41. Liu, J.; Xie, G.N.; Simon, T.W. Turbulent flow and heat transfer enhancement in rectangular channels with novel cylindrical grooves. *Int. J. Heat Mass Transf.* **2015**, *81*, 563–577. [CrossRef]

42. Fessler, J.R.; Eaton, J.K. Turbulence modification by particles in a backward-facing step flow. *J. Fluid Mech.* **1999**, *314*, 97–117. [CrossRef]

MDPI
St. Alban-Anlage 66
4052 Basel
Switzerland
Tel. +41 61 683 77 34
Fax +41 61 302 89 18
www.mdpi.com

Water Editorial Office
E-mail: water@mdpi.com
www.mdpi.com/journal/water